T0202368

BEYOND THE DYNAMICAL UNIVERSE

Praise for *Beyond the Dynamical Universe*

"This book is likely to become a crucial resource for future scientific revolutions, despite (or due to) its daring speculative proposals. Indeed, it offers no less than a complete redefinition of science and explanation, abandoning causality in favor of global consistency. By involving consciousness from the outset, instead of vainly wondering what is its 'material cause', it paves the way to a truly complete view of the world. No aspect of 'what there is' is left aside in this comprehensive book that synergizes physics with philosophy, our knowledge of nature with our knowledge of ourselves."

Michel Bitbol, *CNRS/ENS, Archives Husserl, Paris, France*

"Einstein was worried about the exclusion of the 'now' from physics. Perhaps one should worry less. After reading this book an idea deeply grounded in physics emerges: complementary to our subjective experience, 'presence' and 'passage of time' are universal and fundamental properties of reality."

Marc Wittmann, *Institute for Frontiers Areas of Psychology and Mental Health, Germany*

"This important book drives a well-crafted stake through the heart of the dynamical view of time. The dogma that physics does not need philosophy is another welcome casualty."

Huw Price, *University of Cambridge, UK*

"This book presents a fascinating scientifically informed original metaphysics of nature sure to provoke discussion. And with the price of admission you get a set of wonderfully clear introductions to the cutting edge of modern physics."

William Seager, *University of Toronto Scarborough, Canada*

"From relativity and quantum mechanics to consciousness, Silberstein, Stuckey, and McDevitt take us on an exciting cutting-edge tour of one of the greatest mysteries in science: the nature of time."

Dean Buonomano, *University of California, Los Angeles, USA*

"A tour-de-force on physics and philosophy by a philosopher, a physicist, and a mathematician, *Beyond the Dynamical Universe* is a bold attempt to do away with the standard explanatory paradigm in physics and replace it with a form of blockworld adynamical explanation that might have been inspired by the heptapods in the movie *Arrival*. It is a revolutionary proposal, with consequences for the nature of time and our perception of time, worked out in some detail in separate threads for the non-expert, the philosopher of physics, and the physicist. Well worth a serious read, the book succeeds in being both provocative and instructive on many levels."

Jeffrey Bub, *University of Maryland, USA*

"This book is innovative in form and content. The form—a physicist, a philosopher, and a mathematician contributing parallel and interrelated threads on the same topic—is an exemplary model of interdisciplinarity in the foundations of physics. The content—an adynamical, atemporal approach to solving the long-standing conceptual puzzles about quantum mechanics—is a radical departure from standard modes of physical explanation, and one that just might give us genuine insight into the nature of the quantum world."

Peter Lewis, *Dartmouth College, Department of Philosophy*

"The book is an original and far-reaching attempt to bridge the gap between the physical image of time, presupposing a static view of a universe given in block, and our dynamical experience of passage, based on our perception of events coming into being in succession. To the extent that the essential task of philosophy is to achieve a unified view of the physical universe and of our place in it, this book is an absolute must for scientists and laypeople alike. You simply cannot put it down."

Mauro Dorato, *University of Roma Tre, Italy*

Beyond the Dynamical Universe

Unifying Block Universe Physics and Time
as Experienced

Michael Silberstein
W.M. Stuckey
Timothy McDevitt

Elizabethtown College

OXFORD
UNIVERSITY PRESS

OXFORD
UNIVERSITY PRESS

Great Clarendon Street, Oxford, OX2 6DP,
United Kingdom

Oxford University Press is a department of the University of Oxford.
It furthers the University's objective of excellence in research, scholarship,
and education by publishing worldwide. Oxford is a registered trade mark of
Oxford University Press in the UK and in certain other countries

© Michael Silberstein, W.M. Stuckey and Timothy McDevitt 2018

First Edition published in 2018

Impression: 3

Published in the United States of America by Oxford University Press
198 Madison Avenue, New York, NY 10016, United States of America

British Library Cataloguing in Publication Data
Data available

Library of Congress Control Number: 2017959077

ISBN 978–0–19–880708–7

DOI 10.1093/oso/9780198807087.001

Printed and bound by
CPI Group (UK) Ltd, Croydon, CR0 4YY

Acknowledgements

We would like to thank ten anonymous reviewers at Oxford University Press for their very thoughtful and helpful comments. For a close reading of large portions of the text complete with copious face-saving comments, we owe a special debt of thanks to Ken Wharton, Peter Lewis, Jeff Long, Daniel Peterson, Karen Crowther, Michael Cifone, Ruth Kastner, Tom Lancaster, and Louis Witten. Thanks to Tuyen Le for constructing www.RelationalBlockworld.com and to Alex Sten for work on the figures. Finally, we would like to thank our editor at OUP, Ania Wronski, for her support and guidance.

Contents

Part I

Overview

Overture for Ants

Resolving this problem [the feeling of the passage of time and perception of change], however, will prove to be a formidable task, as our subjective sense of time sits at the center of a perfect storm of unsolved scientific mysteries: consciousness, free will, relativity, quantum mechanics, and the nature of time.

[Buonomano, 2017, p. 266]

Fortuitously, this book addresses Frank Wilczek's challenge,[1]: "To me, ascending from the ant's-eye view to the God's-eye view of physical reality is the most profound challenge for fundamental physics in the next 100 years." We have been working on a "God's-eye" approach to fundamental physics since 2004 so it is gratifying to us that Wilczek issued this challenge just as we began writing a book about our "block universe" approach. Essentially, a philosopher of physics, a foundational/theoretical physicist, and a mathematician will argue that there are many reasons to favor adynamical explanation in the block universe over the currently reigning dynamical explanation in the mechanical universe; thus, Beyond the Dynamical Universe is the title of the book.

More specifically, we claim the entire mechanical paradigm itself is the reason that foundational physics[2] and foundations of physics[3] are at an impasse across the board. The bottom line is that dynamical explanation in the mechanical universe, while it certainly works well in many explanatory contexts in physics, is not fundamental and must be replaced as such in order to develop future physics. We are not suggesting that general relativity (GR) or quantum field theory (QFT) are wrong but that they are merely incomplete. The new paradigm we propose is adynamical explanation in a block universe and we offer a specific example called Relational Blockworld (RBW). In terms of the nature of reality itself, we are proposing a wholesale replacement of the mechanical universe (as fundamental) with the ontology of RBW wherein not time-evolved entities but four-dimensional spacetime "entities"[4] are fundamental. We call this "spatiotemporal ontological contextuality" and according to this idea, for example, an entire quantum experimental setup and procedure from initiation to termination would

---

[1] Nobel Laureate Frank Wilczek: Physics in 100 Years. *Physics Today* 69(4), 32–9 (2016).

[2] Attempts to reconcile quantum field theory and general relativity and/or to unify the fundamental forces of particle physics.

[3] The philosophical study of any area of physics, but typically emphasizing modern physics. We will often use the term "philosophy of physics" so as to avoid confusion with the term "foundational physics."

[4] Spacetime relations and contexts at multiple scales.

Beyond the Dynamical Universe. Michael Silberstein, W.M. Stuckey and Timothy McDevitt, Oxford University Press (2018). © Michael Silberstein, W.M. Stuckey and Timothy McDevitt. DOI 10.1093/oso/9780198807087.001.0001

provide the spatiotemporal context for explaining the corresponding outcomes of the quantum experiment. The new paradigm has profound implications for perennial concerns in:

- The foundations/philosophy of physics.
- Philosophy of science more generally.
- Philosophy of mind.
- Metaphysics.
- Reductionism.
- Emergence.
- The nature of matter, space and time.
- Time as experienced.
- The hard problem of consciousness itself.
- Foundational physics.

We will show how using adynamical global constraints, e.g., the least action principle, the Feynman path integral and the transition amplitude in RBW's modified lattice gauge theory, can resolve some of the current impasses of foundational physics and philosophy of physics. For example, we provide a new approach to quantum gravity and unification that explains dark energy and resolves the black hole firewall paradox. We will explain how contextuality already inherent in general relativity might resolve the dark matter problem without recourse to non-baryonic matter. We resolve the cosmological puzzle of the Big Bang, the flatness problem, the horizon problem, the low entropy problem, the GR paradoxes of closed timelike curves, and conundrums of quantum mechanics (QM) such as the measurement problem and quantum nonlocality. The RBW paradigm also has implications for the experience of time/change and the specialness of the present moment in a block universe. Further, RBW bears on the hard or generation problem of consciousness more generally, as it allows us to reject materialism or physicalism in favor of a kind of neutral monism whence a very natural explanation of temporal experience can be given that doesn't force us to say that time is an illusion, that it's all in the mind/brain, or that it must somehow be put in by hand in foundational physics.

Everyone involved in the philosophy of physics and most involved in foundational physics are hoping that physics will tell them about the nature of reality, together these two communities make up the "natural philosophy" of our age. Philosophy of physics has produced myriad interpretations of QM, e.g., Many Worlds, de Broglie-Bohm, the Two-State Vector Formalism, the Transactional Interpretation, Relational Quantum Mechanics, the Ithaca Interpretation, information theoretic accounts, etc. Decades of research in philosophy of physics has not pared the field of contenders but, quite the opposite, the interpretations of

QM have proliferated. As things currently stand, taken together, both QM and quantum field theory (QFT) are highly predictive and successful, yet there is no consensus as to what they are telling us about the nature of reality. On the relativity side, the debates about possible block universe implications ("eternalism"), whether or not spacetime is relational or a thing-in-itself, etc., also continue without resolution. If inductive inference is the least bit valid and underdetermination persists, there is no magic argument or experiment that will bring unanimity on these issues.

The big picture at the dawn of the twenty-first century on the foundational physics side is that researchers are stymied with respect to the unification of forces in the Standard Model of particle physics and the reconciliation of GR with QFT, that is, quantum gravity. There is not even consensus in foundational physics about whether or not quantum gravity requires unification; for example, string theory does unify gravity with the other fundamental forces while loop quantum gravity does not. The stalemate has the physics community seriously debating whether or not a theory such as string theory or the multiverse needs to be falsifiable to count as a scientific theory! In a sense, some physicists are giving up on the dream of a final theory altogether. Many physicists and science writers lament the lack of progress in foundational physics and worry that it is becoming little more than metaphysics. As if all that wasn't troubling enough, many in both communities now concede that physics (and neuroscience) have no real explanation for time as experienced: the Passage of time, the Direction of time (including some of the physical arrows of time such as the entropic arrow), or Presence, the specialness of the lived present—the Now—all of which are at the heart of everyday human experience if not physics itself.

As Thomas Kuhn would say per *The Structure of Scientific Revolutions*, foundational physics is not in a "normal period" of science, but a "revolutionary" one, and it is struggling to birth a new paradigm. We are living in interesting times for physics and everyone agrees some fundamental axiom of the old paradigm is false, they just do not agree on which one. We believe therefore that the only way forward is to combine the efforts of both these communities and attempt to construct theories that resolve a large number of issues across the board. For example, we believe the attempt to unify GR and QFT, and the attempt to interpret quantum physics should go hand in hand. That is in part what we attempt to do in this book. We think it is time for philosophers of physics to go beyond their usual concerns and start working with people in the foundational physics community to construct new models that attempt to resolve as many anomalies as possible and hopefully lead to new physics. More specifically in terms of our model, we believe that the situation in natural philosophy is portending a Kuhnian revolution per Wilczek's recent challenge. The paradigm being superseded is "dynamical explanation in the mechanical universe," an expression we intend in the broadest possible metaphysical and explanatory sense, not the usual historical designation. The "dynamical universe paradigm" holds that reality is "composed of" fundamental dynamical entities that are time-evolved according to some fundamental equation of motion (dynamical law) from some initial condition. In the dynamical/mechanical universe, conscious experience including the experience of time "supervenes on" these dynamical physical processes. Ironically perhaps, even some proponents of the block universe fully embrace the dynamical universe paradigm.

This book challenges the dynamical universe paradigm in its entirety. Most of the book is devoted to showing that, given Relational Blockworld, many of the current conundrums of theoretical physics and philosophy of physics can be resolved, including the mystery of the experience of time and how that experience relates to the physical universe. Of course, not all readers will buy into our particular form of adynamical explanation in a block universe. Rather, the real potential herein resides in conveying a nascent approach to foundational physics and philosophy of physics. Additionally, we truly hope that even if our specific form of "God's-eye physics" does not work out, the methodology for how to proceed and how to collaborate will catch on. Our model includes not only interdisciplinary interaction but also an attempt to provide a complete ontological and explanatory system that subsumes foundational physics, the philosophy of physics, and conscious experience, as they relate to time and change.

Therefore, we are writing this book for:

1. People working in foundational physics, such as theoretical physicists.
2. People working in the foundations of physics, such as philosophers of physics.
3. Philosophers of mind, cognitive scientists, and metaphysicians interested in the nature and perception of time or the problem of consciousness more generally.
4. Any and all interested graduate students and advanced undergraduates in philosophy, cognitive science, and physics.

While these groups have been engaging one another with increasing frequency, they have very different styles, norms, and mediums of communication even when they are addressing the same questions. Researchers in foundational physics often express frustration with the buzz words and technical jargon of every philosophical tradition. And those are the ones who value philosophy at all. Many theoretical physicists such as Weinberg [Kefalis, 2015], Hawking [Warman, 2011], and Krauss [Pigliucci, 2012] have recently disparaged philosophy as antiquated armchair pseudoscience in need of total replacement by science. Ironically, given the increasing rhetoric about the value and necessity of multidisciplinary interaction, there are schisms everywhere, even within disciplines, such as analytic versus continental philosophy, analytic metaphysics versus philosophy of science, etc. However, in spite of the obvious possibility of failure given the diverse target audience, our goal is to engage and communicate effectively with all these groups; hopefully not as disparate disciplines but ultimately as a multidisciplinary community because it is our contention that the resolution of these issues requires cooperation between all concerned parties. To that end, most chapters have three parts, that is, a main thread, a thread devoted to the philosophy of physics, and a thread devoted to foundational physics.

Most chapters contain a relatively non-technical main thread that is written with virtually no formalism so as to be accessible to advanced undergraduates in physics and philosophy who are not familiar with issues in the philosophy of physics. The main thread is designed to define the problem space, provide the minimally necessary background and our particular solutions to the problems under discussion. Therefore, while the main thread does attempt to communicate with both novice and expert alike, the main thread presents the content of each subject primarily as a vehicle to articulate our specific resolution to the problems under discussion. Therefore we warn the reader that *the chapters are not textbook presentations of these subjects in either style or content*. Indeed, entire textbooks have been written on special relativity, general relativity, quantum mechanics, quantum field theory, quantum gravity, and conscious experience.

Thus, while the main thread is accessible to the advanced undergraduate, it and the other two threads in no way replace a proper introduction to these subjects. The philosophy of physics thread will contain material most people will associate with foundations of physics and the foundational physics thread will focus on material most associate with theoretical physics, hence it will be the thread where the denser formalism is housed. In the overviews for Parts II and III, and throughout the book, we will point readers to the material relevant to their varied interests and highlight the material they might wish to skim. The book *in toto* provides an argument for our thesis so to that end, for each chapter, the most comprehensive reading experience will be engendered by reading first the main thread, then the philosophy of physics thread, and finally the foundational physics thread. These latter two threads are not mere appendices to the main thread; indeed the main thread is not always the longest part of a given chapter. In general, the main thread is designed to:

- Introduce the relevant philosophical concepts and terminology to a physics audience.
- Introduce the relevant physical concepts and terminology to a philosophy audience.

The philosophy of physics thread is designed to:

- Address the philosophical/foundations issues and background mentioned in the main thread more deeply than in the main thread.

The foundational physics thread is designed to:

- Consider formal issues in theoretical physics alluded to in the main thread in more detail than could be done there.

As detailed below, this is our attempt to structure the book for multiple and potentially non-overlapping audiences. But, let us acknowledge that the price of covering so much diverse territory under our new paradigm is that no particular topic is going to get covered in the detail it would in a more focused work. We should also warn the reader that we have made liberal use of quotations. We feel that quoting rather than paraphrasing or merely asserting allows the reader to see precisely what other authors have written without having to spend hours scouring the cited literature. We feel that this is especially important in this context because we are introducing respective fields of study to different and potentially unknown disciplines. For example, we are introducing theoretical physicists to the cognitive neuroscience of consciousness.

An outline of the book is as follows. Part I is an introduction to and overview of the book as a whole. Part II presents the puzzles, problems, and paradoxes of physics that follow from dynamical explanation in the mechanical universe, as well as our resolutions thereof using adynamical explanation in the block universe. After resolving mysteries in general relativity, quantum mechanics, and quantum field theory, we hope our case is strong enough that the reader will accept our invitation to go further "down the rabbit hole" where we use adynamical explanation in the block universe to propose a new path for future physics. We end Part II with possible resolutions of major anomalies such as dark energy, dark matter, and the black hole firewall paradox, as well as address other outstanding concerns related to quantum gravity and unification.

In Part III we deal at length with time and change as experienced. We show how our new re-conception of the physical universe and explanation within it paves the way for a novel kind of neutral monism that deflates the hard problem of consciousness and allows for an explanation of the experience of time and change given a block universe approach to physics. This new way of looking at the relationship between conscious experience and the physical world allows us to avoid eliminativism, reductionism and dualism about phenomenal experience. In the Coda, we acknowledge some remaining problems in foundational physics that need to be resolved, some of which arise because of the new paradigm and some extant ones that await the application of the new paradigm. We will also explore new frontiers in both foundational physics and philosophy of physics that the new paradigm opens up. We realize that much is promised in this book which will rightly make many readers skeptical; we only ask that you suspend judgment about our prophesied end of the dynamical universe until our vision for new physics is made clear.

Finally, anybody who wants the deepest possible appreciation of the issues across all the disciplines, how they relate to one another, and what is fully at stake will have to read every thread as suggested. However, we know many readers will not choose to read the entire book but rather will focus on specific interests. For example, a reader primarily interested in time as experienced in a block universe could profitably read chapters 2, 3, 7, and 8 only, focusing on threads of interest therein. A reader primarily interested in the interpretation of quantum physics could just read threads of interest in chapters 1, 4, and 5. A reader interested primarily in cosmology and quantum gravity could just read threads of interest in chapters 1, 3, 4, 5, and 6. A reader primarily interested in standard topics in philosophy of science such as explanation or topics in analytic metaphysics, such as, the nature of time, mereology, emergence such as could just read threads of interest in chapters 1, 2, and 3, and the philosophy of physics thread for chapter 4. We will provide further guidance in the Part II and Part III overviews.

1
Introduction

1.1 Dynamical versus adynamical explanation

We begin the book with a quotation from Nobel Laureate Frank Wilczek:

A recurring theme in natural philosophy is the tension between the God's-eye view of reality comprehended as a whole and the ant's-eye view of human consciousness, which senses a succession of events in time. Since the days of Isaac Newton, the ant's-eye view has dominated fundamental physics. We divide our description of the world into dynamical laws that, paradoxically, exist outside of time according to some, and initial conditions on which those laws act. The dynamical laws do not determine which initial conditions describe reality. That division has been enormously useful and successful pragmatically, but it leaves us far short of a full scientific account of the world as we know it. The account it gives—things are what they are because they were what they were—raises the question, Why were things that way and not any other? The God's-eye view seems, in the light of relativity theory, to be far more natural. Relativity teaches us to consider spacetime as an organic whole whose different aspects are related by symmetries that are awkward to express if we insist on carving experience into time slices. Hermann Weyl expressed the organic view memorably in his 1949 book *Philosophy of Mathematics and Natural Science* (Princeton University Press, page 116):

The objective world simply is, it does not happen. Only to the gaze of my consciousness, crawling upward along the life line of my body, does a section of this world come to life as a fleeting image in space which continuously changes in time.

To me, ascending from the ant's-eye view to the God's-eye view of physical reality is the most profound challenge for fundamental physics in the next 100 years.

[Wilczek, 2016, p. 37]

We agree with Wilczek and take up his challenge in this book, which we had almost completed when we found his inspirational quotation in *Physics Today*. *We want to point out immediately that the term "God's-eye view" has no religious or theological implications; it is merely a perspective used*

Beyond the Dynamical Universe. Michael Silberstein, W.M. Stuckey and Timothy McDevitt,
Oxford University Press (2018). © Michael Silberstein, W.M. Stuckey and Timothy McDevitt.
DOI 10.1093/oso/9780198807087.001.0001

for computations and modeling Figure 1.1).[1] This perspective is motivated by the "blockworld," the "block universe," or "eternalism," that is, the co-reality or co-existence of the past, present, and future as inferred, for example, from the relativity of simultaneity in special relativity.[2] Explanation associated with this God's-eye view is what we refer to as "adynamical explanation in the block universe." In Wilczek's quotation, the God's-eye view is contrasted with the "ant's-eye view of human consciousness, which senses a succession of events in time." Explanation associated with this ant's-eye view is what we refer to as "dynamical explanation in the mechanical universe."[3] Since our perceptions are formed in time-evolved fashion, we are predisposed to think dynamically and, therefore, we want to understand/explain what we experience dynamically. This dynamical bias is occasionally exploited in science fiction.

For example, in *The Story of Your Life* [Chiang, 2002], Earthlings encounter aliens ("heptapods"[4]) from another planet and attempt to unravel their language

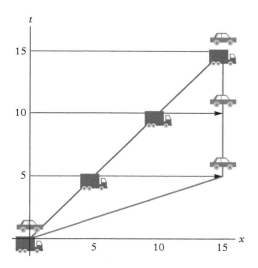

Figure 1.1 *A spacetime depiction of two vehicles racing; their worldlines are shown as colored lines. Time is the vertical direction and space is the horizontal direction. As you can see, while this God's-eye view represents objects moving in space as a function of time, nothing is moving or changing in this view as a whole.*

---

[1] Along these lines, the reader may replace the term "God's-eye view" with the term "block universe view" if they find it objectionable for any reason.

[2] We will explain this implication in chapter 2 for those unfamiliar with the relativity of simultaneity.

[3] As will become clear, we use the term "mechanical universe" much more broadly than merely "Newtonian mechanics." All current physics can be and usually is thought of "mechanically," as we will explain.

[4] This story inspired the movie *Arrival*, released in November 2016.

and their physics. The heptapods, however, experience a block universe, perceiving past, present, and future as equally real. Perhaps they experience the entire block universe, or at least their own worldtubes from birth to death, as one moment like Dr. Manhattan in *Watchmen* [Moore and Gibbons, 1986]. This block universe perspective infects both the language and physics of the aliens, and the humans must overcome this difference in perspective in order to understand their new friends. The following is an episode from the story that nicely illustrates the different emphases in human physics and heptapod physics:

> That day when Gary [the human physicist] first explained Fermat's principle to me [the human linguist], he had mentioned that almost every physical law could be stated as a variational [least action] principle. Yet when humans thought about physical laws, they preferred to work with them in their causal formulation. I could understand that: the physical attributes that humans found intuitive, like kinetic energy or acceleration, were all properties of an object at a given moment in time. And these were conducive to chronological, causal interpretation of events: one moment growing out of another, causes and effects creating a chain reaction that grew from past to future.
>
> In contrast, the physical attributes that the heptapods found intuitive, like "action" or those other things defined by integrals, were meaningful only over a period of time. And these were conducive to a teleological interpretation of events: by viewing events over a period of time, one recognized that there was a requirement that had to be satisfied, a goal of minimizing or maximizing. And one had to know the initial and final states to meet that goal; one needed knowledge of the effects before the cause could be initiated. I was growing to understand that, too.
>
> [Chiang, 2002, pp. 161–69]

The difference in the human and the heptapod physics is only a matter of emphasis since both have access to the same techniques—humans have developed the least action principle which is "meaningful only over a period of time."[5] However, as the story points out, humans prefer to work with the "causal formulation" dealing with "properties of an object at a given moment in time." Indeed, some even consider time-evolved causation *essential* to explanation. This bias is captured nicely by the Tralfamadorians, the blockworld-experiencing aliens in *Slaughterhouse-Five* Vonnegut [1969], who claim the block universe precludes explanation. Here is what one Tralfamadorian says to the protagonist Billy Pilgrim in response to his question "How did I get here?":

> It would take another Earthling to explain it to you. Earthlings are the great explainers, explaining why this event is structured as it is, telling how other events may be achieved or avoided. I am a Tralfamadorian, seeing all time as you might see a stretch of the Rocky Mountains. All time is all time. It does not change. It does not lend itself to warnings or explanations. It simply is.
>
> [Vonnegut, 1969, pp. 85–6]

---

[5] Foundational Physics for Chapter 1 contains the formal details.

The claim that the block universe view cannot provide explanation reflects the bias that explanation must be dynamical and physics certainly shares this bias. As Wilczek said, "Since the days of Isaac Newton, the ant's-eye view has dominated fundamental physics" [Wilczek, 2016, p. 37]. Here is how Sean Carroll described explanation via physics on his blog:

> Let's talk about the actual way physics works, as we understand it. Ever since New-
> ton, the paradigm for fundamental physics has been the same, and includes three
> pieces. First, there is the "space of states": basically, a list of all the possible config-
> urations the universe could conceivably be in. Second, there is some particular state
> representing the universe at some time, typically taken to be the present. Third, there
> is some rule for saying how the universe evolves with time. You give me the universe
> now, the laws of physics say what it will become in the future. This way of thinking
> is just as true for [quantum mechanics] or general relativity or quantum field theory
> as it was for Newtonian mechanics or Maxwell's electrodynamics.
>
> [Carroll, 2012]

The formalism of time-evolved states in a space of states is what Smolin calls the "Newtonian Schema"[6] [Smolin, 2009] and constitutes the "Hamiltonian method" of physics.[7] Strictly speaking, explanation, ontology, and formalism are independent of each other, but obviously some combinations are a better fit than others. Dynamical explanation is consistent with the mechanical universe ontology and these are very amenable to the Newtonian Schema formalism. The mechanical universe ontology has interacting, typically three-dimensional (3D) objects with presumably intrinsic properties moving in space as a function of time. These three concepts fit so well together that many people (outside philosophy) simply conflate them tacitly. This combination is what we mean by "dynamical explanation in the mechanical universe"[8] or what Wharton calls the "Newtonian Schema Universe (NSU)" [Wharton, 2015] and it maps nicely onto our dynamical perceptions, that is, the time-evolved narration of objects moving in space under their mutual influences. Objects moving in space can cause events, for example, a moving rock can break a window, and finding the time-evolved cause of events is what it means to explain those events for most people.

In fact, the reason we use the term "mechanical universe" more broadly than simply "Newtonian mechanics" is because Newtonian "mechanistic thinking" was carried into electromagnetism (e.g., electric and magnetic field perturbations

---

[6] This formalism is not restricted to Newtonian physics. All physics can be done in this fashion, as Carroll points out.

[7] It is not inconceivable that a non-Hamiltonian method of physics that uses time-evolved states could be developed, so we will use Smolin's more general term "Newtonian Schema" for this formalism.

[8] This was shortened to "Dynamical Universe" in the title of the book.

moving through space to mediate force),[9] special relativity (relativistic mechanics), quantum mechanics (e.g., interacting particles with intrinsic properties of mass, charge, and spin), general relativity (e.g., the expanding universe), and quantum field theory (e.g., oscillators linked in space exchanging energy momentum). However, there is another way to explain events that does not require time-evolved narration. This alternative "adynamical explanation" says the reason for events is in accord with a spatiotemporal pattern dictated by some global rule connecting the initial state in the present to the final state in the future (so-called future boundary condition). That is a bit vague, so let's explore the ontology and formalism that are most consistent with this adynamical explanation.

Perhaps not surprisingly, the ontology that is most consistent with this adynamical explanation is the block universe, since we employ future boundary conditions as an explanans[10] rather than an explanandum,[11] as in dynamical explanation. The block universe ontology employs four-dimensional (4D) entities situated relative to each other in spacetime (Figure 1.1). From the God's-eye perspective these 4D entities don't move or change; they're truly adynamical, as described by Geroch:

> There is no dynamics within space-time itself: nothing ever moves therein; nothing happens; nothing changes. In particular, one does not think of particles as moving through space-time, or as following along their world-lines. Rather, particles are just in space-time, once and for all, and the world-line represents, all at once, the complete life history of the particle.
>
> [Geroch, 1978, pp. 20–1]

A worldline is just the plot of a point particle in space and time. For example, approximating a 3D object, like the car or truck in Figure 1.1, as a point particle (0D) leads to the colored 1D worldlines in Figure 1.1. If you don't approximate a 3D object as a point particle, you get a 4D (3D + 1D) worldtube in spacetime instead of a 1D (0D + 1D) worldline. These worldtubes would be the 4D entities in the block universe corresponding to the formalism of classical mechanics. The formalism that is most consistent with adynamical explanation and the block universe ontology is one in which the final state is an input on equal footing with the initial state, and there is some mathematical formulation of the global rule for the spatiotemporal pattern connecting these initial and final states. Wharton calls this formalism the "Lagrangian Schema" [Wharton, 2015] and we call the mathematical formulation of the global rule for the spatiotemporal pattern the "adynamical global constraint." There is a well-developed physics for the Lagrangian Schema alluded to in Chiang's story called the "Lagrangian method"[12] and one of its

---

[9] Indeed, Maxwell believed "all physics research should advance on strict mechanical principles" [Goodstein, 1985, Episode 39].

[10] That which does the explaining.

[11] That which is being explained.

[12] It is not inconceivable that a non-Lagrangian method of physics that uses an adynamical global constraint could be developed, so we will use Wharton's more general term "Lagrangian Schema" for this formalism.

adynamical global constraints is the least (extremal) action principle. This combination of adynamical explanation, block universe ontology, and Lagrangian Schema formalism is what we mean by "adynamical explanation in the block universe." Let us also refer to this as the "Lagrangian Schema Universe (LSU)" as per Wharton.[13]

It is possible, of course, to render both a dynamical explanation in the mechanical universe and an adynamical explanation in the block universe for one and the same phenomenon, as we will see in the following example. And in most cases, the NSU and the LSU both work, although our dynamical experience leads us to assume the NSU actually represents reality while the LSU reflects a mere mathematical convenience. For example, in Foundational Physics for Chapter 1, the differential equations describing the trajectory of a thrown rock produced by the Newtonian Schema and Lagrangian Schema are the same, but they derive from dynamical and adynamical perspectives, respectively. That is, the Newtonian Schema differential equations result from Newton's second law with the force of gravity while the Lagrangian Schema differential equations result from the extremum of the action for paths between the initial and final rock locations with gravitational potential energy. Thus, the dynamical explanation for the broken window would ultimately reside in the rock's initial position and initial velocity given Newton's laws and the force of gravity, while the adynamical explanation would ultimately reside in the initial and final positions of the rock, given the adynamical global constraint that the path connecting those locations is an extremum of the action. Clearly, most people would use the dynamical explanation, regardless of how they set up and solved the corresponding differential equations, that is, "My window is broken because Timmy threw a rock at it!"

But, as we will show in this book, there are cases where dynamical explanation in the mechanical universe is strained, incomplete, or even seemingly inconsistent, while the appropriate adynamical explanation in the block universe has no such issues. Accordingly, the point of this book is that, at least in those cases, the LSU can be viewed as fundamental to the NSU, so when dynamical explanation in the mechanical universe is tortured or possibly inconsistent it should be abandoned in favor of adynamical explanation in the block universe.[14] Hence, the title of the book, *Beyond the Dynamical Universe*.

---

[13] We should point out that the mode of explanation—dynamical versus adynamical—trumps the formalism in making the distinction between the NSU and the LSU designation. Lagrangian and Hamiltonian formalisms produce the same differential equations, as we will see in the rock–window example below.

[14] Keep in mind this isn't a book about the theory of explanation and all its intricacies; therefore, the scope of our contrast case is limited. We acknowledge that the NSU versus LSU modes of explanation as we have characterized them are not exhaustive even in physics. We also concede that there are instances of explanation in the NSU mode that, while not LSU as such, could be construed as acausal or constraint-based, such as the ideal gas law, for example, using a change in pressure to explain a change in temperature and vice versa. The debate about how to view the symmetries of laws or the action is also relevant here. Are symmetries and the conservations laws they give rise to *a priori* constraints on the form of dynamical laws? Or, are they just another way of formally codifying such laws? As we point out in Foundational Physics for Chapter 1, symmetries of the action lead to conservation laws (Noether's

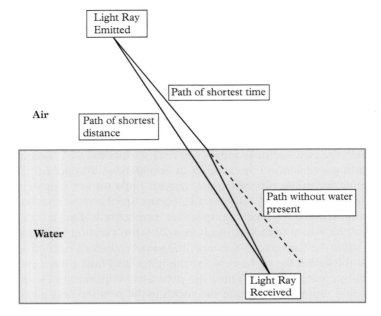

Figure 1.2 *A ray of light emitted in air and received in water takes the path of least time rather than the shortest spatial path. This is called Fermat's principle or the least action principle.*

An easy way to see the difference between the NSU and the LSU is to consider the example of a light ray emitted in air and arriving at a location in water (Figure 1.2). The Lagrangian Schema says that the path taken is the path of least time (Fermat's principle [Feynman et al., 1963, pp. 26–3]). The path of shortest distance (straight line connecting start and end points shown) takes longer, so it's not the path taken. The Newtonian Schema says that the light ray emitted toward the water proceeds without deviation, since nothing is interacting with it, until it hits the water. The water then slows and refracts (changes the direction of) the light according to Snell's Law toward the point where it is ultimately received. While both formalisms yield the same result, we tend to favor the NSU, since the LSU sounds a bit like the light ray intended to go to its ultimate destination and

theorem) and these conservation laws have counterparts in the Newtonian Schema associated with the functional form of forces (or torques in the case of angular motion). In the Lagrangian Schema, the Lagrangian is the fundamental mathematical entity and in the Newtonian Schema, force is the fundamental mathematical entity. Essentially, the Lagrangian is given by the potential and the gradient of the potential is the force, so on the surface neither approach seems to provide an easier path to new physics (new fundamental Lagrangians or new fundamental forces) and arguments rage over which approach should be considered fundamental. Where do we come down on this issue? This will become clear in chapters 4 and 5, but for us the Lagrangian account is fundamental and we will show that the symmetries of the action and thus the conservation laws that result are a consequence of a more fundamental adynamical global constraint.

calculated its path based on the presence of the water. According to the Newtonian Schema, the light ray that was received (perhaps only one of many emitted in all directions) was refracted at the air–water interface; otherwise, it would have continued along the dashed path shown, that is, it didn't adjust its emission direction with the intention of reaching its destination and it certainly doesn't "know" it will encounter water when it is emitted. But, the LSU doesn't attribute any intentionality or prescience to the light ray; rather, it simply relegates dynamical explanation to secondary status *a priori*.

The Lagrangian Schema may strike you as a mere mathematical curiosity; clearly reality is composed of objects that move in space as a function of time under the influence of forces. Since nobody knows what the future holds, how can objects behave in the present in accord with (unknown) future conditions? Consequently, that the light ray takes a path of least time (Fermat's principle) and the rock takes the path of extremum action are merely mathematical "tricks" and certainly have no ultimate explanatory power. But, this reaction is precisely what we warned against, that is, the dynamical explanatory bias. Indeed, if the LSU explanation of the light ray strikes you as endowing light with a sense of purpose and intelligence, you've already tacitly decided that reality *is* objects moving in space under the influence of forces and consequently you have undermined the fundamentality of adynamical explanation in the block universe. This is a very hard bias to overcome; ironically even the many philosophers and physicists who claim to believe in the block universe ontology still adopt the ant's-eye view and assume that fundamental explanation is of the NSU variety. Therefore we must ask that you make every effort to suspend this dynamical bias as you read the book. If you are able to do so, we believe you will be one of the very first to appreciate the depth of Wilczek's challenge.

We admit that a move from the NSU to the LSU at the fundamental level requires a dramatic change in the scientific worldview, so we are presenting a book-length argument for the fundamentality of the LSU, at least in the relevant cases. A change of this magnitude in the scientific worldview (what in the past has been called a "Kuhnian revolution" or paradigm shift) hasn't occurred since we ascended from Aristotle's teleological universe to Newton's mechanical universe in 1687. As with Galileo's astronomical observations in the 1600s that revealed flaws in the teleological universe, some philosophers and physicists today are starting to see red flags concerning the dynamical universe. Exactly how serious these warnings are is a source of much speculation, debate, and disagreement, but consensus in the foundations of physics is that *something* in the current scientific worldview is deeply flawed. As Smolin notes in his chapter entitled "The Unfinished Revolution," the move away from the mechanical universe began with relativity and quantum mechanics, "each required us to break definitively with Newtonian physics" [Smolin, 2006, p. 4]. As he says, this revolution has begun "but has yet to end." He is also absolutely right that "the problems that physicists must solve today are, to a large extent, questions that remain unanswered because of the incompleteness of the twentieth century's scientific revolution."

One problem of course is that there is little consensus on exactly what is wrong with the dynamical worldview and therefore little agreement on where to go next. We would say the revolution remains incomplete because we are still too wedded to the dynamical and mechanical universe picture. In the words of Carlo Rovelli:

> General relativity (GR) altered the classical understanding of the concepts of space and time in a way which . . . is far from being fully understood yet. Quantum mechanics (QM) challenged the classical account of matter and causality, to a degree which is still the subject of controversies. After the discovery of GR we are no longer sure of what is space-time and after the discovery of QM we are no longer sure of what matter is. *The very distinction between space-time and matter is likely to be ill-founded.* . . . I think it is fair to say that today we do not have a consistent picture of the physical world.
>
> [Rovelli, 1999, p. 227, italics added]

We agree with Smolin and Rovelli about the state of play. Unlike Smolin and Rovelli, however, as will become clear going forward, we think *every* axiomatic assumption of the dynamical universe paradigm is wrong at the fundamental level. Namely, throughout this book we will argue that:

1. Dynamical explanation is not always fundamental, as illustrated by phenomena in relativity, quantum physics, and quantum gravity.
2. The universe is not composed of or built up from dynamical, time-evolved entities (e.g., 3D entities with autonomous identities and intrinsic natures). [See chapters 4 and 5 in particular.]
3. Time as experienced, its apparent passage, and the specialness of the present moment are neither strictly "in the head, brain, or mind," nor are they features that must be put into our fundamental physical theories by hand (see chapters 7 and 8).
4. For reasons that will become clear in chapters 7 and 8, the new paradigm outlined herein strongly suggests that physicalism/materialism is false. More positively, this is a universe best conceived in terms of neutral monism wherein so-called material and mental phenomena ultimately comprise a non-dual, non-distinct unity, and not a duality.

To appreciate our alternative to the dynamical universe fully it will be necessary to read the book. However, we believe that many of the basic elements of our alternative paradigm are being floated in various forms and just need to be refined and combined in novel ways. For example, Julian Barbour says in his essay "Reductionist Doubts:"

> According to reductionism, every complex phenomenon can and should be explained in terms of the simplest possible entities and mechanisms. The parts determine the whole. This approach has been an outstanding success in science,

but this essay will point out ways in which it could nevertheless be giving us wrong ideas and holding back progress. For example, it may be impossible to understand key features of the universe such as its pervasive arrow of time and remarkably high degree of isotropy and homogeneity unless we study it holistically—as a true whole. A satisfactory interpretation of quantum mechanics is also likely to be profoundly holistic, involving the entire universe. The phenomenon of entanglement already hints at such a possibility.

[Barbour, 2012]

Here Barbour is focusing on his Machian account of general relativity (GR) and shape dynamics, neither of which we share with him, but his essay is very much in the same spirit as this book, including the invocation of adynamical global constraints in both general relativistic cosmology and quantum mechanics (QM). We will discuss these connections and others throughout the book, but what we are suggesting is that adynamical global constraints in a block universe provide *fundamental* explanation in many cases. Accordingly, four-dimensional, multiscale relationality and contextuality are the fundamental features of reality, *not dynamical entities*.[15]

No doubt dynamical explanation and ontologies are considered fundamental because experience itself is dynamical. Yet, it is only recently that theoretical physics seems increasingly compelled to explain not only the physical arrow(s) of time as such, but time as experienced (the experiential arrow of time). Assuming the physical arrows of time can be explained, for example, the thermodynamic arrow of time, it's not clear that the physical arrows of time are necessary or sufficient to explain time as experienced. Of course, physics has always regarded itself as the fundamental science that ultimately subsumes all phenomena, so perhaps it isn't surprising that a growing number of physicists are increasingly concerned that theoretical physics should subsume and explain the experience that time passes, that change occurs, that time has a direction, and the specialness of the present moment—the Now [Carroll, 2010a; Smolin, 2013; Rovelli, 2017]. Many people such as Smolin believe that the block universe hypothesis greatly increases the difficulty of proffering a satisfactory explanation in this arena. For all these reasons the various mysteries of time have taken center stage in certain quarters of theoretical physics, neuroscience, and cognitive science (see chapters 7 and 8). Consequently, there is a deluge of popular science books that focus on the "many mysteries of time" [Burdick, 2017; Gleick, 2016]. Thus, unsurprisingly, time will be a central concern in this book as well. Ironically perhaps, we'll be arguing that our adynamical God's-eye perspective actually solves many of the problems related to explaining time as experienced.

We claim that completing the twentieth-century's revolution begins by overcoming the human dynamical bias, that is, evolving from the ant's-eye view to

---

[15] We will unpack that dense statement in chapters 4 and 5.

the God's-eye view of physical reality. We will argue that the Tralfamadorian attitude about explanation is unjustified and that adynamical explanation in the block universe is not only possible but often preferable to dynamical explanation in the mechanical universe. We acknowledge that in order for you to overcome this enormous bias and consider the adynamical alternative we must provide an equally large motivation for you to do just that. Thus, a book-length argument is required. And, that argument in a nutshell is that the new adynamical paradigm can solve many of the outstanding problems in theoretical physics and the philosophy of physics. In this book, we hope to persuade the reader that this new paradigm is really what relativity, quantum physics, and experience have been telling us. Thus, this book outlines our attempt to finish the revolution in physics as suggested by Wilczek's challenge. At the end of this chapter, we'll return to a more detailed description of our so-called "Relational Blockworld" alternative, but first we want to say more about the state of play in theoretical physics and philosophy of physics in order to justify the need for our approach.

1.2   The impasse

Physics is composed of several areas of study such as Newtonian mechanics, special relativity (SR), general relativity (GR), thermodynamics and statistical mechanics, Maxwell's equations of electromagnetism, quantum mechanics (QM), and quantum field theory (QFT). Some of these areas of study are subsumed by others. For example, statistical mechanics uses Newtonian mechanics with a large number of particles to obtain thermodynamics. Newtonian mechanics obtains in the limit of small velocity in SR and Newtonian gravity obtains in the limit of weak gravitational fields in GR. In Maxwell's electromagnetism, the magnetic field can be obtained from the electric field using a Lorentz transformation from SR, and SR obtains locally in the curved spacetime of GR. And, QM plus SR gives QFT. Accordingly, all the areas of study are divided into two camps, GR and QFT, which don't seem to accommodate each other fully either formally or conceptually. Thus, physicists have been seeking a theory fundamental to QFT and GR that would reconcile these two pillars of physics.

The theory they seek is called "quantum gravity" (QG) and their study falls into foundational or theoretical physics, which also includes the attempt to unify the fundamental forces (gravity, electromagnetism, weak nuclear, and strong nuclear) and fundamental particles (quarks and leptons) of the Standard Model of particle physics. For example, the electromagnetic force and weak nuclear force combine at high enough energy density and temperature to form the electroweak force. Theories attempting to unify the electroweak and strong nuclear forces at higher energy density and temperature into the electronuclear force are called Grand Unified Theories (GUTs). Theories attempting to unify the electronuclear force and gravity at even higher energy density and temperature are called Super Unified Theories (SUTs). GUTs and SUTs are called "unification" programs and a successful SUT is even called a Theory of Everything (TOE).[16] Some programs of QG also address unification, for example, string theory, while others such as loop quantum gravity do not.

While the boundaries are fuzzy, foundational/theoretical physics is not to be confused with "foundations/philosophy of physics," which is the philosophical study of any area of physics, but typically emphasizing SR, GR, QM, QFT, or QG. Physicists and philosophers participate in both foundational physics and foundations of physics; some fall into both camps. In this first chapter, we will outline some of the concerns and frustrations leading many physicists and philosophers to question at least one or more aspects of the current dynamical worldview concerning physics.

Keep in mind that no one in either camp is suggesting general relativity (GR) or quantum field theory (QFT) is wrong per se, but rather merely incomplete. Everyone agrees that while one or both need to be superseded for reconciliation with one another, for the most part these theories work very well in their domains of discourse. It is just that, for example, after decades of dedicated effort from many brilliant people, foundational physics hasn't produced a consensus theory of quantum gravity (QG) or unification and foundations of physics hasn't produced a consensus interpretation of quantum mechanics (QM) or QFT (these will be covered in chapters 4 to 6). Furthermore, among other things, neither community has a compelling explanation for the origin of the universe, the flatness problem, the horizon problem, the low entropy problem, paradoxes of closed timelike curves, quantum entanglement, the black hole firewall paradox, dark matter, or dark energy,[17] at least not if "compelling" means "consensus."

GR and QFT explain an enormous number of phenomena and underwrite a tremendous number of technical achievements. Tegmark and Wheeler reported, for example, that "according to a recent estimate, about 30% of the U.S. gross national product is now based on inventions made possible by quantum mechanics" [Tegmark and Wheeler, 2001]. Global positioning system (GPS) devices wouldn't work without time adjustments in the GPS satellites per GR. Therein lies the rub. GR and QFT provide tremendous predictive power, so there is no way we're going to scrap them and start from scratch. Yet we can't seem to reconcile them and they leave many nagging explanatory gaps.[18] This is the impasse of foundational physics and foundations of physics leading participants in both camps to start seeing red flags with the dynamical universe.

Exactly what is responsible for these red flags is the source of much disagreement and debate. These concerns and the lack of consensus about how to explain quantum weirdness, such as quantum entanglement and nonlocality (more on those in chapter 4), led to this recent exchange on the radio program *Science*

---

[16] We acknowledge the difficulty a reader new to this field may have with so many unfamiliar acronyms, so we will toggle between acronym and term in this first chapter. If you continue this study, you will need to become familiar with these acronyms.

[17] These issues will be introduced in chapters 3, 4, and 6, respectively.

[18] These explanatory gaps will be explained in chapters 3, 4, 5, and 6.

Friday between the host Ira Flatow, the well-known scientist and science writer George Musser, and the physicist Shohini Ghose:

> IRA FLATOW: And if this tells us nothing, where has that gotten us? I mean, it's more like we're into philosophy then.
>
> SHOHINI GHOSE: Absolutely.
>
> GEORGE MUSSER: And physics often is—I mean, and it should be, foundational physics is at the intersection of metaphysics and physics. It's kind of at that boundary. And so these questions should seem philosophical to us. And in fact, we have to bring all the apparatus not just of experimental and mathematical physics, but of philosophy to try to understand it.
>
> IRA FLATOW: Is physics stuck now, because it can't explain this? There's no unified—I mean, are we at a plateau where we need either new physics or a new idea that'll talk about this spooky action—what's the mechanism, how it works? We know that it works, but we don't know how it works.
>
> GEORGE MUSSER: It's hard to say. I mean, it's something we'll know really only in retrospect, I suppose. On the one hand, there's just a burbling of ideas. It's very exciting to go to conferences and hear people excitedly talk about their ideas and wow, let's go have coffee and talk about your idea. And it's such a great to and fro that's occurring. But I will say that I do get the sense that the physicists need some new input. It could either be an Einstein arising and piecing together things people hadn't seen before that were there but hadn't been appreciated. Or maybe even better, nature has to give us guidance. We need some experiment. We need something that's actually just knocking us on the head and saying, hey, guys, here's a new data point or here's a new experiment or here's a new observation that is really going to guide us forward.
>
> [Flatow, 2016]

When it comes to a lack of consensus, things don't get much better when we move to cosmology and quantum gravity. Witness the recent widely attended conference in Munich at the Ludwig Maximilian University on whether or not, given the criterion of falsifiability, string theory and its alleged multiverse implications count as science, the multiverse being the general claim that there are multiple universes or at least many universes in one with different initial conditions, different values for the constants, and perhaps even different dynamical laws [Greene, 2011]. "The crisis, as Ellis and Silk tell it, is the wildly speculative nature of modern physics theories, which they say reflects a dangerous departure from the scientific method. Many of today's theorists—chief among them the proponents of string theory and the multiverse hypothesis—appear convinced of their ideas on the grounds that they are beautiful or logically compelling, despite the impossibility of testing them" [Wolchover, 2015]. There are many well-known physicists and other interested parties who share these worries about the current state of fundamental physics ['t Hooft, 1997; Woit, 2012; Baggott, 2013], take the following:

the usual heavily-promoted ideology of the past 30 years of fundamental physics research: we must have SUSY, so must have supergravity, so must have string theory, so must have the landscape, so must have a multiverse where we can't predict anything about anything, thus finally achieving success.

[Woit, 2012]

Actually, I would not even be prepared to call string theory a 'theory' but rather a 'model' or not even that: just a hunch.

['t Hooft, 1997, p. 163]

We cannot properly explain either eternal cosmic inflation or string theory here, the two main theories alleged to entail or strongly imply a multiverse of one sort or another, but let us say enough about them so you can understand the skepticism surrounding them. Inflation is a cosmological theory that posits a period of extremely fast, accelerated expansion early in the cosmic history of the universe. The claim is that within a fraction of a second a tiny region of space was blown up and a small region within the inflated bubble since expanded (with the rest of the bubble) to the entire observable universe of today. At inflation's end, the energy that drove this expansion created a hot fireball of particles and radiation which we can, for all practical purposes, equate with the Big Bang [Vilenkin and Tegmark, 2011].

The cosmic inflationary model solves (at least) the following two problems,[19] where "solves" means that if inflation were true, it would allegedly predict these features of the universe. These problems are called "cosmological fine-tuning problems" because they arise from the observation that some of the initial conditions of the universe seem to be fine-tuned to very unlikely or special values such that small deviations from these values would radically alter the state of the universe as it appears now. Here are the two problems solved by inflationary cosmology:

1. The "horizon problem." The universe looks uniform to a very good approximation in every direction. Why is this the case given that the Big Bang happened only 13.8 billion years ago? Widely separated regions of the universe did not have enough time to communicate via an exchange of electromagnetic waves and thereby come to thermal equilibrium. The inflationary model explains this uniformity by claiming that all the far-flung regions were in thermal contact in an extremely small region of space *before* the inflationary period, so thermally speaking, said regions remained similar after the expansion. It is interesting to note that some people speak of cosmic inflation as leading to the Big Bang (i.e., causing or coming before it) and others talk as if the Big Bang comes first, these strike us as very different claims [Greene, 2011].

---

[19] We will provide more details on these problems in chapter 3.

2. The "flatness problem." Why is the density of matter and energy in the universe exactly such as to give rise to the observable flatness of space (the universe)? As we will explain in chapter 3, GR cosmology allows for space to be positively curved or negatively curved to any degree. Residing precisely between all those curved configurations is a single flat-space configuration. Thus, it seems highly unlikely that the universe would not have any curvature. This flatness can be explained by cosmic inflation because any curvature in the small, pre-inflated region would be stretched flat by accelerated expansion of the inflationary period. Since our observable universe resided entirely within that small, pre-inflated region, it is now flat, too.

How does this get us to a multiverse? Because the end of cosmic inflation is allegedly triggered by inherently probabilistic quantum processes, it does not occur uniformly everywhere at once. Inflation presumably ended in what is now our cosmic neighborhood (our "universe") 13.8 billion years ago, but since inflation doesn't end everywhere at the same time, expansion is still perpetually happening in other regions. Thus, we have what is known as "eternal inflation" with other regions or "universes" such as ours constantly being formed. We don't observe these other regions because all such regions are driven away from each other by the continual inflationary expansion, thereby making room for more and more universes to form. Our universe is just one of the many, and the other universes will have different initial conditions, different values for constants, and perhaps different dynamical laws. So, the claim is that cosmic inflation strongly suggests eternal inflation.

If cosmic inflation explains so much, then why the skepticism on the part of some? The first thing to note is that there is absolutely no consensus explanation for why or how cosmic inflation happened in the first place. Keep in mind that cosmic inflation is supposed to solve several cosmological fine-tuning problems. The form of any fine-tuning problem is to explain why some initial condition at the Big Bang exists, why the value of some constant exists, etc., that were it be to be even slightly different, the universe would also be very different now. So whatever the explanation for cosmic inflation is, it had itself better discharge the mystery without creating any new fine-tuning problems or other mysteries. The worry is that so far no such explanation is forthcoming (see chapter 3 for more). Another worry is that the argument from cosmic inflation to eternal inflation has questionable assumptions [Woo, 2017].

Yet another issue raised by early cosmic inflation pioneer Paul Steinhardt and others is that any theory/model such as the multiverse that explains anything and everything explains nothing:

It's a breakdown in the theory. It's as if someone came to you with a theory for why the sky is blue that at first seems plausible, but after some refinements, it produces not just a blue sky, but a purple sky, a polka-dotted sky—you name it. What does it mean to predict something if it predicts everything?

[Steinhardt, as quoted in Woo, 2017]

Rather than multiverse, Steinhardt uses the term "multi-mess." In other words, if a theory predicts everything, it can't be falsified. We will say more on this in chapter 3.

Let us move on to string theory, a theory that suggests a way to unify QFT and GR, though even that claim is debatable. The theory is named for the idea that the different particles of physics can be modeled as the variant behavior of one-dimensional strings and their interactions; for example, the different properties of various particles are determined by the different frequencies strings might take on. More recently the theory has moved beyond strings to include two-dimensional sheets or three-dimensional "branes." Strings can attach at one or both ends of these sheets or branes. Regardless, string theory requires the addition of several other extra spatial dimensions. When the theory focuses on one-dimensional strings, the extra dimensions are quite small and rolled up like tubes. When the theory invokes three-dimensional branes, the extra dimensions are large and the branes (each of which is presumably a universe such as our own) float in said extra spatial dimensions. The geometry or shape of these extra spatial dimensions (i.e., their compactification) is not uniquely determined by the theory. Each different geometry corresponds to different values for constants and perhaps even differences in local dynamical laws. Indeed, each compactification leads to a different vacuum state. At least one state is said to describe our universe in its entirety. The huge number of solutions (10^{500} possible compactifications) at last count, without any perturbative mechanism to select among them, once again raises the worry about a scientific theory failing to make unique and falsifiable predictions. Furthermore, while the theory is perturbatively finite order by order, the perturbation series does not seem to converge [Greene, 2011].

How does all this get us to a multiverse of a sort? Tegmark gives us the answer:

> It's quite common for mathematical equations to have multiple solutions, and as long as the fundamental equations describing our reality do, then eternal inflation generically creates huge regions of space that physically realize each of these solutions. For example, the equations governing water molecules, which have nothing to do with string theory, permit the three solutions corresponding to steam, liquid water and ice, and if space itself can similarly exist in different phases, inflation will tend to realize them all.
>
> [Vilenkin and Tegmark, 2011]

Of course, it is debatable that either eternal inflation or string theory is well motivated or has any confirmation whatsoever, but if we accept that they are, and that at least in concert they both entail a multiverse of a sort, then obviously we no longer have the burden of resolving every fine-tuning problem in physics by deriving values uniquely; probability does all the work for us in the multiverse. Keeping our eye on the ball, among other things, we want an explanation of the Big Bang, the fundamental dynamical laws, the values of constants, and time as experienced. If the multiverse is supposed to be the answer, then of course one would have to believe that string theory and eternal inflation are independently well motivated to begin with and also believe they really do strongly imply or entail a multiverse, all of which has been questioned by Smolin and several others [Smolin, 2006]. Here is what Sean Carroll, a relatively optimistic proponent, says about inflation and string theory:

> Together, inflation and string theory can plausibly bring the multiverse to life. We don't need to postulate a multiverse as part of our ultimate physical theory; we postulate string theory and inflation, both of which are simple, robust ideas that were invented for independent reasons, and we get a multiverse for free. Both inflation and string theory are, at present, entirely speculative ideas; we have no direct empirical evidence that they are correct. . . . What we can say with confidence is that if we get a multiverse in this way, any worries about fine-tuning and the existence of life evaporate.
>
> [Carroll, 2016, p. 309]

Of course, as Carroll suggests, the validity of string theory is ultimately an empirical question and again we are not going to spend much time in this book disparaging dynamical alternatives. However, as Carroll notes, when it comes to explanatory power and confirmation, string theory is at least as bad off as eternal inflation, and may be beyond confirmation or disconfirmation for human beings *in perpetuity* because of the length and times scales involved.

Although we do come down on the side of people who say that scientific theories ought to be falsifiable in principle if not in practice, we have no innate prejudice against either inflation or string theory, or their presumed multiverse implications. Nothing in this book rules out every possible road to a multiverse, we just share in the view that it is not possible to confirm or deny its existence at this time, and explains everything and therefore nothing. We acknowledge that many theorists have suggested various ways at least in principle to test inflation and string theory, as well as the multiverse implications of one sort or another. We don't rule out the possibility that any of this could ever happen. We simply want to establish that the state of play in physics warrants attempting to construct a unifying theory that does make unique predictions, solves multiple outstanding problems, and is falsifiable. In short, we believe we can explain many of the current mysteries in theoretical physics such as certain fine-tuning problems in ways that are falsifiable in principle and don't require a multiverse. As Greene notes:

> Where you come down on the multiverse also depends on your view of science's core mandate. General summaries often emphasize that science is about finding regularities in the workings of the universe, explaining how the regularities both illuminate and reflect underlying laws of nature, and testing the purported laws by making predictions that can be verified or refuted through further experiment and observation.
>
> [Greene, 2011, p. 165]

Let us just testify now that we think this is exactly what separates the scientific enterprise from a *purely* metaphysical one. Any purportedly scientific endeavor that abandons these ideals in principle just simply fails to be science in our book. We are not advocating for strict Popperian "falsifiability" here, and of course we acknowledge that when it comes to the discovery of new theories and paradigms, that initially one need not focus on theory confirmation. For example, Carroll says:

> Modern physics stretches into realms far removed from everyday experience, and sometimes the connection to experiment becomes tenuous at best. String theory and other approaches to quantum gravity involve phenomena that are likely to manifest themselves only at energies enormously higher than anything we have access to here on Earth. The cosmological multiverse and the many-worlds interpretation of quantum mechanics posit other realms that are impossible for us to access directly. Some scientists, leaning on Popper, have suggested that these theories are non-scientific because they are not falsifiable.
>
> The truth is the opposite. Whether or not we can observe them directly, the entities involved in these theories are either real or they are not. Refusing to contemplate their possible existence on the grounds of some a priori principle, even though they might play a crucial role in how the world works, is as non-scientific as it gets.
>
> [Carroll, 2014]

We agree with Carroll in principle, but in the long run there should be serious concern about justifying belief in a theory that goes beyond mere coherence, unity, or the resolution of anomalies. We think discussions in this arena often equivocate and gloss over importantly different claims. For starters we agree that the mere fact some hypothesis of science such as the graviton is unobservable or undetectable by us in practice at the present moment is not alone reason to doubt its existence. After all, we posited the double helix of DNA well before we could detect it. Given the predictive success of the Standard Model of particle physics overall, the existence of the graviton is a reasonable bet. But positing unobservable universes based on highly speculative theories like eternal inflation and string theory, neither of which has anything remotely like the record of the Standard Model, is a whole different ball of wax. For example, famous philosophers Gottfried Leibniz in the seventeenth century and David Lewis in the twentieth century posited the existence of "possible worlds" other than our own in order to solve various metaphysical problems. Were they doing science? We say, as would most, clearly "no." Such speculations are the Paris metre rod of pure metaphysics. We would say the current multiverse craze in theoretical physics must eventually get to confirmation if it is to fare any better and make claims to being science.

Finally, we think Carroll's remarks bring out an important fact. Namely, it is highly unlikely that theoretical physicists would be defending the multiverse as a possible solution to their explanatory gaps if they hadn't been hitting the wall for decades in virtually every endeavor. The two biggest empirical triumphs, the confirmation of the Higgs field and gravitational waves, were predicted several decades ago. Theoretical physics has not made any advances of this sort for many decades. We think this is telling; at the very least, it's sufficient reason to try and look at all the problems in a different light.

So we think it is ironic then that while physics is arguably becoming more like metaphysics in a variety of ways, metaphysics and philosophy of science are becoming much more science-centric, including the focus on QM and QG as witnessed by philosophers of physics such as Nick Huggett and Christian Wüthrich who focus on QG [Huggett et al., 2013]. But, if fundamental theoretical physics is stuck in the practice of metaphysics, relatively speaking, then what can philosophy possibly do to help physics become more like science again? After all, if fundamental physics is largely just more sophisticated metaphysics with a few more empirical constraints thrown in and philosophy is supposed to inherit its legitimacy from being "the handmaiden of science," as Quine suggests, then obviously there is a problem here for both parties. In the words of science writer Jim Baggott, "the issue, then, is not metaphysics per se. The issue is that in fairy-tale physics the metaphysics is *all there is*. Until and unless it can predict something that can be tested by reference to empirical facts, concerning quantity or number, it is nothing but sophistry and illusion" [Baggott, 2013, p. 287].

Of course, at a minimum, one thing that philosophy can do is play the role of physics watchdog, the role that Baggott wishes it would play more often. Take the recent volley between Lawrence Krauss and David Albert and the other crossfire it spawned [Holt, 2012]. It was a wrangle over Krauss' claim that physics was now in a position to answer the ancient question long asked by philosophy, theology, and religion: why is there something rather nothing, or at the very least, how did the universe as we know it come into being from its non-being [Krauss, 2012]? Krauss argues that, given the laws of QFT and the existence of quantum fields themselves in a vacuum state, the theory predicts there could be a primordial quantum fluctuation that gives rise to the Big Bang. This probably would not have been anything more than a dust-up over the meaning of the word "nothing," except Krauss' rhetoric suggested that science was now in a position to dismiss philosophy fully with respect to these ancient questions. Albert made it very clear that such a claim is only plausible if by "nothing" we mean the ground state of quantum fields [Albert, 2012]. The point is that Krauss could not explain either the existence of the laws of QFT or the quantum fields themselves, and these aren't "nothing," obviously.

Other physicists such as Feynman [Storage, 2013], Weinberg [Kefalis, 2015], and Hawking [Warman, 2011] have been saying disparaging things about philosophers for a long time. In fairness to theoretical physicists we should note there are many physicists such as Carroll, Smolin, and Rovelli who do seek out and welcome the help of philosophers of physics. We should also note that there are many brilliant supporters of string theory and the like, and they have cogent rebuttals to many of the preceding concerns. However, when theoretical physicists are leaning on eternal inflation, string theory, and the multiverse, we will say it takes a lot of chutzpah to dismiss philosophy and metaphysics as useless and scholastic. However, this book is not about further disparaging string theory; we merely want to establish that the state of play in foundational physics and foundations of physics warrants seeking out new alternatives based on empiricism and falsifiability.

Going beyond the role of watchdog, we want to advocate for an even stronger claim however. It is time for philosophers of physics and theoretical physicists to do more than just play with current models of quantum gravity (QG). It is time for us, in the spirit of natural philosophy, to collaborate and create alternative models of QG. Per Rovelli:

> As a physicist involved in this effort, I wish that the philosophers who are interested in the scientific description of the world would not confine themselves to commenting and polishing the present fragmentary physical theories, but would take the risk of trying to look ahead.

> [Rovelli, 1999, p. 228–29]

This "looking ahead" not only provides much-needed competition, new blood, and the exploration of a larger possibility space, but hopefully it will generate superior alternatives. In this book, a philosopher, a mathematician, and a physicist hope to provide such an alternative; an alternative to the dynamical universe that we believe has led to the impasse in theoretical physics and foundations of physics. In our model, interpretational issues, formal issues, and theoretical physics are considered collectively, and all inform each other. Our alternative approach to quantum gravity (QG) and unification is falsifiably based on "reference to empirical facts" and resolves the interpretational issues of quantum mechanics (QM) and quantum field theory (QFT), as well as plugging the nagging explanatory gaps regarding the origin of the universe, the low entropy problem, the flatness problem, the horizon problem, closed timelike curves, dark matter, dark energy, and the black hole firewall paradox.

Perhaps physics doesn't just need more watchdogs, interpreters, or even just more experiments to constrain theory; perhaps it also needs alternative theoretical approaches as well. However, we truly hope that even if our specific alternative model for QG is "a washout," the methodology for how to proceed and how to collaborate will catch on. Our model includes not only interdisciplinary interaction but also an attempt to provide a complete ontological and explanatory system that subsumes foundational physics, the foundations of physics, and experience as it relates to time and change. In order to further motivate our particular approach to QG still further, let's look at the dynamical reasoning that led to the current impasse.

1.3   How did we get here?

From very early on, Western thinkers have generally assumed that everything can be explained. Perhaps the cosmological argument for the existence of God is the classic example of such thinking. In that argument Leibniz appeals to a version of the principle of sufficient reason (PSR) which states: "no fact can be real or existing and no statement true without a sufficient reason for its being so and not otherwise" [Leibniz, 1991, §32]. Leibniz uses the principle to argue that the sufficient reason for the "series of things comprehended in the universe of creatures must exist outside this series of contingencies and is found in a necessary being that we call God" [Leibniz, 1991, §36]. While physics dispensed with appeals to God at some point it did not jettison PSR, merely replacing God with fundamental dynamical laws such as those of some future Theory of Everything (TOE) and initial conditions (the Big Bang or some condition leading to it). Furthermore, at least in the theories/models of physics alleging to be relatively fundamental (relativistic quantum field theory, M-theory, loop quantum gravity, causal sets, etc.), the assumption is generally maintained that the fundamental explanation will be a dynamical one; for example, in the case of QFT particles interacting via forces or dynamical laws evolving states of the system. Such theories may deviate from the "manifest image" by employing radical new fundamental entities (wave functions, strings, loops, branes, etc.) evolving in counterintuitive "spaces" (Hilbert, Calabi–Yau, Fock, etc.), but the basic game is always dynamical; it is a story involving boundary conditions plus some system whose state is to be evolved in time, as in the Sean Carroll quotation earlier. In keeping with everyday experience, a very early assumption of Western physics—reaching its apotheosis with Newtonian mechanics—is that the fundamental phenomena in need of explanation are motion and change in time and such explanation will involve dynamical laws.

In the quest for QG and unification, it is the combination of the principle of sufficient reason (PSR) plus the dynamical perspective writ large (dynamical explanation) that has in great part motivated the particular kind of QG or unification being sought. This has led physicists to search for, "a unique final theory of everything, a theory that permits only one universe [as opposed to a multiverse] . . . a unique theory describing a unique world, with all laws and parameters completely fixed by the theory" [Davies, 2006, p. 233]. Such a theory, in order to be deemed a winner, would have to not only allow an in principle *ab initio* derivation of all other physical theories and phenomena but also put an end to most of the fundamental questions such as: why those fundamental laws? Why the Big Bang (initial conditions)? Why those values for the constants? Why that number of spatial and temporal dimensions, etc? The quest for a TOE so envisioned is certainly an offspring of PSR. This is obviously quite a tall order, so we ask, is such an explanatory feat possible? Let us explore that question in a little more depth.

The sentiments expressed by Salmon [1989, p. 181], Rescher [1970, pp. 13–14], and others make clear what is expected of fundamental physics (a TOE) if it is to be truly explanatory. That is, the game of fundamental physics is to provide the answers to why (as opposed to merely how), and this means showing that the fundamental laws, etc., were truly necessitated, meaning that they (and all the other fundamental facts that logically entail everything else) *could not have been otherwise* in some strong metaphysical or logical sense. Of course, this is just a modern-day echo of PSR to which Michael Redhead skeptically responds:

At some stage scientific explanations always turn into descriptions—"that's how it is folks"—there is no ultimate terminus in science for the awkward child who persists in asking why! I do not believe the aim of some self-vindicating *a priori* foundation for science is a credible one.

[Redhead, 1990, p. 152]

The worry is, once we get down to the initial conditions at the Big Bang and whatever the fundamental dynamical laws are supposed to be in our theory of QG, unless they are "self-vindicating," how can those bedrock facts themselves be explained at all, let alone explained dynamically? In response to the failure of theoretical physics to generate this unique final theory, we have seen various theorists embrace one or another version of the multiverse in which there is more than one universe, or many universes in one, each with different initial conditions, values of constants, and perhaps even different dynamical laws. Multiversers claim that one or another theory such as string theory or inflationary cosmology entails a multiverse, but such claims are hotly debated. This is of course in part what leads to the aforementioned worry about falsifiability and making testable predictions.

Some of the other reactions to the lacunae of a unique final theory are even more puzzling. For example, Stephen Hawking advocates "perspectivalism" in his book *The Grand Design* [Hawking and Mlodinow, 2010], and Craig Callender writes:

This radical theory holds that there doesn't exist, even in principle, a single comprehensive theory of the universe. Instead, science offers many incomplete windows onto a common reality, one no more "true" than another. In the authors' hands this position bleeds into an alarming anti-realism: not only does science fail to provide a single description of reality, they say, there is no theory-independent reality at all.

[Callender, 2010b]

This idea was panned by John Horgan in *Scientific American* with the title "Cosmic Clowning: Stephen Hawking's 'new' theory of everything is the same old CRAP" [Horgan, 2010]. Again, it takes a lot of chutzpah simultaneously to make such claims while dismissing and disparaging philosophers.

Note that Leibniz does not have this explanatory problem because he has a necessary being as a first cause (God) at the base of his axiomatic system and a necessary being requires no explanation, dynamical or otherwise, by definition. In any endeavor of axiomatic unification, how can one but fail to explain the fundamental axioms (the most brute facts) unless you could somehow show that said axioms really could not have been otherwise in some strong metaphysical or logical sense? Good luck! Note also that Leibniz takes it for granted that modulo God, the explanation of everything else, is dynamical or causal [Leibniz, 1991, §36]. Again, the point is that the combination of PSR plus the idea that dynamical explanation is fundamental creates real problems for theoretical physics. As we will make clear, in our approach, we avoid the infinite regress of dynamical explanation via a form of self-referentiality based on self-consistency (explained in chapters 3 and 4). In attempting to provide a self-vindicating theory, our approach goes beyond mere "consistency" as rejected by Redhead, it is *self-consistency writ large* provided by a global adynamical organizing principle for the block universe.

As we said, various concerns about time within physics and with regard to the experience of time have taken center stage. Furthermore, these various explanatory worries about time are deeply related to our previous worries about dynamical laws and explanation, as we will see shortly. The overriding concern here is that time as characterized in various physical theories seems to conflict with or, at the very least, fails to explain/underwrite one or another feature of time as experienced. Think again of relativity theory and its seeming block universe implication and the lack of an objectively distinguished present moment or direction in time (see chapter 2). Another such conflict is that all our dynamical laws are time-symmetric and yet, the universe appears to unfold in a very time-asymmetric way. Here is how Sean Carroll puts it:

> The existence of the arrow of time is both a profound feature of the physical universe and a pervasive ingredient of our everyday lives. It's a bit embarrassing, frankly, that with all the progress made by modern physics and cosmology, we still don't have a final answer for why the universe exhibits such a profound asymmetry in time.
>
> [Carroll, 2010a, p. 25]

Carroll is making reference to the temporal symmetry of dynamical laws versus the temporal asymmetry of everyday experience also exhibited by various physical arrows of time, such as that of thermodynamics, which is presumably less fundamental than either classical mechanics or quantum mechanics. Obviously, if time in some appropriate sense isn't fundamental, then the problems alluded to here become that much more difficult, because we then only have the current resources of physics at our disposal to explain why we experience time as having an objectively distinguished passage, an objectively distinguished direction, and an objectively distinguished present moment. Yet, none of those resources seem up to the task; rather they deepen the mystery of time as experienced.

When we get to general relativity (GR) proper and certain accounts of quantum gravity (QG), such concerns about the gap between time as experienced and time in physics only get worse. This is known as the "problem of time:"

> The "problem of time" in quantum gravity is neither a single problem, nor exclusive to quantum gravity—instead, it is a cluster of problems, at least one of which (the "problem of change," discussed below) arises even in classical GR, when it is linked to the interpretation of gauge invariance. Other versions of the problem stem from the disparate way in which time is treated in GR compared to quantum theory. Time, according to quantum theory, is external to the system being studied: it is *fixed* in the sense that it is specified from the outset and is the same in all models of the theory. In GR, however, time forms part of what is being described by the theory. Time in GR is subject to dynamical evolution, and is not 'given once and for all' in the sense that

it is the same across all models [Butterfield and Isham, 1999]. Reconciling these two treatments of time is not possible, and any attempt at formulating a quantum theory of gravity will thus face the problem of time in some form or another.

[Crowther, 2016, p.16]

Essentially, we're talking about time as a coordinate in our physics theory versus time as the experience of change. Saint-Ours reports the following exchange [Saint-Ours, 2015]:

DeWitt: You want a "time." You want to see something evolve.

Kuchař: I do not *want* to see things evolving. I *see* things evolving and I want to *explain* why I see them evolving.

In quantum mechanics (QM), the time coordinate is that of Newtonian mechanics which refers to the unambiguous measurement of time between events, that is, the time measured between events is the same for all observers regardless of their relative motion. This coordinate time mirrors the change that all observers experience "together" and is "external" to the system being measured. This would presumably satisfy Kuchař. In chapter 2 we will see that observers moving relative to each other will actually measure different times between the same events per special relativity (SR). This is the coordinate time of quantum field theory (QFT). While the coordinate time between events in SR differs quantitatively from that in Newtonian mechanics, it still mirrors the change that a set of observers at rest with respect to each other experience "together" and is "external" to the system being measured. Thus, the coordinate time of SR would presumably also satisfy Kuchař. Following QM and QFT, one would assume that a quantum theory of gravity would use the coordinate time from GR. However, space and time aren't "external" to the system being measured in GR, they are part of the system proper. And, since GR is invariant with respect to general coordinate transformations ("generally covariant") its coordinates don't necessarily represent measurement outcomes nor do they necessarily mirror change as experienced by observers. Thus, the time coordinate of GR doesn't necessarily satisfy Kuchař. But, one can use the GR spacetime coordinates to calculate a "proper time" along the worldlines of observers that does provide a measurement of time between events and does mirror change as experienced, at least by individual observers. This would presumably satisfy Kuchař.

Unfortunately, the standard quantization procedure used for QM and QFT requires some sort of globalization, some notion of spatiality, and that means a collection of observers (hypothetical or actual) who agree on proper time, so their synchronized clocks readings could be used to define spatial hypersurfaces. There are certainly GR solutions that accommodate this requirement such as the standard cosmology solutions where a collection of observers who agree on proper time is the set of all observers who measure a homogeneous and isotropic matter–energy distribution (as we will detail in chapter 3). Such a solution might be approximately realized in our universe for those who are rest with respect to the cosmic background radiation. The proper time in these solutions is typically used as a coordinate time and has spawned the QG approach called "causal dynamical triangulation"(see chapter 6). This QG program would presumably satisfy Kuchař.

The complaint from some about such an approach to QG is that it places undue emphasis on a particular coordinate system in a theory most widely lauded for its general covariance. Thus, the "canonical" approach to quantizing gravity borrows from a more general Hamiltonian formulation of GR called the "Arnowitt–Deser–Misner (ADM) formulation." In the ADM formulation of GR, one carves the GR spacetime into spatial hypersurfaces that are time-evolved per general coordinates so as to obtain a Hamiltonian formulation of GR. Of course, you might expect that this approach will lead to a problem precisely because GR is invariant under general coordinate transformations and you would be right. In this formulation, Einstein's equations say the Hamiltonian H = 0. Promoting H to an operator and plugging this into Schrödinger's equation as in ordinary quantization gives $\hat{H} \mid \Psi\rangle = 0$, that is, the Wheeler–DeWitt equation, so that $\frac{\partial \Psi}{\partial t} = 0$. Apparently, both coordinate time and any hope for including a representation of the experience of change in QG have disappeared. [This is an oversimplified presentation of "the problem of time": for more substantial overviews see Isham, 1992; Saint-Ours, 2015; Crowther, 2016; Weinstein and Rickles, 2016].

Some people such as Barbour [1999] just accept that what the frozen formalism is telling us is true, namely, that reality is timeless and changeless. His approach to QG is based on the timeless Wheeler–Dewitt equation $\hat{H} \mid \Psi\rangle = 0$ void of time evolution. Barbour takes this picture seriously and argues that configuration space is fundamental and each point in that space contains a 3D snapshot of some configuration of the universe. Here is how Smolin describes Barbour's view: "That's all there is. There's one quantum universe, described by one quantum state. That universe consists of a very large collection of moments. Some are more common than others. Some, indeed, are vastly more common than others" [Smolin, 2013, pp. 85–6].

Of course, one needn't accept this interpretation of the Wheeler–DeWitt equation or even accept the canonical quantization method itself, as there are other approaches to QG (chapter 6). Asher Peres notes, for example, "the problem of time" can be avoided in the quantization of GR "provided that one degree of freedom is kept classical,[20] so that it can be used as a clock" [Peres, 1997, p. 3]. In particular we show you how our Relational Blockworld approach resolves the "the problem of time" in chapter 6, and it is related to Pere's point.

Though Smolin does not put it this way, there is a certain irony in the fact that physics, historically driven both by PSR and dynamism, keeps being led to the block universe of SR or perhaps even worse, its GR–QM extension leading to "frozen time." Again, change is merely a redundancy of the representation in

---

[20] Classical physics is the non-quantum physics: general relativity, Newtonian mechanics, and Maxwell's equations. When we use the term "classical context," "classical reality," or "classical spacetime" we are referring to the model of objective reality obtained using classical physics.

GR because time evolution is a gauge motion, and hence does not correspond to any change at all in the "physical" degrees of freedom in the theory.[21] As Earman puts it:

> Taken at face value, the gauge interpretation of GR implies a truly frozen universe: not just the 'block universe' that philosophers endlessly carp about—that is, a universe stripped of A-Series change or shifting 'nowness'—but a universe stripped of its B-Series change in that no genuine physical magnitude (= gauge invariant quantity) changes its value with time.
>
> [Earman, 2003, p. 152]

As Barry Dainton notes, Earman's value judgment in this regard is open to debate:

> According to Belot and Earman, the doctrine that time is not physically fundamental is stronger and more radical than the doctrine that 'temporal becoming' is unreal. This is questionable. As far as departures from our ordinary ways of thinking are concerned, it is passage (or 'becoming') that matters: the (metaphysical) thesis that time does not pass, that all moments are equally real, is truly astonishing when first encountered; almost impossible to believe.
>
> [Dainton, 2001, p. 43]

We think Earman has a point here in that the Wheeler–DeWitt equation does not describe any temporal evolution of the quantum state whatsoever, hence his remark about the B-series. So perhaps in terms of physical arrows of time, Wheeler–DeWitt is more troublesome and counterintuitive. For example, what is the correct interpretation of probability in the context of a wave function for the universe that does not evolve. What could universal timeless probabilities possibly mean? However, with regard to the various facets of the experiential arrow of time, we don't see that Wheeler–DeWitt is any more problematic than the standard block universe model. That is, we think these two models are experientially equivalent in that nothing in experience would tell you if you were living in one of these worlds as opposed to the other. Therefore, with respect to the worry about time in fundamental physics versus time as experienced, we think both issues must be addressed.

Lee Smolin goes in the opposite direction about time; his book, *Time Reborn*, advocates an approach to QG in which time (in some sense) is fundamental (see chapter 6 for details):

> This picture emerges from a class of approaches to quantum gravity in which space is not taken as fundamental, but time is. These approaches posit a fundamental quantum structure—one that doesn't need space to define it. The idea is that space emerges, just as thermodynamics emerges from the physics of atoms. Such approaches are background-independent, because they do not assume the existence of a fixed-background geometry. Rather the primitive notion is of a graph or network,

---

[21] Gauge invariance also reflects relational structure, as we will show in chapter 4.

defined intrinsically, without reference to space. . . . By taking time as fundamental, these approaches differ from older background-independent approaches positing that spacetime—all together, as in the block universe—must emerge from a more fundamental description in which neither space nor time is primitive. These include loop quantum gravity, causal sets, and some other approaches to string theory.

[Smolin, 2013, p. 178]

His motivations for this view are first to explain time as experienced. He assumes that if the appropriate physical notion of time is somehow fundamental in physics, then we will have an explanation for time as experienced. Second, he believes that if time is fundamental, even more fundamental than laws or constants that can therefore change over time, then we can solve the explanatory gaps of the dynamical paradigm when applied to the universe as a whole. He believes many of these things can be easily explained if some sort of global time and change are fundamental to physical reality. Though he does not frame it this way, Smolin is advocating for a fundamentally neo-Whiteheadian process conception of reality.

As Smolin sees it, if we take seriously that the universe as a whole must be explained as the ultimate system in a complete cosmology, and we assume the standard dichotomy between initial/boundary conditions and eternal, universal dynamical laws, and we assume that time is either an illusion or at best emergent, then we are left with the impossible task of explaining why the fundamental facts could not have been otherwise. Smolin believes that "the result is that a coherent spacetime emerges only in models in which [a global] time is assumed to be real" [Smolin, 2013, p. 189]. As a result of these and other considerations, Smolin concludes that "space may be an illusion, but time must be real" [Smolin, 2013, p. 192]. Again, he goes even further to claim that time and change are more fundamental than dynamical laws in that the laws and the values of constants can change over time. Now you can see one reason why the worries about dynamical explanation in the mechanical universe when applied to the universe as a whole and the worries about the nature and explanation of time are deeply connected.

In Smolin's diagnosis [2013], it is precisely PSR plus the assumption that dynamical explanation is fundamental when applied to the entire universe that leads to the troubles of string theory. String theory initially adopts these assumptions then fails to uniquely constrain the geometry of the various dimensions uniquely, the values for various constants, etc. Then, string theorists attempt to turn this vice into a virtue with the multiverse interpretation which lead to skepticism on the part of many, as we discussed earlier.

To be fair, many of the aforementioned worries about string theory or the multiverse are not unique to them, but are perhaps even worse in alternative accounts of QG:

All these approaches [to quantum gravity] have their share of problems and challenges; especially, each has at best remote and tenuous connection to experiment, and so there are only weak empirical constraints on theory construction and choice.

[Huggett et al., 2013, p. 244]

The confirmation and empirical constraint worry is a universal problem for all current accounts of QG, but regarding what Nick Huggett, Tiziana Vistarini, and Christian Wüthrich call the "truly iconoclastic" accounts of QG, such as causal set theory and causal dynamical triangulation, things are even worse.

> Members of this family tend to offer programmatic schemes rather than full-fledged theories. They gain in attraction as more conventional approaches [string theory and loop quantum gravity] fail to produce a complete and coherent quantum theory of gravity.
>
> [Huggett et al., 2013, p. 244]

We will discuss the state of various accounts of QG in chapter 6, but many of them, in addition to being perhaps beyond confirmation or disconfirmation in principle, are quite nascent in general and incomplete as Huggett suggests.

It must be said that while we share Smolin's skepticism about string theory in particular and the multiverse more generally, Smolin's neo-Heraclitean solution is just a multiverse in time rather than space or spacetime and is therefore no less of a cop out if one were hoping for that unique final theory. So while the multiverse gives us multiple universes or sub-universes with varying constants and perhaps laws, Smolin's universe gives us changing constants and perhaps laws on a single time-line that vary over time as the universe unfolds. Either way, it constitutes a rejection of the mechanical conception of laws and constants as universal and unchanging. Just like the multiverse in space or spacetime, Smolin's multiverse in time only pushes the fundamental WHY questions back one step. For example, will we need meta-laws to explain why the first-order laws and other fundamental facts change over time and so on ad infinitum? How can we call something a law or a constant if it changes in time? Again, this seems rather an abandonment of these essential mechanical notions.

Most theorists probably embrace neither Barbour's *timeless universe* nor Smolin's *time reborn*; however, they do embrace dynamism or the mechanical paradigm even when they profess to accept the block universe. For example, Maudlin pushes further than simply insisting on an asymmetry between the past and the future; he insists on a particular understanding of this asymmetry, as is revealed in the following.

> But the passage of time connotes more than just an intrinsic asymmetry: not just any asymmetry would produce passing. Space, for example, could contain some sort of intrinsic asymmetry, but that alone would not justify the claim that there is a 'passage of space' or that space passes. The passage of time underwrites claims about one state 'coming out of' or 'being produced from' another, while a generic spatial (or indeed a generic temporal) asymmetry would not underwrite such locutions.
>
> [Maudlin, 2006, p. 109–10]

What makes Maudlin's position interesting and novel is revealed in this passage; though he commits himself to eternalism,[22] he claims that there is something special about the "passing" of time that brute asymmetries alone cannot account for. Here is how Carroll characterizes both Smolin and Maudlin:

> There has, predictably, been some pushback. Tim Maudlin, a philosopher, and Lee Smolin, a physicist, have argued vociferously that time is real, and that the passage of time plays what we might call a generative role: It indeed brings the future into existence. They think of time as an active player rather than a mere bookkeeping device. Both researchers have been developing new mathematical tools and physical models to buttress their views. Maudlin's novel approach focuses on the topology of spacetime itself—how different points in the universe are sewn together. Whereas traditional topology uses regions of space as fundamental building blocks, Maudlin takes worldlines (paths of particles through time) as the most basic object. From there, time evolution seems like a central feature of physics. Smolin, in contrast, has suggested that the laws of physics themselves are evolving with time. We wouldn't notice this from moment to moment, but over cosmological time scales, the parameters we think of as fixed may eventually take on very different values.
>
> [Carroll, 2015]

Obviously, Maudlin's attempt to make the passage of time fundamental in some sense is much less radical than Smolin's. As we will see in chapter 2, Smolin advocates for presentism (only the global present is real) a global-moving Now that is a complete rejection of the block universe. Maudlin clearly wants dynamical laws and past events to *generate* or *bring about*, in some sense, future events. But since he accepts the block universe in which from a God's-eye view all events are "just there," however, the sense of production here must be very subtle. And indeed, as we will discuss at length in chapters 2, 7, and 8, it turns out Maudlin's real concern is over the direction of time.

So it is significant that one of the battle lines in foundational physics and QG is whether or not in some sense time is fundamental as Smolin and Maudlin allege, is an illusion as Barbour claims, or emerges from what is fundamental (i.e., the true theory of quantum gravity), as many others allege (see Carroll, 2016 for more details). The latter, "positing that spacetime—all together, as in the block universe—must emerge from a more fundamental description in which neither space nor time is primitive" [Smolin, 2013, p. 178]. Indeed, as Smolin says, the same battle line exists for the question of space. Different theories of QG start with or without space and/or time at bottom (either in the relativistic sense, Newtonian sense, or everyday phenomenological sense) and attempt to derive them.

Finally, aside from all the other issues, we should also remind the reader that none of the current theories of QG in any way resolve the interpretative issues of QM or QFT. And, as Sean Carroll notes about QFT,

---

[22] "Insofar as belief in the reality of the past and the future constitutes a belief in the 'block universe', I believe in the block universe" [Maudlin, 2006, p. 109].

We know the Core Theory [quantum field theory] isn't the final answer. It doesn't account for the dark matter that dominates the matter density of the universe, and neither does it describe black holes or what happened at the Big Bang. . . . Astrophysics needs more than the Core Theory, but our everyday experience is well within its domain of applicability.

[Carroll, 2016, p. 189]

The point is that, for all practical purposes, QFT is the Theory of Everything (TOE) for the physics of "everyday experience" and while it is a highly successful and predictive theory, it inherits all the interpretational worries of non-relativistic QM and adds a good deal more that are unique to it (more on that in chapter 5). It seems to us that a real contender for QG ought to at least shed light on these concerns or at the very least a good interpretation ought to have implications for accounts of QG.

1.4   God's-eye physics to the rescue

In this section, we primarily want to provide a conceptual introduction to our particular adynamical ontology and formalism called Relational Blockworld (RBW) [Stuckey et al., 2006, 2008, 2015; Silberstein et al., 2008, 2013]. As the following passage suggests, essentially, we need to start thinking like a heptapod:

Imagine that I am a God-like being who has decided to design and then create a logically consistent universe with laws of nature similar to those that obtain in our universe. . . . Since the universe will be of the block-variety I will have to create it as a whole: the beginning, middle and end will come into being together. . . . Well, assume that our universe is a static block, even if it never 'came into being', it nonetheless exists (timelessly) as a coherent whole, containing a globally consistent spread of events. At the weakest level, "consistency" here simply means that the laws of logic are obeyed, but in the case of universes like our own, where there are universe-wide laws of nature, the consistency constraint is stronger: everything that happens is in accord with the laws of nature. In saying that the consistency is "global" I mean that the different parts of the universe all have to fit smoothly together, rather like the pieces of a well-made mosaic or jigsaw puzzle.

[Dainton, 2001, p. 119]

The overriding assumption in physics is that dynamical explanation is fundamental, but taking the block universe seriously opens up the possibility that adynamical global constraints might be fundamental both for individual systems within the universe and for the universe as a whole. When we provide a dynamical explanation for why one event followed another, the answer is independent of future boundary conditions; indeed, it is independent of conditions anywhere else other than those of the 3D hypersurface in the immediate past. In contrast, an adynamical answer provides a global constraint for the system as a whole, given input anywhere in the past, present, and/or future of the event in question. As we will elaborate, our fundamental ontological entities are "spacetimesource elements" and they are 4D in nature. Furthermore, in our model the insight of Einstein and Minkowski that space and time cannot be separated is upheld even for foundational physics and is further extended to matter itself, hence the expression "spacetimesource."[23]

A change in the Western scientific worldview from dynamical explanation to adynamical explanation brings an end to the reign of the mechanical universe writ large. As we said, a change of this magnitude last occurred when Newton's mechanical universe replaced Aristotle's teleological universe. That many people equate "mechanical universe" (ontology) with "Newtonian mechanics" (theory) isn't wrong. Newtonian "mechanistic thinking" was carried into electromagnetism, SR, QM, GR, and QFT, as we elaborated earlier. That is exactly what we are claiming is responsible for the underlying impasse in theoretical physics and the foundations of physics. In hindsight, the mechanical universe ontology should have been left with Newtonian mechanics; it should not have been assumed in any of these later theories.

Richard DeWitt, among others in the foundations of physics community, foresaw just such a Kuhnian revolution resulting from the mysterious "Bell-like influences" of QM which we will detail in chapter 4:

> In the past, fundamental new discoveries have led to changes—including theoretical, technological, and conceptual changes—that could not even be imagined when the discoveries were first made. The discovery that we live in a universe that, deep down, allows for Bell-like influences strikes me as just such a fundamental, important new discovery. . . . If I am right about this, then we are living in a period that is in many ways like that of the early 1600s. At that time, new discoveries, such as those involving Galileo and the telescope, eventually led to an entirely new way of thinking about the sort of universe we live in. Today, at the very least, the discovery of Bell-like influences forces us to give up the Newtonian view that the universe is entirely a mechanistic universe. And I suspect this is only the tip of the iceberg, and that this discovery, like those in the 1600s, will lead to a quite different view of the sort of universe in which we live.
>
> [DeWitt, 2004, p. 304]

QM weirdness is just one example of the outstanding mysterious phenomena that we will explain in this book using RBW. Many subsequent chapters will detail the physics and ontology of our Relational Blockworld approach, but the primary goal of this chapter was to provide the background for what is to come and motivate the need for the paradigm shift and the approach we are recommending. We believe cracks in the NSU exist in both relativity and quantum physics. For example, in GR we have the puzzle of the origin of the universe, the flatness problem, the horizon problem, the low entropy problem, and the paradoxes of closed timelike curves (discussed in chapter 3). In QM, we have the measurement problem and

---

[23] The term "source" is used for the source or sink of particles in quantum field theory and it is meant to convey something similar here.

the mystery of quantum entanglement (discussed in chapter 4). And, in QFT we have numerous technical issues (discussed in chapter 5) and the black hole firewall paradox (discussed in chapter 6). We, on the other hand, propose to take seriously the possibility that the LSU might be fundamental. What we will argue in this book is that many of the problems in physics are the result of the ant's-eye dynamical bias of the NSU and that a new God's-eye adynamical paradigm per the LSU can eliminate those problems, as we alluded to earlier. So, in summary, foundational physics and foundations of physics are both at an impasse.

We believe the cause of the impasse in both camps is one and the same: universal adherence to dynamical explanation in the mechanical universe. Recall the light ray example above. In the NSU one might say "Wow, of all the directions the light ray could have been emitted, it was emitted in precisely the right direction to end up at its place of reception. That's almost miraculous!" Or one might ask, "Of all the colors that could have been emitted, why was the light emitted of that precise color?" Or, "Why was any light emitted at all?" In the NSU, explanation resides in time-evolved or narratable stories, and as Wilczek said, "things are what they are because they were what they were" in this storytelling form of explanation. Thus, the current situation is explained as the result of the situation immediately preceding it via the equations of the Newtonian Schema. Of course, one can then ask why the previous situation and the situation prior to that and so on until finally arriving at some beginning point asking "why were things that way and not any other?" However, if this beginning point has no prior "situation" and can't otherwise be explained, as with the beginning state of the universe (Big Bang), it is by definition inexplicable, and thus mysterious, in the NSU. Similarly, universal conditions at very early times, such as the flatness of space, the homogeneity of temperature, and the low entropy, make the NSU story of cosmology seem "miraculous" or "fine tuned."

In the spatiotemporally global perspective of the LSU, conditions at any point in spacetime are that way because of conditions at all the other points in spacetime. Explanation is globally self-referential; conditions are self-consistent per the adynamical global constraint or least action principle of the Lagrangian Schema. Thus, conditions at any point in spacetime are no more mysterious than at any other point, providing one can cook up a rigorous LSU explanation for the phenomena in question. We will discharge the cosmological fine-tuning problems in just such a manner in chapter 3. Thus, on this way of looking at explanation, asking "why *all* the points and conditions of spacetime?" is moving beyond empirical science and therefore beyond the realm of physics.

Another element of NSU storytelling is the dynamical "character" of the story, that is, a mediating object defined by its intrinsic properties moving through space as a function of time. In electromagnetism, this mediating object is the electromagnetic field and, as it turns out, a charged particle will both create an electromagnetic field and then interact with it. This self-interaction creates inconsistencies for the theory [Feynman, 1964] which ultimately led to attempts to describe electromagnetic interaction without mediating fields in a process called "direct action." Of course, without a mediating/causal/time-evolved entity, the

Newtonian Schema is undermined. However, the lack of a mediating entity is no problem for a least action principle, so direct action was naturally cast in the Lagrangian Schema.

In RBW, having taken seriously the possibility of the LSU as fundamental, we have come to realize there are many cases where direct action can resolve problems. While there are many viable NSU interpretations of QM, there are quantum phenomena that strongly suggest the possibility of direct action. As we will see in chapter 4, the existence of a time-evolved object moving from the emission event to the detection event in a quantum experiment can cause a problem because there are quantum experiments where the existence of unmeasured properties (definite properties that exist without interactions) seems to be forbidden. This lack of "counterfactual definiteness" makes the existence of a time-evolved quantum causal entity difficult to define and leads to many "quantum mysteries." So, at the fundamental level of objective reality, we will employ direct action and the Lagrangian Schema to connect a quantum emission event to its associated detection events thereby eliminating the offending assumptions leading to mysteries of quantum mechanics (QM), such as the measurement problem and quantum entanglement/nonlocality (chapter 4). For example, the reason for mysteriously correlated QM outcomes associated with entangled states is simply the adynamical global constraint relating emission and detection events. End of story.

This dramatically alters the view of quantum field theory (as explained in chapter 5) which changes our view of unification (as explained in chapter 6) and results in a deviation in the geometry of spacetime established per general relativity (GR). In chapter 6, we will use that deviation (our brand of quantum gravity) to fit the supernova distance moduli data without need of mysterious dark energy. We will also use the contextuality already inherent in GR to fit galactic rotation curves, X-ray cluster mass profiles, and the angular power spectrum of the cosmic microwave background without need of exotic new non-baryonic dark matter.

This lack of time-evolved quantum causal agents also changes the understanding of black hole evaporation from a dynamical story to an adynamical emission-detection couple. In the NSU version of black hole evaporation, one partner of a virtual particle pair is captured by the black hole while its partner escapes to infinity as "Hawking radiation." This leads to the so-called black hole firewall paradox (explained in chapter 6) whereby the escaping member of the pair must be entangled with its captured partner, but it must also be entangled with all previously emitted Hawking radiation to avoid the black hole information paradox. This dual entanglement is not allowed in quantum mechanics; thus, the black hole firewall paradox is widely considered to present the most severe example of the incompatibility of quantum physics and general relativity, so its resolution should strongly hint at quantum gravity, if not outright require it.[24]

---

[24] There is widespread disagreement as to whether or not all of the assumptions leading to this paradox are reasonable, but we're not here to argue for or against those assumptions. We are simply explaining the paradox as it follows from a particular dynamical perspective and resolving it with our adynamical approach.

We agree with this sentiment noting that the black hole firewall paradox provides a stark contrast between the dynamical NSU and the adynamical LSU, where it is a non-starter in our direct action version of black hole evaporation whereby Hawking radiation is viewed as a tunneling process (keeping in mind the use of direct action). Thus, in our approach to quantum gravity an emission event inside the black hole is linked to a detection event outside the black hole per tunneling without requiring entangled virtual particles (replaced by direct action) or subsequent entanglement between different detection events (replacing quantum particles escaping to infinity).

Regarding the tension between time as experienced and time as characterized in various theories of foundational physics, as counterintuitive as it sounds, we will show that our particular account of the block universe actually provides a dissolution of that tension and an explanation for the various facets of the phenomenology of time. In our attempt to heal the divide between time as experienced and time in foundational physics, we are going to reject all the views canvassed earlier. These views assume that:

1. There must be some notion of time in fundamental physics that explains or corresponds to key features of time as experienced.
2. Failing that, time as experienced must be explained as being in the brain or mind only.

We reject both of these assumptions. We can only "get away with this" because RBW allows us to, indeed forces us to, reconceive both the physical and the mental, and the relationship between the two. In short, the third option is neutral monism, which holds that so-called mental/subjective/internal experience and so-called physical/objective/external features of reality, such as those pertaining to time, are in fact not essentially different; they are non-dual, non-distinct features of a more fundamental neutral reality.

Thus, time as experienced is neither in the head (the subject) nor the external world (the object) as those are conceived in the Cartesian picture–temporal experience is fundamentally relational. It is the self-consistency relation or cut between subject and object in the field of the neutral that allows for the experience of time. This relation or structure is not *in* anything nor located any*where*; it is the ground for such concepts. Given neutral monism, the conscious subject/objective world are two sides of the same coin; therefore, the dynamical character of consciousness and the world are two sides of the same coin. Consciousness or subjectivity is not some utterly distinct internal process that imposes or projects time and change onto a static universe as if the mind were just some virtual reality machine that we are stuck behind and the "physical," "external" world some thing-in-itself that only physics manages to describe with any objectivity. No, in our view the very existence of the co-dependent, co-determining conscious minimal self/world-in-space-and-time unity is yet another adynamical global constraint, you simply cannot have one without the other. All of this is spelled out in detail in chapters 7 and 8.

To put it simply, the problem of time as experienced is a subset of the "hard problem" or "generation problem" of consciousness. The generation problem is this: if matter, whatever it is, is fundamental and it is inherently non-conscious or lacking in subjectivity, then how do dynamical material processes *generate* conscious experiences such as the ones related to time? Neutral monism in general rejects the fundamentally dualist assumption that fuels the generation problem and the RBW brand of neutral monism goes a step further and rejects the dynamical paradigm all together, and that includes conscious experience. Therefore the explanation for subjectivity in general and time as experienced in particular isn't ultimately going to be a dynamical or causal one as that is understood in the NSU or causal mechanistic paradigm.

To state the obvious on this view, natural science, conceived as being about an external physical world minus all conscious experience, is not explanatorily or ontologically fundamental. The world is really not spacetime plus matter with experience as a mysterious add-on, but actually the world is "spacetimesource-subjectivity." As Martin Heidegger said, "human life does not happen in time but rather is time itself" [Heidegger, 1925, p. 169].

As philosopher of physics Craig Callender notes, philosophers such as Merleau-Ponty would say that Smolin and others are misguided for exactly this reason:

> French philosopher Maurice Merleau-Ponty argued that time itself does not really flow and that its apparent flow is a product of our "surreptitiously putting into the river a witness of its course." That is, the tendency to believe time flows is a result of forgetting to put ourselves and our connections to the world into the picture. Merleau-Ponty was speaking of our subjective experience of time, and until recently no one ever guessed that objective time might itself be explained as a result of those connections. Time may exist only by breaking the world into subsystems and looking at what ties them together. In this picture, physical time emerges by virtue of our thinking ourselves as separate from everything else.
>
> [Callender, 2010a, p. 65]

As we will discuss in chapter 8, the suggestions made here about the explanation for time as experienced are not merely phenomenology, they also have grounding in the recent cognitive science of temporal experience. For example, the cognitive neuroscientist Marc Wittmann who studies time perception wrote the following in his recent book, *Felt Time*:

> Because I have a body, I perceive the passing of time.
>
> [Wittmann, 2016, p. 133]

> Temporal experience, self-consciousness, and the perception of bodily states and feelings are tightly bound to each other; they cannot be experienced separately.
>
> [Wittmann, 2016, p. 135]

Research on consciousness inevitably shows that our concepts of self, time and body are interrelated. Presence means becoming aware of a physical and psychic self that is temporally extended. To be self-conscious is to recognize oneself as something that persists through time and is embodied.

[Wittmann, 2016, p. 104]

Do not misunderstand us, we are not saying that every problem related to some physical aspect of time, such as the problem of time in quantum gravity, is resolved by neutral monism. This is the problem to which Callender is actually alluding, but as he suggests when he invokes Merleau-Ponty, the form of our solution to the problem of time and our explanation for time as experienced are quite similar. With regard to the latter, making a cut in the neutral "field of experience" between what is subject and object is the ultimate in "breaking the world into subsystems." The point that Wittmann is making is that being embodied agents/subjects in an environment is a concrete instantiation of the phenomenology of breaking the world into subsystems consisting of subject and object. Embodiment is part of the explanation for how and why this cut gets made, and that cut is behind the experience of time's passage. As we will learn, the bottom line is that the answer to the questions why is there experience (subjectivity) at all, and why is experience spatiotemporal (objectivity), is the same. Again, there is much to unpack here, and it all happens in chapters 7 and 8.

In summary, we hope that this book is a milestone toward making Wilczek's prediction come true:

I predict that, in 100 years, Weyl's vision—which, in essence, goes back to the Greek philosophers Parmenides and Plato—will be fully vindicated, as the fundamental laws will no longer admit arbitrary initial conditions. "What is" and "what happens" will be understood as inseparable aspects of a single, trans-temporal reality.

[Wilczek, 2015]

Since the adynamical explanation we propose employs a block universe, in chapter 2 we will next introduce the standard argument for the block universe from special relativity (SR). Philosophy of Physics for Chapter 2 has a novel variant on that argument designed to plug any loopholes, and many possible loopholes are considered therein. The reader who is happy to accept the block universe implication of SR for the sake of argument and is familiar with the timelike/spacelike structure of spacetime may skim chapter 2 and proceed to chapter 3.

Keep in mind what we told you in Overture for Ants, the main thread of each chapter provides the minimally essential content for proceeding with our argument. Overviews for Parts II and III explain what the minimal expectations for each chapter are and what information is provided in the main thread, the philosophy of physics thread, and the foundational physics thread for each chapter. Given the diversity of our intended audience, we leave it to each reader to decide what they need to take from each chapter.

Foundational Physics for Chapter 1

In the main thread, we talked about the Newtonian Schema and the Lagrangian Schema approaches to physics. We housed these approaches in the Newtonian Schema Universe (NSU) and Lagrangian Schema Universe (LSU), respectively, with their dynamical and adynamical explanatory methods, respectively. In this foundational physics thread, we will provide the Newtonian Schema and Lagrangian Schema for a rock breaking a window, alluded to in the main thread, in an effort to clarify the distinctions between the NSU with its Newtonian Schema physics and dynamical explanation and the LSU with its Lagrangian Schema physics and adynamical explanation. We refer to the 2D depiction of events in Figure 1.3 throughout.

Let us start with the familiar Newtonian Schema analysis. Newton's second law says the sum of the forces equals the time rate of change of momentum:

$$\sum_i \vec{F}_i = \frac{d\vec{p}}{dt} \tag{1.1}$$

With constant mass m, the time rate of change of the rock's momentum $\vec{p} = m\vec{v}$ is $\frac{d\vec{p}}{dt} = m\frac{d\vec{v}}{dt} = m\vec{a}$. This gives the familiar form of Newton's second law, $\vec{F} = m\vec{a}$. There is only one force acting here, that of gravity, so we have $F_x = 0$ and $F_y = -mg$ where g is the acceleration of gravity near the surface of Earth. Eq. 1.1 then gives us two differential equations, one in the x direction and one in the y direction, which must be solved:

$$F_x = 0 = m\frac{d^2x}{dt^2} \tag{1.2}$$

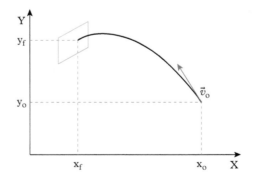

Figure 1.3 *The trajectory (black curve) of a rock thrown from (x_o, y_o) at time $t = 0$ with initial velocity $\vec{v}_o$ arriving at (x_f, y_f) in the window (green box) at time $t = t_f$.*

and

$$F_y = -mg = m\frac{d^2y}{dt^2} \tag{1.3}$$

The general solutions to these second-order differential equations are

$$x(t) = x_o + v_{x_o}t \tag{1.4}$$

and

$$y(t) = y_o + v_{y_o}t - \frac{1}{2}gt^2 \tag{1.5}$$

respectively. To create a particular solution for our specific situation, we must provide two pieces of information for each of Eqs. 1.4 and 1.5. We could, for example, specify the initial position and initial velocity, that is, the initial position with components $x(0)$ and $y(0)$ gives the otherwise arbitrary constants x_o and y_o, respectively, directly. And, the initial velocity with components $\frac{dx}{dt}(0)$ and $\frac{dy}{dt}(0)$ gives the otherwise arbitrary constants v_{x_o} and v_{y_o}, respectively, directly. And, the other hand, for example, we could provide the positions at two different times, that is, the initial position $(x(0),y(0))$ and the final position $(x(t_f),y(t_f))$, say. Either way, the dynamical explanation for why the window is broken would be along the lines of "The rock was thrown with just the right initial velocity from the thrower's hand location," or "Timmy threw a rock at it!" That is, the dynamical explanation of the NSU would ultimately reside in the initial position and initial velocity, given Newton's second law with $F_x = 0$ and $F_y = -mg$. Now let's look at the Lagrangian Schema analysis and adynamical explanation.

The action is

$$S = \int L(q,\dot{q})dt \tag{1.6}$$

and demanding that $\delta S = 0$ with fixed end points, that is, $\delta q = 0$, gives the Euler–Lagrange equation

$$\frac{d}{dt}\frac{\partial L}{\partial \dot{q}} - \frac{\partial L}{\partial q} = 0 \tag{1.7}$$

For us, the Lagrangian is

$$L = \frac{1}{2}mv^2 - V \tag{1.8}$$

where the potential $V = mgy$ with the path connecting the initial and final locations, that is, $(x(0),y(0))$ and $(x(t_f),y(t_f))$. Since we are working in more than 1D, we need the higher-dimensional Euler–Lagrange equations

$$\frac{d}{dt}\frac{\partial L}{\partial \dot{q}_i} - \frac{\partial L}{\partial q_i} = 0 \qquad (1.9)$$

one equation for each i. If we let $q_1 = x$ and $q_2 = y$, we have $\dot{q}_1 = v_x$ and $\dot{q}_2 = v_y$. Eqs. 1.8 and 1.9 then give

$$m\dot{v}_x = 0 \qquad (1.10)$$

and

$$m\dot{v}_y + mg = 0 \qquad (1.11)$$

which are exactly Eqs. 1.2 and 1.3, respectively, given $\dot{v}_x = \frac{d^2x}{dt^2}$ and $\dot{v}_y = \frac{d^2y}{dt^2}$. In general, the Euler–Lagrange equations reproduce Newton's equations because essentially the Euler–Lagrange equations relate force $m\vec{a}$ and the potential V, that is, $m\frac{d^2\vec{r}}{dt^2} = -\vec{\nabla}V$, just as Newton's laws. So, you're right back with the same differential equations, but the adynamical explanation is different. The adynamical explanation for the broken window is that the rock's trajectory is an extremum of the action for paths connecting $(x(0),y(0))$, and $(x(t_f),y(t_f))$. In other words, the adynamical explanation of the LSU would ultimately reside in the initial and final positions, given the adynamical global constraint $\delta S = 0$.

Finally, we want to remind you of Noether's theorem, that is, symmetries of the action result in conservation laws. In our example, we see that $x \to x + c$ with c an arbitrary constant does not change the action, since L contains only the derivative of x. In general, when L is not a function of q_i, Eq. 1.9 gives

$$\frac{d}{dt}\frac{\partial L}{\partial \dot{q}_i} = 0 \qquad (1.12)$$

which means

$$\frac{\partial L}{\partial \dot{q}_i} = p_i \qquad (1.13)$$

is a constant. In this case, $p_i = p_1 = mv_x$: momentum in the x direction is conserved. The Newtonian Schema version of this conservation law comes from Eq. 1.2 which can be written as

$$\frac{dp_x}{dt} = 0 \qquad (1.14)$$

Likewise, when the sum of torques in a particular direction is zero, angular momentum in that direction is conserved and the Lagrangian is rotationally invariant about that direction.

Part II

Adynamical Explanation: Physics

Part II Overview

Part II focuses on the mysteries of physics resulting from dynamical explanation in the mechanical universe and our resolutions thereof via adynamical explanation in the block universe. This includes our proposed adynamical approach to future physics. All the chapters of Part II are broken into three threads to serve the needs of our diverse audience. Keep in mind that the book is not "the main thread plus appendices." The threads serve entirely different purposes as we will describe below to guide the reader. When we talk about "the novice," we mean the advanced undergraduate in philosophy or physics who may not be familiar with the issues from foundational physics and foundations of physics that we are presenting. In most cases, these readers are familiar with the physics but not the concerns we raise, since most introductory physics textbooks ignore such issues. Thus, our presentation in each chapter is not a textbook presentation of the broader subject in which the concerns are housed.

In the main thread of chapter 2 we supply a short, conceptual introduction to special relativity and a typical argument from physics for block universe adapted from DeWitt [2004, pp. 213–19]. This main thread is written at the level of the novice and the physicist may complain that it is missing the computations, so those have been placed in Foundational Physics for Chapter 2. The philosopher may complain that the main thread argument ignores many assumptions and counterarguments, so those have been placed in Philosophy of Physics for Chapter 2. In subsequent chapters, the reader need only understand what is meant by a block universe with its spacelike versus timelike structure and that the block universe is strongly supported by physics and philosophy.

The main thread of chapter 3 introduces general relativity (GR), Big Bang cosmology, and closed timelike curves conceptually, with their associated puzzles, problems, and paradoxes. As with the main thread of chapter 2, this main thread is written at the level of the novice. The reader who wants to see the associated formalism is directed to Foundational Physics for Chapter 3. The philosophical nuances such as the status of the block universe argument in GR and debates about the Past Hypothesis have been placed in Philosophy of Physics for Chapter 3. In subsequent chapters, the reader need only appreciate how dynamical explanation in the mechanical universe is responsible for all these mysteries in GR and that adynamical explanation in the block universe resolves all these mysteries easily.

The main thread of chapter 4 introduces some of the major mysteries and interpretational issues of quantum mechanics (QM) at the level of the novice. These mysteries and issues include: "quantum superposition," "quantum nonlocality," "Bell's inequality," "entanglement," "delayed choice," "the measurement problem," and "no counterfactual definiteness." As with all other chapters in this book, chapter 4 is not intended as a textbook introduction to these concepts, so it is not written in textbook fashion. Instead, in the next section of the main thread we introduce these mysteries via experiments in the spirit of N.D. Mermin— some have actually been carried out while others are gedanken experiments in accord with QM predictions. We believe this is the simplest way to introduce these mysteries to the novice as they refer directly to empirical facts and require no QM formalism to explain. All the relevant formalism for this chapter is provided in Foundational Physics for Chapter 4. After introducing the major mysteries and interpretational issues of QM, we show how Relational Blockworld (RBW) resolves them. The metaphysical underpinnings of our interpretation of QM, such as "contextual emergence," "spatiotemporal ontological contextuality," "adynamical global constraints," etc., are provided in the philosophy of physics thread. Also in Philosophy of Physics for Chapter 4, we situate our RBW interpretation with respect to retrocausal accounts, and explain that RBW is a realist, psi-epistemic account of QM. In subsequent chapters, the reader need only understand that all of these mysteries and interpretational issues of QM result from dynamical explanation in the mechanical universe and are dispatched using our brand of adynamical explanation in the block universe.

A brief introduction to particle physics and quantum field theory (QFT) is presented in the main thread of chapter 5. The impasse of unification described in chapter 1 is historically reviewed in the main thread of chapter 5 showing that the dynamical paradigm pervades the development of particle physics and QFT. A new adynamical path for unification is therefore proposed based on RBW. The main thread contains some introductory physics, that is, simple harmonic motion with the associated matrix multiplication and a bit of relativistic kinematics, and a presentation and discussion of the Central Identity of Quantum Field Theory, so the novice may find this main thread a bit denser than the previous main threads. As with the main thread of chapters 2, 3, and 4, entire textbooks are written on this topic, so no single chapter could adequately cover this material and that is not our intent. Rather, we are simply trying to argue that dynamical explanation in the mechanical universe led to the impasse regarding unification in particle physics per QFT, and RBW's adynamical approach provides an entirely new view of unification. That and the basic idea of QFT are all the reader need understand from chapter 5 on. In Foundational Physics for Chapter 5, we show how classical field theory is related to QFT and introduce gauge fields per standard QFT. For the interested specialist, in Philosophy of Physics for Chapter 5 we use RBW to resolve interpretational issues of gauge invariance, gauge fixing, the Aharonov– Bohm effect, regularization, and renormalization, and we largely discharge the problems of Poincaré invariance in a graphical approach, covering inequivalent representations, and Haag's theorem.

Chapters 1–4 present a relatively simple, non-technical argument at the level of the novice in the main threads that dynamical explanation in the mechanical universe leads to many mysteries throughout physics that disappear using adynamical explanation in the block universe. Philosophical and mathematical details for the specialist and technically curious are placed in parallel threads in philosophy of physics and foundational physics, respectively, but our goal for the novice and specialist alike is the same: to convince them of the explanatory superiority of the Lagrangian Schema Universe (LSU) over the Newtonian Schema Universe (NSU), at least in these troublesome cases. Given the power of the LSU to resolve the mysteries of current physics, we began in chapter 5 by extrapolating it to unification and quantum gravity. We now say to the reader what Morpheus told Neo in *The Matrix*: "You take the blue pill, the story ends. You wake up in your bed and believe whatever you want to believe. You take the red pill, you stay in Wonderland, and I show you how deep the rabbit hole goes." The "blue pill" represents safe adherence to the NSU and the "red pill" represents a speculative foray into the LSU.

For the reader who takes the "red pill," the main thread of chapter 6 motivates the need for quantum gravity (QG), and introduces the RBW approach to QG, unification, dark matter, and dark energy with minimal formalism: the curve fits and details of our modified Regge calculus and modified lattice gauge theory approaches are displayed via graphs and figures. Nonetheless, the novice will certainly find this main thread to be denser than the others. Our advice to the reader is not to worry if you don't fully grasp our modified Regge calculus and modified lattice gauge theory. Just understand that buying into the LSU has the potential to provide a new path for future physics. For the reader who actually wants to understand how we fit astronomical data associated with dark matter and dark energy, an overview of the computational details has been placed in Foundational Physics for Chapter 6. In the philosophy of physics thread for chapter 6, RBW's taxonomic location with respect to other discrete approaches to QG is detailed, it is argued that the search for QG is stymied by the dynamical paradigm across the board, and it is argued that an adynamical global constraint as the basis for QG in the block universe provides a self-vindicating unification of physics. Again, the key message from chapter 6 is that the LSU provides an entirely new approach to future physics, just as Wilczek suggests.

While this concludes the physics covered in the book, the bravest and most curious reader will likely want to proceed all the way to the bottom of the RBW rabbit hole and check out RBW's approach to consciousness and time as experienced in Part III.

2

The Block Universe from Special Relativity

Why are these two things inconsistent with each other? I felt that I was fa-cing an extremely difficult problem. I suspected that Lorentz's ideas had to be modified somehow, but spent almost a year on fruitless thoughts. And I felt that this puzzle was not to be easily solved. But a friend of mine living in Bern (Switzerland) [Michele Besso] helped me by chance. One beautiful day, I visited him and said to him: 'I presently have a problem that I have been totally unable to solve. Today I have brought this 'struggle' with me.' We then had extensive discussions, and suddenly I realized the solution. The very next day, I visited him again and immediately said to him: 'Thanks to you, I have completely solved my problem.' My solution actually concerned the concept of time. Namely, time cannot be absolutely defined by itself, and there is an un-breakable connection between time and signal velocity. Using this idea, I could now resolve the great difficulty that I previously felt. After I had this inspir-ation, it took only five weeks to complete what is now known as the special theory of relativity.

Nobel Laureate Albert Einstein [Stachel, 2002, p. 185]

2.1 Introduction

Newtonian mechanics is based on absolute time and space, that is, that every-one measures the same time and distance between the same events, regardless of their relative motion. This means that Newtonian mechanics is invariant with respect to Galilean coordinate transformations and there is an unambigu-ous notion of absolute simultaneity; everyone agrees on the collection of events happening everywhere in space at any given time. In contrast, Maxwell's equa-tions of electromagnetism are invariant under Lorentz coordinate transformations whereby observers in motion relative to each other can measure a different time and distance between the same events (as we will see in this chapter). Accord-ingly, observers moving relative to each other disagree as to what events are happening simultaneously (relativity of simultaneity). At the end of the nine-teenth century, these two theories were our best theories of physics, so this

Beyond the Dynamical Universe. Michael Silberstein, W.M. Stuckey and Timothy McDevitt,
Oxford University Press (2018). © Michael Silberstein, W.M. Stuckey and Timothy McDevitt.
DOI 10.1093/oso/9780198807087.001.0001

contradiction was a problem at the time. In 1905, Einstein solved this problem by introducing Lorentz-invariant special relativistic mechanics which showed Newtonian mechanics is only valid for small velocities (small relative to the speed of light). In that sense, special relativity (SR) can be thought of as reconciling or unifying mechanics and electromagnetism. Minkowski later constructed a four-dimensional geometric "spacetime" model for SR called "Minkowski spacetime." These developments are what lead us to the block universe implications of SR.

Therefore, in this chapter, including all its threads, we will explain why so many people are convinced that SR strongly suggests a block universe. The path to that explanation starts with time dilation ("moving clocks run slow") and length contraction ("moving objects shrink") per SR. The derivation of time dilation and length contraction from the postulates of SR is given in Foundations of Physics for Chapter 2, but that derivation isn't required for the novice to appreciate the paradox. Let us explore what those two consequences of SR imply.

If you and observers at rest with respect to you (hereafter simply "you") see someone else moving, you should see them aging more slowly than you. But, motion is relative, so they (and observers at rest with respect to them) say you're the people moving and therefore they should see you aging more slowly than they are. Similarly, if you see them moving, you should measure their meter sticks to be shorter than yours, while they say you are moving and therefore your meter sticks are shorter than theirs. Here in the main thread we will explain how the relativity of simultaneity resolves this paradox in SR.

In turn, it is the relativity of simultaneity that is the key premise in the inference to a block universe. In the philosophical (foundations of physics) literature, the debate over whether or not relativity really does entail or strongly suggest a block universe is long running. This will be the subject of Philosophy of Physics for Chapter 2 immediately following the main thread here.

In that thread, we will attempt to provide a much more detailed discussion of the relationship between relativity and the block universe. We will also provide a more detailed argument for the block universe from relativity that we hope is relatively free of loopholes. As with the computations in Foundational Physics for Chapter 2, the philosophically nuanced arguments in Philosophy of Physics for Chapter 2 are not necessary for the novice to proceed if they are willing to accept the block universe implication of SR. But, keep in mind, however, that the main thread is designed for the widest possible audience to get up to speed on the key issues and technical concepts, the philosophy of physics thread is designed to provide the most philosophical detail and conceptual background, and the foundational physics thread is where to find formal rigor and mathematical proofs for those who want them. While this setup is designed to speak to our many diverse audiences, it does mean that a reader may not fully appreciate the significance of the main thread until they have read at least the philosophy of physics thread.

So, read what you must in chapter 2 to allow yourself to accept the block universe premise. Starting in chapter 3 we will see how the Lagrangian Schema

Universe (LSU), that is, adynamical explanation in the block universe, allows us to start resolving several problems in both foundations of physics and foundational physics. Dirac long ago pointed out a formal advantage of Lagrangian Schema physics over Newtonian Schema (Hamiltonian) physics due to relativity:

> Quantum mechanics was built up on a foundation of analogy with the Hamiltonian theory of classical mechanics . . . Now there is an alternative formulation for classical dynamics, provided by the Lagrangian . . . The two formulations are, of course, closely related, but there are reasons for believing that the Lagrangian one is the more fundamental. In the first place the Lagrangian method allows one to collect together all the equations of motion and express them as the stationary property of a certain action function . . . There is no corresponding action principle in terms of the coordinates and momenta of the Hamiltonian theory. Secondly, the Lagrangian method can easily be expressed relativistically; while the Hamiltonian method is essentially non-relativistic in form, since it marks out a particular time variable as the canonical conjugate of the Hamiltonian function.

[Dirac, 1933, p. 1]

By the time the reader has read chapter 6 the full meaning and import of this insight will become apparent.

2.2 Abridged introduction to special relativity

We start with the postulates of SR [NobelPrize.org, 2016]:

1. The Principle of Relativity—The laws of physics are the same in all inertial frames of reference.
2. The Constancy of Speed of Light in Vacuum—The speed of light in vacuum has the same value c in all inertial frames of reference.

It is the second postulate that leads to time dilation and length contraction. The simple mathematical explanation is left to Foundational Physics for Chapter 2, but you can intuitively see that something about space and time measurements has to change if everyone measures the same speed of light even when they are moving relative to each other. For example, suppose you and I are at rest with respect to each other and I throw a baseball to you.[1] If the baseball moves away from me at 20 miles per hour (mph), then it is moving toward you at 20 mph. If I get in a car and drive toward you at 10 mph, holding the ball rather than throwing it, the ball and I are moving toward you at 10 mph. If I instead throw the baseball to you at 20 mph relative to me while in the car that is already moving

---

[1] While there are three authors here, for simplicity sake we shall adopt the royal "I" in what follows.

toward you at 10 mph, then the baseball is moving toward you 20 mph faster than I am. That means the baseball is moving toward you at 30 mph. Simple, right? Now replace the baseball with a flashlight that I shine toward you. The light beam moving relative to me at c replaces the baseball moving relative to me at 20 mph. Therefore, it seems that when I get in a car and drive toward you at 10 mph and shine the flashlight in your direction, you should measure the speed of the light beam to be c *plus* 10 mph. However, SR says you will measure the light beam to move at speed c, just as when I was standing still relative to you. How can that be?

SR says that you are making an assumption about spatial distance and temporal duration in the measurement process that isn't true. You are assuming the spatial distance between events is measured the same by everyone, regardless of their relative motions. Likewise, you are assuming that the temporal duration between events is measured the same by everyone, regardless of their relative motions. That is the assumption of Newtonian physics and it is compatible with our everyday experience, but SR says it is not spatial distance and temporal duration between events *per se* that are the same for everyone; rather, it is the *combined* spacetime distance between events that is the same for everyone. The mathematical method for converting spatial distance and temporal duration from one frame moving relative to another frame is the Lorentz transformation.

In the example that follows, we will relegate the Lorentz transformations to the Foundational Physics for Chapter 2 and exaggerate the temporal differences to make the effects more pronounced. It will then be clear how the relativity of simultaneity allows for time dilation and length contraction when motion is relative. However, this explanation will most likely violate your sensibilities, so we will tell you that these effects have been experimentally confirmed in accelerators with high speed particles many times. Indeed, if SR were wrong about this, particle accelerators would not work as designed.

2.3 Relativity of simultaneity

In order to show how the relativity of simultaneity resolves the apparent contradiction that I see your clocks running slow and your meter sticks to be short while you say the same about mine, consider the following four events adapted from the DeWitt/Mermin version of this story [DeWitt, 2004, pp. 213–19]:

- Event 1: 20-year-old Joe and 20-year-old Sara kiss
- Event 2: 20-year-old Bob and 17.5-year-old Kim kiss
- Event 3: 22-year-old Bob and 20-year-old Alice kiss
- Event 4: 25.6-year-old Bob and 24.5-year-old Sara kiss

The girls and the boys agree on the facts contained in these four events. Further, Joe and Bob see the girls moving in the positive x direction (Figure 2.1), so the girls see the boys moving in the opposite direction at the same speed (Figure 2.2). Who is actually moving?

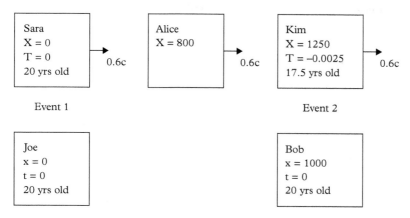

Figure 2.1 *The girls are triplets and the boys are twins. In the figures that follow, we exaggerate the time differences for effect. The temporal origin corresponds to being 20 years old and –0.0025 s corresponds to being 2.5 years younger than 20 years old, that is, 17.5 years old, and so on. In Event 1, 20-year-old Joe and 20-year-old Sara kiss and in Event 2, 20-year-old Bob and 17.5-year-old Kim kiss. The boys and girls agree on their ages in those events, so Events 1 and 2 occur at the same time for the boys and they conclude the girls are not the same age. Also, the boys measure the distance between Sara and Kim as 1000 km, not 1250 km as the girls measure it, so the boys say the girls' distances are length-contracted.*

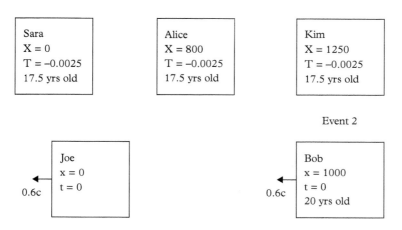

Figure 2.2 *The girls are triplets and 20-year-old Joe and 20-year-old Sara kiss (Event 1) and 20-year-old Bob and 17.5-year-old Kim kiss (Event 2). The boys and girls agree on those facts, so the girls don't believe Events 1 and 2 occur at the same time, that is, Event 2 occurred 2.5 years before Event 1.*

The answer to this question is central to the block universe perspective. According to the first postulate of SR, there is no way to discern absolute, uniform motion (the laws of physics are the same in all inertial frames of reference), so either perspective is equally valid. The girls are correct in saying it is the boys who are moving, and vice versa. This is equivalent to saying there is no preferred, inertial frame of reference in the spacetime of SR. Finally, we throw a monkey wrench into this scenario—the boys are twins and the girls are triplets. This establishes simultaneity for each set, that is, events are simultaneous for the boys if the events occur when the boys are the same age; likewise for the girls.[2] Now let's look at the implications of these four events.

Looking at Figure 2.1 we have the boys' perspective. Since the boys say Events 1 and 2 are simultaneous, they must conclude that the girls are not triplets, that is, Kim is 2.5 years younger than Sara. Second, they conclude that the girls' meter sticks are length-contracted because the girls say the distance between Sara and Kim is 1250 km while the boys say it is only 1000 km.

Of course the girls disagree. They have celebrated birthdays together for many years and they know for a fact that they're the same age. Event 2 is definitely not simultaneous with Event 1, it happened 2.5 years before Event 1 (Figure 2.2). The event simultaneous with Event 1 is Event 3 as far as the girls are concerned (Figure 2.3). Thus, they conclude that the boys are not twins, that is, Bob is two years older than Joe. Second, they conclude that the boys' meter sticks are length-contracted because the boys say the distance between Joe and Bob is 1000 km while the girls say it is only 800 km. This is how the relativity of simultaneity allows for the boys to claim the girls meter sticks are too short and the girls' to claim the boys' meter sticks are too short. Again, motion is relative (first postulate), so both the girls and the boys are correct in their assertions; therefore, simultaneity is relative. That is, whether or not the boys are twins or the girls are triplets (i.e., whether or not they are the same age) is relative to one's frame of reference. So, we have seen how the relativity of simultaneity resolves the paradox that each set of siblings sees length contraction in the other set's measurements. What about time dilation?

The girls see clearly that the boys are aging more slowly because the girls aged 2.5 years between the time Bob and Kim kiss (Figure 2.1) and the time Bob and Alice kiss (Figure 2.3), while Bob only aged two years. Again, this conclusion is based on the fact that the girls are the same age, something with which the boys obviously disagree. Rather, they say it is the girls who are aging more slowly because the boys aged 5.6 years between the time Joe and Sara kiss (Figure 2.1) and the time Bob and Sara kiss (Figure 2.4), while Sara only aged 4.5 years. Again, this conclusion is based on the fact that the boys are the same age, something with which the girls obviously disagree. So, what is the consequence of this relativity of simultaneity?

---

[2] Of course, per SR, it is possible for one of the girls/boys to age more slowly than the other girls/boys and therefore to have simultaneous events not correspond to having the same ages. We are just trying to keep the presentation simple.

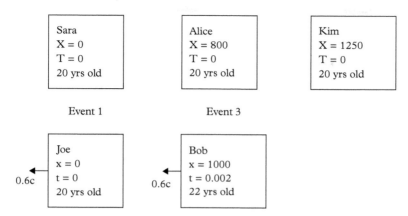

Figure 2.3 *The girls are triplets and 20-year-old Joe and 20-year-old Sara kiss (Event 1) and 20-year-old Alice and 22-year-old Bob kiss (Event 3). The boys and girls agree on their ages in those events, so assuming the girls are triplets, they say Events 1 and 3 occur at the same time and the boys are not the same age. Since 17.5-year-old Kim and 20-year-old Bob kiss (Event 2), the girls aged 2.5 years between Events 2 and 3 while Bob only aged 2 years, so the girls say the boys are aging more slowly than they are. Also, the girls measure the distance between Joe and Bob as 800 km, not 1000 km as the boys measure it, so the girls say the boys' distances are length-contracted.*

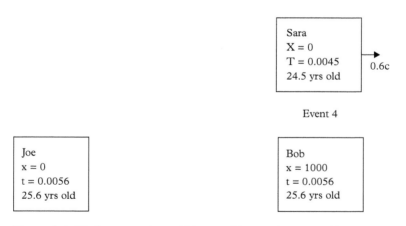

Figure 2.4 *The boys are twins and 20-year-old Joe and 20-year-old Sara kiss (Event 1) and 25.6-year-old Bob and 24.5-year-old Sara kiss (Event 4). The boys and girls agree on their ages in those events, so assuming the boys are twins, the boys aged 5.6 years between Events 1 and 4 while the girls only aged 4.5 years. Therefore, the boys say the girls are aging more slowly than they are.*

The relativity of simultaneity renders the view known as "presentism" highly suspect, or at least the most common version of that view. Global Presentism is the belief that everyone shares a unique, "real" present while their common past states no longer exist and their common future states are yet to exist; that is, only the global present is real. Per presentism, everyone could agree with a statement such as "Sam is surfing in California while Jonathan is shoveling snow in New York." If Sam is 25 years old, the 24-year-old version of Sam no longer exists, is no longer "real," and the 26-year-old version does not yet exist, will not be "real" for another year. There is a sense that we share in a "real" present moment with everything else in the universe and this attribute of "realness" moves along all worldlines synchronously. In fact, what we mean by "the universe" is vague unless everyone agrees on a spatial surface of simultaneity in spacetime. But, as we just saw, the relativity of simultaneity is contrary to this commonsense notion of presentism: the girls and boys do not agree on what constitutes "the present," the boys say they are twins and the girls disagree while the girls say they are triplets and the boys disagree. So, let's bring this all together to infer block universe from the relativity of simultaneity.

We just need to know who "coexists" at any given time; we just need to define the spatial frame which defines "the universe" at any given time. Well, unless we can touch things that are not "real," Joe and Sara "coexist" at Event 1 and Bob and Kim "coexist" at Event 2 (Figures 2.1 and 2.5). But, Joe and Bob "coexist" at Events 1 and 2 (they are twins) so we see that all four characters involved in Events 1 and 2 "coexist." This means 20-year-old Sara "coexists" with 17.5-year-old Kim and 20-year-old Sara "coexists" with 20-year-old Kim. Thus, 20 year-old Kim "co-exists" with 17.5-year-old Kim and that means the future is as "real" as the present for 17.5-year-old Kim, while the past is as "real" as the present for 20-year-old Kim: block universe.

As far as the block universe implications for the boys, notice that at Event 3 we have 20-year-old Alice "coexisting" with 22-year-old Bob while she "coexists" with 20-year-old Sara (Figures 2.3 and 2.5). However, 20-year-old Sara also "co-exists" with 20-year-old Joe at Event 1. Since 20-year-old Bob "coexists" with 20-year-old Joe, 20-year-old Bob "coexists" with 22-year-old Bob. That means the future is as "real" as the present for 20-year-old Bob, while the past is as "real" as the present for 22-year-old Bob: block universe.

In conclusion, the boys say the girls' clocks run slow and the girls' meter sticks are short, while the girls say exactly the same thing about the boys' clocks and meter sticks. These seemingly incompatible facts can hold because the boys and girls disagree as to which events are simultaneous—Events 1 and 2 are simultaneous for the boys, Events 1 and 3 are simultaneous for the girls (Figure 2.5)—the boys believe they are twins and deny the girls are triplets and vice versa. Thus relativity of simultaneity implies the past, present, and future must "coexist": reality is a block universe. We highly recommend Nova's program, *The Fabric of the Cosmos: The Illusion of Time* for another explanation of the block universe. Although,

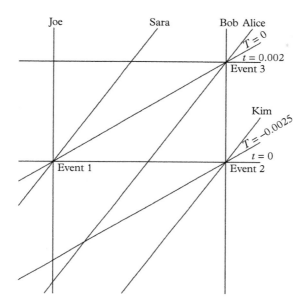

Figure 2.5 *This spacetime diagram shows surfaces of simultaneity for the boys and girls. Events 1 and 2 are simultaneous for the boys (both occurring at t = 0) while Events 1 and 3 are simultaneous for the girls (both occurring at T = 0).*

as we explain in chapters 7 and 8, block universe does not entail that "time is an illusion." What we just presented here our version of the classic argument from SR to the block universe. However the argument can be challenged in a number of ways. Philosophy of Physics for Chapter 2 will provide another version of the argument and attempt to address most of the possible objections.

Philosophy of Physics for Chapter 2

2.4   Introduction

There is a long-standing debate in metaphysics and the foundations of physics about whether relativity entails the block universe, that is, the reality of all events, past, present, and future. The argument usually proceeds from the relativity of simultaneity (RoS), as we saw in the main thread. The short answer is "no," the RoS doesn't *entail* the block universe: extra premises must be added, assumptions must be made, etc. This isn't at all surprising since the metaphysical implications of any empirical observation, scientific theory, or formalism are always under-determined. In the following we provide an extended argument from relativity for the block universe, including rebuttals, that we hope is as complete as possible. This argument and almost all of the material of this section originated in the chapter: "In Defense of Eternalism: The Relativity of Simultaneity and Block World" in *Space, Time, and Spacetime—Physical and Philosophical Implications of Minkowski's Unification of Space and Time* [Peterson and Silberstein, 2010]. But first we want to provide some general remarks about the state of the debate.

As will become clear from this discussion and that covering of the experiential arrow of time in chapters 7 and 8, the consensus is that nothing in science or experience has ruled out metaphysical alternatives to the block universe, such as presentism (the claim that only the present is real) or the growing-block model (the idea that the past and present are fixed but the future is open somehow; it "grows" from the past and present), and the "moving-spotlight" theory which says that something "illuminates" the present moment and distinguishes it as such. Presentism is a dynamical model of time in that there is a succession of present moments; the growing-block model is dynamical in that while the present and past are real, the future comes into being. The moving-spotlight theory is compatible with the claim that it is a block universe in the sense that all events are "just there," but only the illuminated ones are present, as with Weyl's conscious mind crawling along the worldtube. There are certainly hybrid views and modified versions of each of these views. There is the "growing–glowing" block universe, as Dainton calls it [Dainton, 2001], in which only the global wave-front of the present (the edge of the block) is fully real and/or only it has conscious observers. The hope is that if only the edge is fully real or has conscious observers then that would explain why the present moment feels special: it is the experience of "creation" (what we will call Presence in chapter 7). Another example is one could have a kind of presentism without any dynamism: only the present is real but there is no succession, there is only one moment. This is the experience of the character Dr. Manhattan in *Watchmen* [Moore and Gibbons, 1986]. For example, he experiences the entire block universe at once as one moment. We will deal with the relationship between conscious experience and physics in chapters 7 and 8, where we will deal with Presence, Passage, Direction, and the like. The consensus is that these alternative views are empirically or experientially equivalent and thus nothing about our experience rules any of these views out. Many people also claim that these metaphysical views about time are also metaphysically equivalent. We will take issue with the latter claim in this thread. What we can say now about these metaphysical alternatives about the nature of time is that if one accepts the arguments offered in this chapter then relativity does rule out any form of dynamical presentism and any growing block model in which the future is truly not real.

However, it is certainly also true that the block universe view has enjoyed near consensus acceptance among physicists and philosophy for decades, precisely because of relativity. According to Dainton the consensus may be starting to change:

> Nonetheless, it is a striking development that scientists working at the forefront of physics and cosmology are starting to take a serious interest in the different metaphysical conceptions of time that philosophers have been debating for centuries (if not longer). At the very least, it means that philosophers who combine an interest in temporal metaphysics with a respect for contemporary physics and cosmology have more room to maneuver. As we wait for further scientific developments, we no longer need to work on the assumption that the block universe is the only conception of time with respectable scientific credentials.
>
> [Dainton, 2016, p. 78]

While we doubt that the consensus has changed much regarding the block universe for either philosophers or physicists, as we discuss in chapters 7 and 8, most of the movement away from the block universe has been the worry on the part of a few foundational physicists who believe fundamental physics is incomplete if it can't explain time as experienced. Indeed, many hope to reduce the experiential arrow of time to some physical arrow of time. The consensus is that the block universe posit greatly magnifies this incompleteness worry regarding the experiential arrow. Again, the experiential arrow of time will be the focus of chapters 7 and 8. Here we note that the two main scientists Dainton has in mind are Lee Smolin and George Ellis. We will return to a discussion and critique of their views at the end of this thread, as well as the Maudlin view we alluded to in chapter 1.

As will become even more clear in what follows, one can escape the inference from RoS to block universe in a number of ways. The most obvious way is just to refuse to be a realist about the Minkowski interpretation of relativity or about relativity more generally. One can also try to add some *metaphysical* resources to the theory that provide a preferred frame, but as Miller notes there is a well-known problem with that move:

> Thus what STR tells us is that it is in principle impossible to *determine* which plane is the metaphysically privileged one. But it does not tell us that no plane is in fact metaphysically privileged . . . But this contention is in striking tension with the claim that although there is a privileged hyper-plane, we could never know which it is, since its being metaphysically privileged does not entail its also being empirically or physically privileged.
>
> [Miller, 2013, p. 353]

On the other hand, one can add or pick out some physical feature of the universe that constitutes a preferred frame such as the cosmic time assumed in GR. For example, several interpretations of quantum mechanics such as Bohmian mechanics seem to demand a preferred frame [Zimmerman, 2011, ch. 7], collapse interpretations seem also to require a preferred frame, etc. But again, as the arguments for a preferred frame or against the block universe more generally are typically motivated by features of the experiential arrow of time, the trick is to show or make believable that whatever physical feature one selects to play, this role corresponds with/is the cause of that experience. We will discuss the problems therein in this thread and in chapters 7 and 8.

In what follows, we will consider these and other ways to avoid the blockworld implication of the relativity of simultaneity (RoS), but what we hope to show is that if one is a realist about the Minkowski interpretation of relativity and finds no plausible overriding mechanism for adorning the structure of Minkowski spacetime, then the block universe implication is as persuasive as most any explanation in all of metaphysics. But again, and more importantly, we believe that the best reason to believe in the block universe is not RoS alone, or even SR plus GR alone, but is instead the work that the block universe and adynamical global explanation can do more generally in resolving problems in the foundations of physics, foundational physics, and metaphysics. We hope the book is a testament to this and therefore it is the book as a whole that constitutes our argument for the block universe.

This thread focuses almost exclusively on the inference from special relativity (SR) to the block universe. This discussion stems from two competing notions of time. The first, originally suggested by Heraclitus, is called "presentism."[3] Though we will later clarify the presentist position in more definite terms so that it can be made relevant to a more thorough and modern treatment of presentist/eternalist debate, a good starting definition for presentism (see figure 2.9) is the view that only the present is real; both the past and the future are unreal.[4] This view is close to, but not exactly the same as, "possibilism," (see figure 2.10) which states that the future is unreal while both the past and the present are real (i.e., the growing block picture). Both these stances claim adequately to capture the manifest human perception of time. We tend to view ourselves as occupying a unique temporal frame that we call the present that always moves away from the past toward an uncertain future.

However, with the advent of relativity, a different stance whose primary ancient proponent was Parmenides of Elea provided a viable alternative to Heraclitean presentism. This new stance, called "eternalism," (see figure 2.8) was translated into the language of relativity by Hermann Minkowski in 1908 and suggests that time and space should be united in a single, four-dimensional (4D) manifold.

---

[3] Recent defenders of presentism include Bourne [2007], Craig [2001], and Craig and Smith [2007], whom we take to be our primary presentist opponents for the purposes of this discussion.

[4] What is meant here by "real" is the topic of great debate (see Dorato, 2006b and Savitt, 2006b, for more on this issue), and we will later clarify our criteria for reality in such a way that many of the vagaries that arise from an imprecise definition of "real" are dismissed.

Thus arose the notion of a 4D "block universe" (BU) in which the past, present, and future are all equally real. Two arguments by Putnam [1967] and Rietdijk [1966] allegedly show that SR with its RoS implies that only the BU perspective can obtain.

First, we examine the basic structure of the RoS eternalism argument suggested by Putnam, Rietdijk, and more recently Stuckey, Silberstein, and Cifone [Silberstein et al., 2006, 2007, 2008] (hereafter SSC), and present our own interpretation or version of the argument for eternalism. Following our proposal, we suggest various points of contention that presentists and possibilists might exploit or have exploited in seeking to either refute eternalism or collapse the presentism/eternalism dichotomy. We have compiled a reasonably exhaustive taxonomy of possibilities that the presentist or possibilist could take to avoid the argument from RoS for BU. After elaborating our own version of the argument, we respond to each counterargument and show that these objections do not dismiss RoS's problems for presentism.

2.5   The argument from the relativity of simultaneity

2.5.1   General outline and definition of terms

Before presenting our RoS argument against presentism, we first provide a general outline of RoS arguments for eternalism and give preliminary definitions of some relevant terms. The general form of the arguments against presentism utilized by Putnam, Rietdijk, and SSC goes as follows:

1. Define presentism.
2. Define the term "co-real."[5]
3. Show that the consequences of the definition of "co-real" and RoS contradict presentism.
4. Conclude that presentism is false from the conjunction of 1 and 3.
5. Conclude that eternalism is true from the rejection of presentism.

To begin with, we must provide our own definitions for the terms that form the foundation of our revamped version of the RoS argument. The first term to be defined is "presentism." Presentism is a kind of realism that takes as real only those events[6] which occur in the present. For instance, since we are sitting next to

---

[5] The term "co-real" appears only in the SSC papers, but since these present the most recent incarnation of the RoS argument against presentism, we follow their terminology here. It should be noted that Rietdijk does not provide an analysis of reality in his paper, and while Putnam does discuss some basic assumptions about reality that are necessary for his argument to go through, they are not argued for or supported in any great detail.

[6] We use the term "events" here to bypass any concerns that may arise due to the identity of individuals like those raised by French and Krause [2006] or issues of endurance and perdurance. Such

our friend Joe who is currently reading a paper, the event of his reading a paper and the event of our writing this thread are both real while the event of Joe's leaving to eat dinner is not real because it has not happened yet and the event of our leaving to eat lunch is not real because it has already happened. In terms of simultaneity then, one can define presentism as the view that the only real things are those which are simultaneous with a given present event. Eternalism, by contrast, is the view that all events past, present, and future are equally real. Thus, Joe's reading, our typing, Joe's leaving for dinner, and our leaving for lunch are all equally real despite the fact that one of these events has already occurred while another has yet to occur. Eternalists hold that all events are equally real, regardless of whether or not said events are simultaneous.

There are two elements then that are important for establishing both presentism and eternalism: reality and simultaneity. The debate presupposes that there is a unique (non-equivocal) sense of the term reality that both sides share. The dispute therefore is over whether or not present events have some ontologically privileged status qua their property of "existing at some time t where t is in the present." To this end we will first minimally characterize the terms "reality" and "simultaneity" for use in the context of our revamped argument. Before beginning, we should emphasize that we are being purposefully vague with our first characterization of reality here so as to determine reality's most general non-equivocal properties which we will build upon later in this thread. Two events which "share reality" as we characterize it share a single, unique feature (i.e., the same ontological status with respect to realness); this uniqueness, we believe, is the absolute minimal criterion an event would have to satisfy for it to be considered "real" in any meaningful sense of the word.

To improve out understand the minimal sense of reality at work here, we define two separate principles: the "reality value" and "reality relation." "Reality values" or "R-values" can be thought of as representing the ontological status of any given event. Within spacetime, every event can be assigned an R-value that denotes its ontological status, and there is a one-to-one and onto mapping of possible R-values onto ontological statuses. In the interest of defining reality generally, we will not attempt to enumerate how many R-values exist, but one could easily take reality to be binary and thus assert that, for any event, if its R-value is 1, that event "is real," and if its R-value is 0, that event "is not real." One could use higher values like 2 and higher to denote other states, such as "possibly real," "real in the future," etc., but, as previously stated, we will not attempt to enumerate all such possible R-values here.[7] It should be pointed out that our uniqueness criterion

---

issues as identity and endurance/perdurance, while interesting, need not directly bear on this debate, and so we invoke events that are assumed to be of infinitely small extension and duration (as such they should be fully understood only in terms of their identifying coordinates) to bypass such debates. We are not committed to the claim that such events are in some way the atomistic components of what exists in spacetime; rather, we simply invoke them to avoid begging the question on issues like identity and endurance/perdurance.

[7] See Appendix B for a more nuanced view of R-values and possible objections to the RoS argument that one might raise based on our naive characterization of R-values described here.

on reality translates into this system simply as the claim that every event has a single unique R-value. This seems intuitive since an event with an R-value of both 1 and 0, in our scheme, would be both real and unreal, which would be a contradiction.

Our other sense of reality as expressed in the "reality relation" will be essential to our discussion of co-reality. The reality relation can be recast as the idea of "equal reality" and exists between any two or more events that can be considered "equally real." Translated in terms of R-values, a reality relation exists between any two events that must have the same R-value. For instance, if events A and B are equally real, then the R-value of event A is the same as the R-value of event B. One should notice here that our definition of "equally real" does not assume that two equally real events are both "real;" equally real events A and B may have whatever R-value you please as long as the R-values are the same for both A and B. This relation explains what a presentist means when she says, "The present is the only thing that is real" since the presentist will hold that events in the future and the past will have different R-values from events in the present.[8] Thus, our purposefully limited characterization of the "equally real" relation has been defined so as to be useful in a definition of co-reality.

As for simultaneity, if it is possible for one to construct a hyperplane of simultaneity between any two or more events, then these events are said to be simultaneous. Such simultaneous events are required to be spacelike separated events that appear to be simultaneous in some subluminal inertial reference frame. Lightlike and timelike separated events cannot have a hyperplane of simultaneity constructed between them. Also, at least in Minkowski spacetime, a hyperplane of simultaneity may be drawn between any two spacelike separated events, meaning that the spacelike separation of events A and B is necessary and sufficient for their simultaneity (in some frame anyway).

Combining the criteria of equal reality ("equally real" means that two events have the same R-value) and simultaneity ("simultaneous" means that two events

---

[8] To reiterate, what we have characterized here is the minimal position a presentist must take with regard to a characterization of reality. It might be objected that, at this point, we have not actually defined "what reality is." We will cash out a richer notion of reality later in the thread so that we are careful not to beg the question against critics like Savitt and Dorato; for now, we are characterizing reality only to a minimal degree in an attempt to determine the properties of the "co-reality" relation, and as such we need only endorse the minimal sense of reality that bears upon our discussion of co-reality. The presentist might object to our characterization of her conception of reality, but to refuse the characterization of reality we have provided here would be to take an anti-realist stance since a non-unique or equivocal conception of reality would make the idea of "reality" a useless concept for the purposes of this debate. Thus, the presentist cannot argue against our minimal characterization of reality and remain a committed presentist, and the same goes for the eternalist. In the words of Dolev, if one denies this minimal ontological assumption then "neither the tensed nor the tenseless view has the final word in the metaphysics of time." The presentist could argue against us on the grounds that it is relations, perhaps, that are fundamentally real and not events; this, however, would simply lead us to re-atomize our spacetime such that these relations become the fundamental ontic units which assume R-values and the relation of "equally real" connects two such lesser relations. Therefore, even if one makes an argument that forces us to change the fundamental ontic units of our setup, our basic characterization of R-values and "equal reality" can stand unadulterated.

are spacelike separated such that a hyperplane of simultaneity can be constructed between the two events) gives us the relation of "co-reality," which refers to, as the name suggests, two events that are equally real and "simultaneous." The presentist perspective can be restated in terms of this "co-reality" as the stance that "simultaneity between events is a necessary and sufficient condition for the reality (i.e., for both events sharing the R-value 1 corresponding to "real") of these events if at least one of these events occurs in the "present." For the presentist, any two spacelike separated points are thus co-real as we have defined "co-reality." Our restatement of presentism in terms of co-reality here is the assumption that we alluded to in step 1 earlier.

Our previous examples should make our notion of co-reality more explicit. For instance, the presentist says Joe's paper reading and our thread typing are co-real events because they are spacelike separated, meaning that there exists some frame in which these two events are simultaneous. However, our thread typing and our leaving for lunch are timelike separated, so there is no sub-luminal frame in which these two events are simultaneous and they are therefore not co-real. These two criteria of reality and simultaneity as we have defined them are necessary and sufficient for our use of "co-real," and so we turn next to our RoS argument that utilizes this notion of "co-reality" to reveal the contradictory nature of presentism when combined with relativity.

2.5.2 RoS argument

One could argue that, having already defined "co-reality" as we have, the RoS argument has already been made for us: any two spacelike separated points are equally real, and spacelikeness is not transitive (i.e., A and B could be spacelike separated and B and C spacelike separated but A and C timelike separated), so we must conclude that any two events (timelike, lightlike, or spacelike separated from each other) are equally real. The RoS argument here, however, is a bit more nuanced than the argument just proposed, and it makes it easier for one to determine which definitions and assumptions about reality play what role in the argument. As such, we hope the reader will bear with the exposition for this longer argument.

Consider the following situation: our friends John and Josephine stub their toes at the same time in my stationary reference frame.[9] The event of John stubbing his toe is labeled A in Figure 2.6 and the event of Josephine stubbing her toe is labeled B in Figure 2.6. At a later time (but again, simultaneously in our rest frame), both Josephine and John shout in pain from stubbing their respective toes. John's shout of pain is labeled A' while Josephine's shout of pain is labeled B' in Figure 2.6. I note that in my frame, both toe-stubs occur at time t1 in Figure 2.6. Thus, events A and B are simultaneous and co-real as per the previously established criteria.

---

[9] We are assuming that these "toe-stubs" in this example are the kind of events described in footnote 5 for the reasons stated in that footnote.

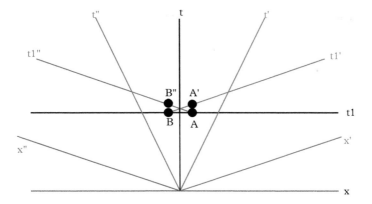

Figure 2.6 *RoS proof spacetime diagram.*

Now, some time before this the alien battle-cruisers P and D pass each other directly over our heads. The primed axes refer to the frame for battle-cruiser P and the double-primed axes refer to the frame for battle-cruiser D. Both of these battle-cruisers tell a different story from ours. For battle-cruiser P events B and A' occur at the same time, and thus B and A' are equally real per co-reality. For battle cruiser D, however, events B' and A occur at the same time, and thus B' and A are equally real per co-reality. We now introduce the symbol "r" to stand for "shares an R-value with" or "is equally real with." The following three statements are true (at least from someone's perspective):

- ArB
- BrA'
- B'rA

From the previously established criteria for equal reality, we can establish two important facts about co-real events α, β, and γ. First, if $\alpha r \beta$ is true, then $\beta r \alpha$ is true since R-values are unique. Thus, the operator "r" is commutative. This fact must be true since equal reality is an equivalence relation.[10] The second

---

[10] One might object that, for the historicists/possiblists (growing block from here on) in particular, the "co-real" relation is not an equivalence relation. For instance, right now the Norman Invasion is "real" to us because it is in our past, and so the growing block theorist would want to say that such an event is as real as our writing this book; however, at the time of the Norman Invasion, we were not yet born, so we were "not real" at that time. The equal reality relation only holds one way. However, one can respond to this claim by citing the fact that the equal reality of simultaneous events is an equivalence relation in the growing block even if the "equal reality" relation in general is not. Two events that happen at the same time must be equally real if it is temporality alone that bestows metaphysical status on events. The above argument only necessitates the treatment of "equal reality" as an equivalence relation for cases where the two "equally real" relata are spacelike separated and thus simultaneous. In such a case, equal reality is an equivalence relation even for the growing

important fact about equal reality is that the co-real operator is transitive, even across frames. Simply put, assuming the co-real relation is an equivalence relation, then it must also be transitive as well. That means that if $\alpha r \beta$ is the case and $\beta r \gamma$ is the case, then $\alpha r \gamma$ must also be the case. This follows directly as consequence of our definition for equal reality.[11] Thus, applying the properties of transitivity and commutativity to the above relations, we arrive at the result that:

- ArA'
- BrB'

Generalizing from this result, one can conclude that a prior event (the stubbing of a toe) is as real as a later event (a shout of pain). If the first event (A, for instance) occurs in the "present," then A' occurs in the future and the RoS argument suggests that the future is as real as the present. Likewise, if A' occurs in the present, then A occurs in the past and the RoS argument suggests that the past is as real as the present. Both of these conclusions contradict the presentist assertion that the present is real while the past and future are not since past, present, and future must share the same ontological status by the above argument. Since presentism in conjunction with relativity and our other basic assumptions leads to a contradiction, presentism must be false given our assumptions. Finally, since variations of this argument would answer equally well anyone who would argue that only the past is real or only the future is real, the only conclusion left for a realist is that eternalism must be correct since both presentism and possibilism must be discarded. We have thus achieved our goal of constructing a rigorous argument for eternalism from RoS in the tradition of Rietdijk, Putnam, and SSC though our argument provides a more detailed analysis of the assumptions about the nature of "is real" that go into this kind of argument.

2.6   Presentist points of contention

There are several points in this argument for eternalism that presentists (or anti-realists, for that matter) could attack or have attacked. The point of this section is to provide a basic taxonomy of points of contention presentists utilize or could utilize to disagree with both the argument presented above and eternalism in general.

---

block. Thus, the fact that equal reality is not necessarily an equivalence relation in general does not mean that equal reality is not an equivalence relation in the case of simultaneity; in fact, the opposite is true.

[11] This feature of co-reality is perhaps not intuitive, but a simple conceptual argument can show why equal reality, as we have defined it, must be a transitive property. If two events A and B are co-real in a given frame, this means that they share an R-value. Likewise, co-real events B and C must also share a unique R-value. Since the uniqueness criterion on reality implies that the R-value shared by A and B must be the same R-value shared by B and C, it then follows that A and C must have the same R-value as well, and thus they must be equally real.

2.6.1 Deflationary objections: No presentist/eternalist distinction

The first attack on the RoS argument which works equally well on any argument trying to prove or disprove eternalism is that there is, in fact, no metaphysical or empirical distinction between the views supported by presentists and those supported by eternalists. This collapse of the dichotomy between presentism and eternalism is most ardently argued for by Savitt [2006b] and Dorato [2006b] in recent papers. Both of these papers utilize semantic arguments to suggest that the distinction between presentism and eternalism boils down to a difference in definitions for "real" which translates, in various contexts, to differences in tensed versus tenseless existence claims. These two authors claim that presentism and eternalism are both essentially either vacuously true when viewed with the proper definition of existence (e.g., to say that the present is the only thing that "exists now" is tautological since "now" is defined in terms of the present) or analytically false when viewed with the improper sense of existence (e.g., to say that the present is the only thing that "exists tenselessly" is to ignore the past and future that are assumed in the phrase "exists tenselessly"). These two authors go on to attack various defenses of eternalism that rely on modality and various other semantic considerations, leading them to the conclusion that the problem posed by the presentist/eternalist debate is truly a non-starter by way of a "Wittgenstein-like" or "Austin-like" deflation.

In an earlier paper, Dorato [2006a] brings in various other semantic arguments against eternalism specifically in an attempt to show how eternalism is as problematic as presentism. The first contention Dorato raises is against the eternalist perspective that "the past, present, and future are all real at the same time," which he views as meaningless since one cannot say anything about the relationship between the past, present, and future at a given time since all three temporal regions cannot be simultaneous. There must be a temporal separation between the past, present, and future for them to be well defined, so any statement about how the past, present, and future interact at a given time collapses this distinction and thus becomes meaningless. The second argument against eternalism on semantic grounds is that an eternal truth like "event A takes place at time t" may be timeless, but the object of this statement, event A, is not necessarily as timeless as the statement about it. Dorato thus believes that eternalism confuses the following two statements:

1. "X is the case at t" is an eternal truth.
2. X exists eternally.

Thus, since eternalism makes this error, it is somewhat a nonsensical position to hold. These two linguistic objections to eternalism, as well as the much larger objection that there is no metaphysical presentist/eternalist dichotomy, will be addressed later.

2.6.2   Compatibilist and incompatibilist objections

Two other groups of people who reject the RoS argument for BU are the compatibilists and incompatibilists. Compatibilist philosophers of time attempt to hang presentism on a given relativistic invariant (like the fact that all inertial frames agree on the ordering of timelike events, or "proper time").[12] Incompatibilists, on the other hand, invoke some preferred frame or other entity with which to adorn Minkowski spacetime in hopes that this new frame will provide a suitable place to hang presentism and becoming. These positions constitute a shift in the definition of "co-reality" as it was presented previously. Both compatibilists and incompatibilists would reject our definition and propose another, though various compatibilists and incompatibilists will propose differing versions of "co-reality." There are essentially two ways philosophers can and do object to the RoS argument:

1. Reject our characterization of simultaneity in our definition of co-reality (redefine simultaneity, compatibilist and incompatibilist objection).

2. Reject our characterization of reality in our definition of co-reality (reject transitivity of co-reality, compatibilist objection).

Option 1 can and has been justified on several different grounds. It has most famously been argued that either a) simultaneity is not a suitable criterion for reality because the present refers to only the "here and now," not simply the now, or b) simultaneity is relative to some preferred foliation of spacetime.[13] Objection a) is raised most famously by Stein [1968, 1991] in his response to Putnam, and objection b) has been raised by various philosophers and physicists who have rather disparate views as to what the preferred foliation of spacetime is and whence it issues.[14] We will address both of these objections to the RoS argument individually in the following sections.

Compatibilist option 2 is typically raised either by those like Savitt [2006a] and Dolev [2007] who believe that an argument for a transitive reality has not and cannot be convincingly made especially within the framework of SR or by antirealists (including solipsists) who believe that the phrase "reality" should only pertain to one's own frame (or, worse yet, only to oneself). The first of these objections is then the only one particularly relevant to the presentist/eternalist debate because an anti-realist would no sooner be a presentist than an eternalist. The transitivity of "is co-real with" is objected to on this view precisely because

---

[12] It should be noted that we do not disagree with the compatibilist assertion that to be real something must be "real in all frames:" in fact, we embrace this idea, and it is a central aspect of our definition for reality that frame-invariant properties like timelike separation are necessarily "real" features of spacetime.
[13] The first of these objections, a), is a compatibilist objection while the second objection, b), is an incompatibilist objection.
[14] See Cifone [2004] for specific examples of proposed preferred foliations to spacetime.

it leads to the view that presentism is wrong. Thus, it seems like any presentist interested in saving her stance would object to the transitivity of co-reality implied by our definition of reality as many before her have chosen to do.

2.7 Response to objection

2.7.1 Defining terms: Establishing a metaphysical presentist/eternalist distinction

Dorato and Savitt claim that there is no metaphysical or empirical distinction between the eternalist and presentist perspectives by critically examining the words "is," "exists," and "real" used in several definitions of reality and in doing so point out the shoddy conclusions that linguistic sloppiness engenders in the presentist/eternalist debate. Our goal in this section is to provide an original definition of reality which supports a metaphysical/empirical distinction between the presentist and eternalist positions. Such a reasonable definition is sufficient to counter Dorato's and Savitt's deflationary claims.[15]

Our definition of reality relies upon two concepts: "definiteness" and "distinctness." For an event to be real, we posit, the event must be both definite and distinct. We take a definite event to be one which is meaningfully determined. A useful example of the distinction between definite and indefinite can be found in the standard account of quantum superpositions.[16] With respect to a particular variable like spin in the x-direction, a pure-state quantum system may be in an eigenstate or a superposition of eigenstates. If there exists a multitude of systems in the same eigenstate, an x-spin measurement on any of these systems will always yield the same value. Thus, we say that an eigenstate of x-spin is property-definite with respect to spin in the x-direction. However, if the system is in a superposition with respect to x-spin, different systems prepared in the same x-spin superposition may give different x-spin values when measured. There is no way to predict the value of the x-spin of such a superposition after measurement given any information about the system prior to measurement, and as such, the superposition of x-spin is said to be property-indefinite with respect to x-spin. Generalizing from our characterization of property definiteness, we define event-definiteness as definiteness with respect to at least one property. Thus, if an event is property-definite with respect to at least one property, we say it is event definite and thus "real."

---

[15] For an alternative response to such a deflation by way of logical and linguistic analysis, see pages 14–17 of Sider [2001].

[16] We are not claiming that quantum superpositions are unreal or non-existent simpliciter; rather, we are providing an example in an instrumental spirit of how a property might be indefinite and thus suggesting how one might generalize from this example to form an idea of general indefiniteness. This indefiniteness, if made general and applicable to all properties, would make an event effectively unreal. However, superpositions themselves are by their very nature in a determinate state with regard to some property, so they are obviously not wholly unreal.

We should note here that our event-definiteness criterion is an objective criterion of a system and as such, unlike property-definiteness, a system must be indefinite with respect to all of its properties to be considered indefinite qua system. Therefore, quantum superpositions are not objectively indefinite, for there exists some property with respect to which this superposition is definite by the very nature of superpositions; it is only the x-spin value of such a superposition that is indefinite. If a given event is definite with regard to any property, it is taken to be objectively definite and thus may be real (as long as it meets our distinctness criterion as well, that is).

It should also be pointed out that event-definiteness is a frame-independent property of events in the universe, though different observers may disagree about the state of a given system (as Rovelli points out in his paper on relational quantum mechanics [1996]), they will all agree about whether or not it is definite simpliciter. One might take issue with our assertion of the frame-independence of definiteness; for instance, some postulate that quantum collapse is hyperplane-dependent, and thus an observer in one frame will see a quantum system as having some definite property "y" while an observer in another frame might observe y to be indefinite. However, even if collapse is so dependent, the fact remains that each of these observers will observe there to be some definite property "z," and thus, by our definition, one must take the quantum superposition to be definite qua system. That is, there is no frame of reference from which one can observe the quantum system in question to be without any definite properties. Therefore, our definition of definiteness directly implies the kind of transitivity we exploit in our RoS argument.[17]

The other criterion for an event to be real is that it must be distinct. A distinct event must be in some way different from other distinct events (like Leibniz, call it the discernability of non-identicals). Such a criterion for the distinctness of events is different from a criterion that requires the distinctness of particles. While it may be that two completely indistinguishable particles can both be distinct, the issue of concern here is the reality of events, and it is the case that two completely indistinguishable events cannot be distinct per the identity of indiscernables; or if you prefer, two completely indistinguishable events cannot be numerically distinct. This criterion of distinctness may be viewed as a more pragmatic concern (we have no reason to take event B to be numerically distinct from event A if all of B's properties are identical to A's). Such a criterion of reality keeps one from treating as real two (allegedly distinct) "events" that might seem to be different but are truly one and the same event—the differences are purely perspectival as in the Lorentz transformations of SR. For example, as per Newton's third law of

---

[17] We should point out here that presentists who claim that there simply are no past or future events can be treated as taking such event as indefinite on our picture here, since a non-existent event cannot have any definite properties. Thus, our account of definiteness provides a criterion for reality that explains this possible presentist stance.

motion, there is no need for us to count as distinct both the event of a car hitting a wall and the event of the wall hitting the car; they are simply two different ways of viewing the same singular event.[18]

Having established these two criteria for reality, does there appear to be a difference between the presentist and eternalist positions? The answer is "yes" because the distinctness and definiteness of the past and future are not analytic. The presentist claims that past and future events lack both/either definiteness and/or distinctness simpliciter while the eternalist says all events past, present, and future possess both definiteness and distinctness simpliciter. The first fact to note about the future is that it is unknown to us. One might even be tempted to say that it appears indefinite since it seems (at least on some stochastic accounts of quantum outcomes) that there is no way for us to know the future (in principle) no matter how much we know about the present. Such stochastic accounts of objective quantum indefiniteness (as opposed to subjective quantum indefiniteness for deterministic interpretations) should not be confused with what we will call O- (objective) indefiniteness and S- (subjective) indefiniteness more generally. O- and S-indefiniteness are best understood as a different kind of indefiniteness entirely which will be made clearer by an appeal to the idea of "Newton's god" (NG), an entity in the fifth or higher dimension "looking down" at her spacetime "sensorium."

Depending on whether the future is O-indefinite or S-indefinite, NG would observe different things as she looked down on her "sensorium." If the future and past are S- indefinite only, NG would physically see[19] the past, present, and future—all spacetime, a 4D BU. NG would see events in the past, present, and future—a static multi-colored marble of worldlines/worldtubes, if you will. If the future and past are truly O-indefinite, however, NG would not be able to see the future or past from her fifth-dimensional perch, but only a continually temporally evolving present. If the future is truly O-indefinite, it does not matter whether NG is observing us flipping a coin or measuring the spin of an electron with stochastic outcomes; either way, she will not observe the future outcome, and likewise if the future is merely S-indefinite then in both the classical and quantum case NG will observe the future outcome. In the O-indefinite case, NG

---

[18] One might well wonder what purpose introducing "distinctness" as a criterion here serves above and beyond the work already done by definiteness. Distinctness is important in this discussion because it allows for nuances within possible presentist positions. We believe there may be presentists who concede that some future events are determined in that they have some definite property, yet who may still reject that the future and present are "equally real." They could do so by way of distinctness, claiming that there are an infinite number of events (one of which will be actual, the rest of which will not be) which are all "definite" in some sense but are indistinguishable. The future would thus be definite but not distinct, and so the presentist could write it off as unreal. For the purposes of this discussion, it is in our interests to give as many reasonable possibilities to the presentist as we can, and so we have included distinctness in our discussion.

[19] When discussing what NG "sees" we are only invoking the traditional physical sensory modalities of this entity. We make no claims about other ways of knowing or omniscience that one in NG's position might be able to employ by means other than perception.

may be able to predict the outcome just as any one of us may be able to predict the outcome of a coin flip, but NG will not be able to observe this future outcome.

Savitt would perhaps respond as follows: the presentist has to admit that the past was O-definite. Similarly, the future will be O-definite. But that's exactly what the eternalist says, however: the past was O-definite, the present is O-definite, the future will be O-definite. Past events aren't O-definite now because there are no past events now that could have definite properties. But let's break things down into NG looking at a particular timeslice of the universe. According to the eternalist, at each timeslice, what events are O-definite will be the same, but, according to Savitt, what events are O-definite will vary from timeslice to timeslice. So if we imagine NG looking at timeslices in particular, we can see the difference between the two views.

The eternalist, presentist, and possibilist positions become clear and distinct given this characterization of O- and S-indefiniteness. Eternalists believe that the future and past are only S-indefinite; though beings within spacetime may not be able to observe the past or the future, a being outside of spacetime would be able to easily observe them. Thus, NG sees a 4D BU when she looks "down on" the universe. The presentist, on the other hand, holds both the past and future as truly O-indefinite and thus believes that NG would see an evolving 3D timeslice of the universe when she looks "down on" her "sensorium."[20] Finally, the possibilist takes the future to be O-indefinite but the past S-indefinite only, thus leading to the belief that NG would see a growing BU when she looks "down" on the universe. Diagrams of these various NG perspectives (Figures 2.8–2.11) may be found in at the end of Philosophy of Physics for Chapter 2.

Another way of viewing our "Newton's god" argument is in terms of "where" time is in the presentist picture compared to the eternalist picture. In the presentist picture, NG is still constrained by time. The fact that NG is removed from the strictures of the universe does not entail her separation from some notion of time in which she must still continue to exist. It is possible, then, for NG to remove herself from space without removing herself from time on the presentist picture. On the eternalist picture, however, NG is free from the strictures of temporality. It is unclear what the character of the 5D universe NG inhabits is (the fifth dimension could be conceived as some sort of second-order time, a fourth-order space, or some phenomenology of dimensions we do not experience—think of the "Bulk Beings" in Christopher Nolan's film, *Interstellar*, for example); however, the point is that NG is free from time as well as space as it exists in the BU since the two are inextricably linked, and thus time has the same ontological status as space. The eternalist does not have to argue that time behaves the same way as space does, simply that time and space are inextricably linked, which is a stance that the presentist rejects.

There may be some who believe that NG is not a suitable tool for dealing with the presentist/eternalist distinction; in particular, one might find our NG

---

[20] On some presentist views, she might even see a point. See Stein [1968] for more on this.

question-begging since a God's-eye point of view might somehow allegedly violate basic tenets of SR; however, one must note that by hypothesis NG is removed from the 4D manifold (spacetime) that she observes. Such a being would be constrained to see a spacetime that conforms to special relativity (SR) even though this "God-frame" itself would not so conform. SR can only make claims about perceptions of spacetime from within spacetime, and since this "God instead of god-frame" is outside of spacetime, this relativistic objection does not obtain. Even without positing the existence of NG or even a position from which NG could look, we have already shown that the presentist/eternalist distinction can be stated in terms of the separability of space and time, and so if this objection to NG as question-begging is simply that one cannot remove oneself from space without removing oneself from time as well, then the objection has already conceded our point to us. Using our novel argument for the eternalist position, Dorato's two previous objections to eternalism can be ignored as well. Nowhere in our argument do we claim that the past, present, and future are all "simultaneous," nor is there any confusion between eternal truths about existence and the eternal persistence of events. First, an appeal to some sort of "second order" time is completely unnecessary for our formulation of the eternalist position, and as such the accompanying language of the "past, present, and future existing simultaneously" has been discarded. As noted earlier, NG's frame need not necessarily be conceived as some sort of second order time; further, it is merely a thought experiment to show that Dorato/Savitt-type arguments are dependent on verificationism of a sort SR need not entail.

It would be absurd to argue, therefore, that two perspectives as different as these are, are in fact, metaphysically and empirically equivalent in principle. In other words, the claim to metaphysical, empirical, or experiential equivalence assumes that the ant's-eye view is the only view. For this reason, Dorato's and Savitt's grander claims must be dismissed. The most these two authors can suggest is that a better definition of reality is necessary before the presentist/eternalist debate can be undertaken, and so, with such a definition provided, Dorato's and Savitt's deflationary claims can be ignored. Dorato and Savitt are right to point out concerns with definitions of terms (such as "real") in arguments such as ours, but generally speaking this is the most that linguistic analysis can contribute to the presentism/eternalism debate. The most such appeals can do is determine that certain positions in the debate are "unspeakables" or that the language used must be clarified for the debate to proceed.

2.7.2 The transitivity of reality

Our new definition of an event's reality as a combination of definiteness and distinctness also has implications for the second compatibilist objection to the RoS argument, namely that there is no good reason why reality or the "is co-real with" relation ought to be transitive. The first response to this claim is that any relativistically invariant relational property must be transitive across all reference frames.

For example, consider the property of "lightlikeness along direction x."[21] Any two events that are lightlike separated in some direction share this property, and all observers in all frames will agree that two events are lightlike separated if they are so due to the fact that the speed of light in a vacuum is a universal constant. Thus, if event A is lightlike separated from event B and event B is lightlike separated from event C in the same direction, then event A must be lightlike separated from event C (in this same direction). This deduction is true even if one adds different relativistic frames into the equation. For instance, if event A is lightlike separated from event B in direction x in a frame moving with velocity v and event B is lightlike separated from event C in direction x in a frame moving with velocity u where u is not equal to v, it is still the case that event A and event C are lightlike separated in a frame moving with velocity w no matter what the value of w[22]. Thus, from this simple example, one can see that a relativistic invariant quantity is transitive across inertial frames.

There are two other relativistic invariant properties aside from "lightlikeness" that we would like to discuss now. The first of these is number. All observers, no matter their frame, will agree on the number of events that occur. Thus, no matter what frame an observer is in, it will never be the case that she will see an event take place that another observer does or could not see. Though observers may disagree about some of the properties of an event, no observer will see a "novel" event; that is, there is no event simpliciter that one can only see if one is in a certain reference frame. This means that the very existence, the very definiteness of an event-as-such must be a relativistic invariant, and thus as per our pre-established criterion, definiteness must be transitive across frames.

Another relativistic invariant is the spacetime interval between two events. This separation is defined by the spacetime metric as

$$s^2 = (ct)^2 - x^2 - y^2 - z^2 \tag{2.1}$$

where "s" is the spacetime interval, "c" is the speed of light, "t" represents time, and "x," "y," and "z" are spatial coordinates in 3-space. Because the interval between events is an invariant, it is always possible for observers in different frames to distinguish between different spacetime events in a consistent manner. Because of this, no observer will confuse two events that are seen as distinct in another frame. Thus, the invariance of the spacetime interval implies that distinctness is a relativistic invariant. Thus, as per our pre-established criterion, distinctness also must be transitive across frames.

---

[21] The "x" in "along direction x" in this property should be a four-dimensional vector pointing from one event to the other. We include this condition to rule out the following, non-transitive case: consider a light beam shot out from a spaceship at A, reflected off of a mirror at B, and returned to the ship at C. A and B are lightlike, B and C are lightlike, but A and C are timelike. However, this non-transitivity arises from the fact that the direction of the light is changed at B, and so the vector x shifts at this point. We thank Gordon Belot for bringing this objection to our attention.

[22] Within relativistic limits, of course.

Now, since reality in our formulation has definiteness and distinctness as necessary and sufficient conditions and since both definiteness and distinctness are relativistic invariants, it follows that reality, the conjunction of definiteness and distinctness, should also be a relativistic invariant. Finally, as has already been established, any relativistic invariant must be transitive across frames, and therefore our "equal reality" relation must be transitive across frames. This argument suggests that, as a logical consequence of SR combined with our definition for reality, the frame-independent reality of the universe must obtain. This logic provides more than sufficient reasoning to support objectivity in our co-reality definition, and so the weight now falls on the shoulders of Savitt and the presentists to explain why "is real for" should not be transitive if they want to continue pushing this point.

Someone might object as follows: why assume that the "is real for" relation must be relativistically invariant? What is real for me might not be real for you. We are happy to admit that "what seems real to you may not be what seems real to me," but making reality a frame-relative notion doesn't seem to leave us with anything like the notion of reality that we have relied on for years. Another way of framing this is that the metaphysical status we assign to an event should be a property of that event alone (or perhaps a relation between the event and some specifiable background if we are structural realists), not a relation between the event and some observer. People who take this line need to provide some argument for why a more permissive definition of reality or objectivity is desirable. After all, we have given such an argument for the more restrictive definition. we are not sure what such an argument would look like or why we ought to consider it compelling.

2.7.3   Against the point presentist

There have been several arguments against the "here, now" presentist as Stein[23] presents him. This variety of presentist holds the present to consist of a single point in spacetime and defines the "now" as both temporal and spatial. There have already been several excellent responses to Stein's view, most notably those provided by Cifone [2004] and Petkov [2006]. We will here reiterate and rephrase Cifone's and Petkov's points to show that the "point" presentists (see figure 2.11), as they are traditionally called, do not hold a viable position.

The first argument against point presentism comes from Cifone. As previously discussed, it is easy enough to see how anti-realism can be reduced to a form of point presentism, but the opposite seems true as well. Point presentists can be taken to be essentially solipsists since what exists at only one point (presumably,

---

[23] Bourne [2007] points out that Stein was not assuming a "common-sense" notion of simultaneity when he attempted to redefine simultaneity within relativity as the "here" and the "now." It seems that Stein's original point was not so much that simultaneity had a different nature than previously thought but rather that the conception of simultaneity that comes to play in everyday discourse has no currency in SR.

the point where the point presentist currently exists) is all that exists. This is not an argument in itself, and there are ways around point presentist solipsism, but these views are almost equally bad. If there is more than one "point present" in the world (i.e., if he rejects solipsism), what is required for a point to be "the present?" Is there some "present-maker" that defines the present, that selects it out from all possible "presents?" If there is, what would such a "present-maker" be? What is more, if there are a large number of "presents" that all compose reality, why do none of them agree with each other? For if the present is only a single point, it follows that multiple "nows" will not count other "nows" as real. There will be no agreement among different observers in different frames, let alone different observers in the same frame, as to what constitutes reality. Thus, it seems that the point presentist loses all semblance of self-consistency when he explains his position and runs the risk of having his position collapsed into absurdity.

Perhaps most damning to the point presentist, however, is Petkov's response. Petkov points out that a point presentist reduces reality to a single, zero-dimensional point. If point presentism is the case, he asks, why does the universe appear to be four-dimensional, as evidenced by the aforementioned 4D spacetime invariants? The universe defended by presentism which lacks the 4D manifold in favor of a 3D universe seems unable to support or explain phenomena like length contraction and time dilation, but it appears nearly impossible to reconcile a zero-dimensional view of spacetime with such phenomena. Such a view, Petkov argues, reduces to solipsism. After all, consider two observers, A and B. If A and B are distinct observers, any observation event by observer A will not be real to observer B since only observer B's "here and now" are real to him. This solipsism leads to the loss of realism that Cifone [2004] points out. Petkov also claims that only a 4D view is supported by SR by refuting the 3D picture of the world as well. His argument is that the phenomena of length contraction and time dilation, both of which allow different observers to hold ontologically distinct and correct beliefs about the 3D properties of an object, cannot be as completely described by a 3D worldview as by a 4D block universe view. He compares the situation to looking at a 2D plane; one can certainly describe the plane as a series of lines in the x-direction for different, constant values in the y-direction, but this "complete" description of the phenomenon does not change the fact that it is a 2D plane and not a 1D line that is being described. If a 3D world is inadequate, then it stands to reason that lower dimensional representations of spacetime would likewise be inadequate. Thus, the 0D description of the world presented by the point presentist must be incorrect. If one is to believe in the point presentist as a viable alternative to the eternalist and the traditional presentist, the point presentist must provide physical support for a 0D universe or else abandon his view.

Before leaving point presentism, however, there is one perspective similar to Stein's that advocates changing the definition of simultaneity in order to save the presentist from the RoS argument. This more recent shift is presented by Bourne [2007] and ought to be addressed here since it is a challenge to the notion of simultaneity we employ, a challenge that adheres to the logic that Stein originally used when proposing point presentism (see footnote 23). Bourne argues that simultaneity is absolute within spacetime. According to Bourne, the notion of absolute reality does not translate into the language of relativity because no one can determine whether or not two events are simultaneous by observations within a frame. He turns simultaneity on its head in presentism, not by defining "what is real" by "what is present" but rather "what is present" by "what is real." Bourne appeals to a linguistic analysis in terms of conjunction, instead of observables in the world as the basis for reality and thus simultaneity. In short, Bourne's reinterpretation of simultaneity insists that simultaneity is absolute by ruling out the possibility of determining simultaneous events (or, it seems, reality) by observation alone.

Bourne's reinterpretation of simultaneity shows to what extremes presentists must go to rescue their philosophy of time from the RoS argument. By the time Bourne is finished with simultaneity, there is nothing resembling the common-sense notion of simultaneity left. Not only is simultaneity dictated as absolute without empirical evidence or verification (for surely one cannot appeal to physical grounds for such an argument), but simultaneity has now also been removed from the realm of science altogether. There is no longer any observation that can determine if two things occur at the same time! Not only does this assertion fly in the face of common-sense views of simultaneity, it also poses dire consequences for science and human knowledge when combined with presentism. If Bourne's simultaneity gives us no access to a distinctively "real" character for "real" events, how can any empirical evidence help in determining which things are real and which things are not? Does linguistics then pose a better means to come to truths about the natural world than science does for Bourne? If we are planning on choosing a metaphysics of time that best accounts for the phenomena at hand without making any wild metaphysical claims, it seems clear that Bourne's reinterpretation of simultaneity does not save presentism since even the claim that the past, present, and future are all equally real is a more conservative claim than that simultaneity and reality are both phenomena to which no one has empirical access.

It is, however, possible that one can reinterpret Bourne's claims about the simultaneity in physical terms; such a reinterpretation of Bourne's simultaneity would necessitate a preferred foliation of spacetime.[24] Though we will not address Bourne's revised notion of simultaneity directly any further since he does not explain his simultaneity in terms of preferred foliations of spacetime in any satisfying way, we will address preferred foliation presentists generally in the next section.

2.7.4 Preferred foliations in spacetime

A slightly tougher objection to RoS is raised by those suggesting that spacetime has a preferred foliation. Such a foliation would run counter to current beliefs not

---

[24] Bourne explicitly endorses such preferred-foliation presentists in his book, though he does so in a different section from the one in which he advocates his radical revision of simultaneity.

only about eternalism but about relativity as well, for one of the chief tenets of relativity as it is traditionally interpreted[25] is that there exists no preferred reference frame. The good news for the eternalist is that there is very little physical evidence[26] to support such a preferred foliation, but such preferred foliations may be postulated. Assuming that such a foliation is found, then, does our RoS argument for BU still follow?

The first response to the preferred foliation objection is that no preferred foliation theory as it currently stands, even if it were proven to be true, provides the necessary physical mechanisms that would be necessary to explain why such a frame would be preferred. Until physical motivation for a preferred frame is provided, one cannot abandon the RoS argument; modulo perhaps the previous discussion of Shape Dynamics. Perhaps there is some way in which the "now" transforms as it goes into other frames. Perhaps the "now," though it is dependent on its preferred spacetime foliation, is still present or still has metaphysical influence on other frames. Until physical motivation for a preferred reference frame is provided, one simply cannot know these things. After all, we do use CMB ("cosmic time") as a pragmatic preferred frame in physics but it does not impugn BU any more than proper time does. In a purely relativistic context the claim that the Big Bang occurred 14 billion years ago is completely frame-dependent, there are other possible, equally valid choices to be made. The point is that none of these invariant features internal to SR changes the fact that M4 unadorned has no resources to construct an absolute and objective preferred frame and that RoS implies the reality of all events. On our view, one can always conventionally define a preferred frame such as cosmic time; however, unless one can show that a preferred frame such as a physical mechanism is the cause of physical effects like Lorentz contraction and time dilation (as opposed to mere relativistic effects), a pragmatic-preferred frame of this sort does not negate BU.[27]

---

[25] Other non-standard interpretations, like the Lorentz interpretation, yield the same results as the Minkowski 4D spacetime (M4), no preferred frame interpretation of spacetime, so it should be pointed out that it is not the physical results of SR that are threatened by the preferred frame but rather the currently held understanding of SR which is under fire. See Appendix A for more information on the rejection of the geometrical SR interpretation.

[26] There are those who claim that at the end of the day, a correct theory of quantum gravity or a correct interpretation of quantum mechanics (such as Bohmian mechanics) might yield an absolute preferred frame. While technically true, recent work by Silberstein [2007] and Monton [2006] suggests that: a) an absolute preferred frame is not a likely consequence of future theorizing in either case and b) even if these preferred-foliation theories do pan out as expected, they will run into all the problems outlined in this section.

[27] Another objection to such a move comes from John Mather, winner of the 2006 Nobel Prize in physics with George Smoot for their discovery of the blackbody form and anisotropy of the cosmic microwave background (CMB) radiation, in a talk given at Swarthmore College in October 2007. In his talk, Mather suggested that there may be many "preferred frames" provided by the CMB depending on how the source of the CMB is moving. If there are, in fact, a multitude of "preferred frames," any idea of "reality" that could be grounded in CMB would be useless for presentism because our uniqueness criterion would be violated. There would be many "real" frames that one could choose. It should also be noted that Mather himself does not believe that the CMB frame should be treated as anything more than a useful frame for doing calculations; that is, like the proper time frame, the CMB

Callender's objection [Silberstein, 2007] to the preferred foliation view, however, is perhaps stronger. Callender proposes a problem he calls the "coordination problem." The idea is that even if there is a preferred reference frame[28] there is no reason to believe that this reference frame would provide anyone with a suitable "now" upon which to base presentism. One must in some way prove that the physical preferred frame is precisely the same as the metaphysical preferred frame posited by the presentists. How would one be able to make such an association? And perhaps more importantly, even if it were possible for one to argue that the physical and metaphysical preferred frames were, in fact, one in the same, how would this alter the presentist's conception of the present?

The presentist must explain why changing one's velocity should cause one's views about oneself to be more or less in line with "reality." When one gets in a car and drives to the store, for instance, one has changed their inertial frame; is one now closer to the "real" frame or farther from it? Either way, one doesn't experience the immediate world differently, nor does one perceive any differences in oneself, yet one's ontological status has changed. What, then, is the basis for calling such a velocity shift a "shift into (or out of) delusion" since one notices no difference in oneself when one speeds up or slows down? The other problem for the preferred frame presentist is a related concern: if the preferred frame is what is "real" but one experiences the world in exactly the same way whether one is in the preferred frame or not, why should one care about "reality?" What makes reality a meaningful concept to one if it is not linked with any physical, psychological, or epistemological change? For a preferred frame presentist, reality has no important implications other than to save presentism. Again, reality becomes distantly removed from our experiences, and though we may be able to convert all of our dimensions, temporal and spatial, into our "real" dimensions according to the preferred frame, these real dimensions will be no more important to our lives than our dimensions according to any other frame.

In the end it seems like the preferred foliation proponent is providing a view that is perhaps as inimical to the presentist as to the eternalist. One of the major reasons why presentists hold the position they do is that it seems to agree with human manifest experience of time. If this experience were hung on some preferred frame due to CMB radiation or preferred frames as posited by some Bohmians and collapse theorists, it would be possible for a "now" to exist that was completely alien to human experience. Does the phrase "now" even have any meaning when it has been removed from human perceptions of time? The burden falls to the presentists here to prove that a meaningful "now," a physical preferred foliation of spacetime, and an identical metaphysical preferred foliation of spacetime are all compatible, and since no such reconciliation of all three of these ideas has been provided by the presentist camp, we are forced to conclude

---

frame is not "real" in some special way, but is rather just a helpful tool for physical calculations [Mather 2007, personal communication].

[28] Specifically, Callender is concerned with a preferred reference frame that might emerge from robust violations of the locality principle in Bohmian mechanics (and other modal interpretations) or preferred frames required for instantaneous collapse in some collapse accounts of quantum mechanics.

with Saunders [2002] that the burden of proof in the presentism/eternalism debate lies entirely on the shoulders of presentists instead of with the eternalists
because there is nothing obvious in the resources of M4 alone to be a preferred
frame to ground presentism, at least nothing not ad hoc, merely pragmatic, or
perspectival.

2.7.5   The spatial presentist: Absurdity in incompatibilist presentism?

Having answered the presentist objections of the RoS argument in turn, we would
like to propose another argument along the same lines as the RoS argument
which, we believe, should serve as a pre-emptive criticism against incompatibilist presentist arguments to come. Suppose that there exists a new kind of realist
called a spatial presentist. The spatial presentist believes not that all events occurring simultaneously are real but that all events that occur in the same place are
real. Perhaps there is a sphere (infinitesimally small, for our purposes) that the
spatial presentist has set aside, following which he claims that "the only things
that are real are those in this sphere." One might ask, then, what would be real
after the creation of the sphere at an event A in the above diagram, which shows,
from relativistic considerations, what will be real.

From Figure 2.7 it is clear that we are left in a situation directly analogous
to the temporal presentist situation previously established in our RoS argument,
for the spacetime diagram shows a property we will call the relativity of same
position or RoSP. One can simply rotate our Figure 2.6 and make an RoSP argument to disprove spatial presentism in the same way that the RoS argument
disproves temporal presentism. The arguments are completely symmetrical in
the same way that RoSP is symmetrical with RoS, but what does this show? Only
that if an incompatibilist presentist of the non-spatial variety wants to assert that
temporal presentism and temporal presentism alone is correct by proposing some
new feature of spacetime, she must be careful that her argument and mechanism
establish presentism but do not allow for spatial presentism.[29] This is yet another
burden that the incompatibilist presentist must carry. The symmetry between
RoS and RoSP suggests that incompatibilist presentists must establish a physical
basis for temporal asymmetry so that spatial presentism does not become as viable
and defensible a position as presentism itself, for reconciliation between spatial
and temporal presentism must lead to point presentism, which we previously
discussed.

Presentists might respond that time is distinct from space in that time passes
and space does not. Thus there is no motivation for spatial presentism. That is,
maybe some presentist trick might be adaptable to the spatial context, but why

---

[29] This is, of course, assuming that the presentist in question is not a point presentist or some new
form of presentist who wishes to tie the conception of the "now" together with some more evolved
conception of the "here."

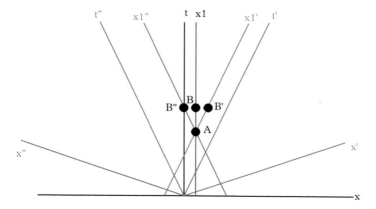

Figure 2.7 *Spatial presentist argument.*

do that? As Savitt would say: "Spatial-presentist talk might be coherent, but who would want to talk that way?"

We are simply placing the additional burden on presentists that they must prompt any response to the RoS argument in a way that wouldn't be equally justified as a response to the RoSP argument. While it is true that temporal passage is a feature (a problematic one, for reasons we have discussed) that applies to RoS but not RoSP arguments; however, temporal passage, as we have shown, isn't enough to rescue presentists from RoS arguments, which means that, to take advantage of this strategy, presentists need to 1) come up with a flaw in the RoS argument and 2) link that flaw to temporal passage so that it doesn't apply to the RoSP argument as well. Therefore we have still established an extra burden for the presentist here. How hard it will be to satisfy we cannot say.

2.8 Conclusion

Though the traditional formulations of the Putnam, Rietdijk, and SSC's RoS argument for the block universe may be ill-defined in certain parts that leave the argument open to attacks by philosophers of language and presentists, we have reformulated the argument with more specific definitions that make eternalism the likely victor over presentism. Thus, the task before the presentist in defending herself has become even grander; she must:

1. Find a way to dispel the RoS argument.
2. Show why presentism is more likely than eternalism.

3. Integrate temporal asymmetry as fundamental to her argument lest her argument run the risk of establishing an obviously false view (spatial presentism) as well as it establishes her temporal presentism.[30]

It is clear from the foregoing that the most common presentist argument that "space and time are not perceived to act in the same way" is not sufficient to shoulder the weight of a full presentist defense, and thus a more developed presentist argument addressing all of our concerns must be proposed before presentism can escape from the jaws of the RoS argument. Even the retreat into the position of Savitt and Dorato that there is no significant difference between presentism and eternalism seems a difficult one to hold in light of definitions for definiteness and distinctness like the ones we have provided. Thus, in conclusion, we echo Saunders in stating that while eternalism in itself may not have been deductively established by our arguments, the burden falls upon the presentist to show why eternalism is not much more probable. There is only one other compelling argument against the block universe: the argument from the experiential arrow of time. The claim is that time as experienced is not compatible with the block universe. That argument will be addressed in chapters 7 and 8.

Appendix A: Against the dynamical interpretation of special relativity

Various people have defended a dynamical interpretation ("constructive" in Einstein's language) of SR of late (e.g., Brown, 2005a). In the following passage Callender claims the latter interpretation is a potential problem for the RoS argument for BU:

> In my opinion, by far the best way for the tenser to respond to Putnam et al. is to adopt the Lorentz 1915 interpretation of time dilation and Fitzgerald contraction. Lorentz attributed these effects (and hence the famous null results regarding an aether) to the Lorentz invariance of the dynamical laws governing matter and radiation, not to spacetime structure. On this view, Lorentz invariance is not a spacetime symmetry but a dynamical symmetry, and the special relativistic effects of dilation and contraction are not purely kinematical. The background spacetime is Newtonian or neo-Newtonian, not Minkowskian. Both Newtonian and neo-Newtonian spacetime include a global absolute simultaneity among their invariant structures (with Newtonian spacetime singling out one of neo-Newtonian spacetime's many preferred inertial frames as the rest frame). On this picture, there is no relativity of simultaneity and spacetime is uniquely decomposable into space and time. Nonetheless, because matter and radiation transform between different

---

[30] We would like to note at this point that there is an obvious reason why spatial presentism has never caught on in the philosophy of time: it does not agree with our perceptions of reality. However, if one wants to dismiss spatial presentism on these grounds but remain a presentist, one's workload is not lessened since one must now conclusively prove a link between experiences and reality.

frames via the Lorentz transformations, the theory is empirically adequate. Putnam's argument has no purchase here because Lorentz invariance has no repercussions for the structure of space and time. Moreover, the theory shouldn't be viewed as a desperate attempt to save absolute simultaneity in the face of the phenomena, but it should rather be viewed as a natural extension of the well-known Lorentz invariance of the free Maxwell equations. The reason why some tensers have sought all manner of strange replacements for special relativity when this comparatively elegant theory exists is baffling.

[Callender, 2007, p. 3]

See also Brown [2005a] and his essay [Brown and Pooley, 2006] for a more developed argument for this stance.

First, Brown [2005a, p. vii] himself is clear that he is not defending either ether or a preferred frame, unlike Lorentz himself. We grant that SR is neutral about the ontology of spacetime, but we think there are good reasons for preferring the kinematical over the dynamical interpretation, though we cannot pursue them here.[31] We do want to note that we are not convinced of Callender's claim that the dynamical interpretation of SR necessarily negates the RoS. At least in the case of Brown, who again, does not claim to be defending absolute simultaneity, while his arguments may lead to spacetime relationalism, they do not obviously entail the falseness of RoS as such. So until someone provides a cogent argument from spacetime relationalism to the falsity of the RoS, our argument remains intact.

Second, even granting an absolute frame, Brown's dynamical interpretation does not obviously save the presentist since she must still face some of the problems raised above. For example, even if there is an absolute spacetime and a universal moment of the present, there is no reason to believe, as per Callender's objection discussed earlier, that such a present lines up with human experience of the present. What is more, as long as Lorentz contractions and dilations exist, one observer A traveling at relativistic velocities with respect to another observer B must entertain two options: either A's clocks are running fast and B's clocks are synched up with the true present moment, or B's clocks are running fast and A's clocks are synched up with the true present moment, or both A's or B's clocks are running fast with respect to the true present moment. There is no empirical way to determine which of these options obtains, and so, once again, an appeal to a preferred spacetime foliation untethers our notion of the present from our experiences of the world around us. So if Brown and Callender want to escape the BU inference, they need to do more than claim that relativistic phenomena need to be redescribed as follows: it is not that an observer moving at relativistic velocities sees a different present—there is only one present in this spacetime. Rather, moving clocks actually slow down, so they cease to be a good guide to how fast time is actually passing.

---

[31] See Michel Janssens "Drawing the line between kinematics and dynamics in special relativity" in the Phil. Sci. archive (reference number 3895) for good arguments favoring the kinematical interpretation. See also Petkov [2007] where the kinematic interpretation of Minkowski spacetime realism has consequences not easily or obviously accounted for by the dynamical interpretation.

Finally, if we are to take seriously the implication that quantum mechanics (our best theory of matter) is to SR what statistical mechanics is to thermodynamics, then had not quantum mechanics better be able to explain (in some robust sense of the word) the key features of SR such as Lorentz invariance? Obviously, this condition has not been met and merely interpreting Lorentz invariance to be restricted to dynamical laws only hardly does the trick.

Appendix B: Objection to RoS argument from a preferred frame or time as fundamental

As you recall from chapter 1 in his book *Time Reborn: From the Crisis in Physics to the Future of the Universe* [2013], Smolin rejects the block universe altogether in favor of a kind of presentism in which fundamental physics has a global moving Now. Indeed, change and time themselves are fundamental and even the laws of physics themselves can change over time. We mentioned some of the problems with Smolin's view in chapter 1, but we will return to his view momentarily. Ellis' rejection of relativity is much less radical [Ellis, 2014]. He has a glowing-growing block model but the growth of the block is not global, it is frame dependent. He has made a nice argument, maybe as nice as is possible, assuming there is something to his "Crystalizing Block Universe" in dealing with delayed choice experiments, which is one of his main motivations (more on delayed choice experiments in chapter 4). He has replaced "meta time" with "proper time" so as to connect a physical (universal) Present to all individual Nows. He has two internal problems that we can only briefly mention here.

First, he states that he has to develop an initial value formulation for his global Present (surface S). This will be tricky when a given S can possess both timelike and spacelike subsets, as with an inhomogeneous (realistic) spacetime. For example, there is no well-defined means of determining distance along a path that varies between timelike and spacelike. Thus,, we imagine some difficulties finding a dynamics for the evolution of S.

Second, S is irregular precisely because some people reside on/near localized massive bodies M while others do not (M is the source of inhomogeneity in the "inhomogeneous spacetime"). Suppose one twin (Joe) moves near M leaving his twin (Bob) far from M. what happens when Joe moves from M to shake hands (metaphor) with Bob? Then you will have two twins touching each other that do not reside on the same S (they will be two different ages). Ellis doesn't mention this at all, but might respond as he does regarding closed timelike curves: "This would require the fundamental worldlines to intersect; but if the fundamental worldlines intersect, the density diverges and a spacetime singularity occurs." Of course, no singularity occurs when two people shake hands, so this cannot be the resolution. It seems he is ruling out the possibility of two fundamental observers of different ages ever shaking hands. Fundamental observers should include humans moving around the universe (otherwise how does his Present connect with our individual Nows, as he claims early on?). So, he has to account for the fact that it is possible in general relativity (GR) to have twins part company and rejoin to shake hands at different ages. In that case, which twin is "real?" That is, which twin's Now coincides with the Present? For these sorts of reasons and given the fact that relativity viewed realistically and unadorned really is incompatible with the growing-block model, we suspect that anybody really serious about defending presentism or growing block in the face of relativity is going to have to be much more radical in their rejection or transformation of relativity, which brings us back to Smolin.

In Smolin's case, he reformulates GR in terms of Shape Dynamics (SD). He claims that you can have a Present/Now (a preferred frame), that is both objective and universe-wide (global-now) given the "new" formulation of GR known as "Shape Dynamics" developed by Julian Barbour and company:

> In a word, in general relativity size is universal and time is relative, whereas in shape dynamics time is universal and size is relative. Remarkably, though, these two theories are equivalent to each other, because you can—by a clever mathematical trick that isn't necessary to go into here—trade the relativity of time for the relativity of size. . . . The physical content of the two descriptions will be the same, and any question about an observable quantity will have the same answer.
>
> [Smolin, 2013, p. 170]

Smolin claims you can incorporate absolute simultaneity into GR and it makes exactly the same empirical predictions, even with a single universal present. The trick is it relativizes size rather than time. Observers moving relative to one another will all agree on when two events occur but disagree on how large the relevant objects are. Here is how Gryb describes SD on his tutorial website:

> Shape Dynamics is a new theory of gravity that, for most situations, is completely indistinguishable from General Relativity, but which is unlike General Relativity in that it is based on a completely different symmetry principle and notion of time. In Shape Dynamics, there is no notion of relative simultaneity: simultaneity is absolute. Instead, it is local spatial scale that is relative. Despite these distinctly different features, it can be proved that, for a specific choice of local scale, one can almost always reproduce a spacetime that solves the Einstein equations.
>
> [Gryb, 2014]

While SD is essentially a new interpretation of GR, as Gryb says:

> More pragmatically, given that no attempt to find a completely adequate theory of quantum gravity has been successful, it is perhaps useful to investigate different formulations of the classical theory (in this case, one with a different symmetry principle) which could lead to insights into the quantization. The alternative symmetry principle offered by Shape Dynamics provides a new theory space for quantum gravity and differs from standard GR on a global level. This could change the whole structure of the quantum theory.
>
> [Gryb, 2014]

Here is one way Gryb's suggestion could be implemented, essentially a Hamiltonian quantization of GR with a preferred frame:

> The final move is to adopt a re-description of gravity in terms of a formalism that features a notion of preferred slicing. One attractive possibility along these lines is suggested by the shape dynamics formalism originally advocated by Barbour and collaborators (Barbour (2003b, 2011); Anderson et al. (2003, 2005)) and then brought into modern form in Gomes et al. (2011). Within this formalism, the principle of local (spatial) scale invariance is introduced with the consequence of favouring a particular notion of simultaneity. This selects a unique global Hamiltonian and thus allows for relational quantization to be applied. Shape dynamics is based upon a re-codification of the physical degrees of freedom of general relativity via exploitation of a duality between two sets of symmetries. Whereas general relativity is locally time reparametrization invariant and spatially diffeomorphism invariant; shape dynamics is globally time reparametrization invariant, spatially diffeomorphism invariant, and locally scale (i.e., conformally or Weyl) invariant. In the class of spacetimes where it is possible to move from one formalism to the other (those that are 'CMC foliable') the physical degrees of freedom described by the two formalisms are provably equivalent, they are merely clothed in different descriptive redundancy.
>
> [Evans et al., 2016, pp. 10–11]

While there is no lengthy discussion of SD and presentism versus the block universe in the literature that we could readily find, we can say that SD is certainly compatible with both. And, while Smolin may see SD as an opportunity to introduce a preferred frame into GR in a natural way that would negate the block universe implications of relativity, others such as Barbour do not. For example, Evans, Gryb, and Thebault in their paper "Psi-Epistemic Quantum Cosmology?" invoke SD in a central way but in the context of a block universe with time-symmetric quantum mechanics in the Price–Wharton family (see Evans et al., 2016). Note that this discussion thus far has been about SD in the context of the block universe, a preferred frame, a moving Now, etc., that is, about the passage of time. Researchers such as Koslowski, Mercati, and Barbour have, however, also suggested that SD might help explain the Direction of time (that there is an objective difference between moving toward the past and moving toward the future). They suggest that rather than the Direction of time or thermodynamic arrow of time emerging from the initial singularity at the Big Bang, given SD, it could be driven by gravitational clumping [Becker, 2015]. We will also deal with Direction in chapters 7 and 8.

As will be clear throughout this book, it is important not to conflate several distinct issues pertaining to time such as the passage of time (time flows or is flux-like as with presentism) versus the Direction of time, you could have one without the other as we will make clear in chapters 7 and 8. Nor are either passage or Direction necessarily incompatible with a block universe. Maudlin for example, whom we mentioned in chapter 1, is a blockworlder who wants both passage and Direction. Many find his view harder to pin down, as Hoefer notes:

Is Maudlin defending, in any sense, the correctness of the A-series view of time [a moving Now]? Here things get a bit murky. Maudlin fully accepts the B-series, and accepts "the block universe"—if by this we mean that spacetime is a four-dimensional whole with no ontological distinction between parts lying to the future of us vs parts lying to the past. But after defending time's passage in chapter 4 of [Maudlin, 2007], Maudlin returns to the A-series momentarily in his concluding remarks. He points out that rather than there being one A-series, there is an infinity of them, one for each moment that may be chosen as "the present" and thus serve as the divider between past and future. This is simply a terminological point that can be readily conceded. Maudlin claims that by having the B-series, he gets all these A-series 'for free.' But this begs the question: what sort of "now" or present is Maudlin laying claim to, to serve as the divider between past and future? Which spatial structure represents the present?

[Hoefer, 2011, pp. 75–76]

As Hoefer notes, it isn't clear what Maudlin has in mind here, whether he wants a preferred foliation of spacetime based on something intrinsic to relativity (such as cosmic time) or quantum (many people claim that the Bohmian mechanics interpretation of quantum demands a preferred frame), or whether Maudlin wants to add something extra to relativity to get passage or Direction. Maudlin also seems primarily moved by the experiential arrow of time so we will return to all this in chapters 7 and 8. Maudlin does, however, say this:

The passage of time is an intrinsic asymmetry in the temporal structure of the world, an asymmetry that has no spatial counterpart. It is the asymmetry that grounds the distinction between sequences which run from past to future and sequences which run from future to past.

[Maudlin, 2007, p. 108]

Maudlin gives us more insight in other works. If we understand him correctly, He believes ontological status should be inferred from mathematical structure in physics [Maudlin, 2011]. In physics, one of the most fundamental mathematical structures is the differentiable manifold of spacetime. A differentiable manifold is based on a topological space and a topological space is based on set theory. In particular, a topological space is a set and collection of its subsets (called "open sets") that satisfy three axioms (the specifics of which do not concern us). Thus, the most basic mathematical entity for the concept of spacetime is an open set. Maudlin rightly points out that the open sets in the spacetime manifold don't correspond to anything of physical significance, so he suggests trying to find a mathematical structure fundamental to open sets that do have physical significance. He has succeeded in doing precisely that by using directed line segments to define open sets which satisfy the three axioms of a topological space. When the topological space in question is that underwriting the spacetime manifold, the directed line segments provide the orientation for the lightcone structure and therefore the geometry of spacetime. That is, the ordering structure of the line

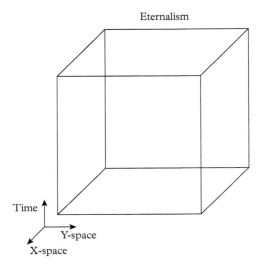

Eternalism

Time

Y-space

X-space

Figure 2.8 *Eternalist perspective on spacetime.*

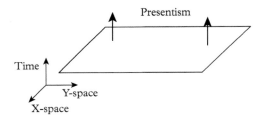

Presentism

Time

Y-space

X-space

Figure 2.9 *Presentist perspective on spacetime.*

segments underwrites a directed time for globally hyperbolic spacetimes. Accordingly, spacetime can be understood to rest fundamentally upon directed time with space being an emergent concept.

So both Smolin and Maudlin agree that time is fundamental and space emergent, but they have very different notions of time. However again, Smolin's primary concern is passage and presence and Maudlin's is direction. Smolin is rejecting the block universe and Maudlin is not. For reasons we will make clear in chapters 7 and 8, our suspicion at the end of the day is that anyone who wants a robust notion of passage or direction would be better off just jettisoning the block universe as Smolin does.

The point of this preliminary discussion is not to assess Smolin's program, Ellis' program, or Maudlin's program—after all they are quite nascent—but to provide a snapshot of the state of the debate at the present moment. Formally

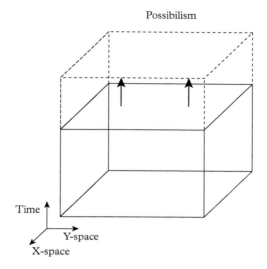

Figure 2.10 *Possibilist perspective on spacetime.*

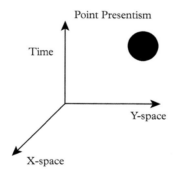

Figure 2.11 *Point presentist perspective on spacetime. This perspective, idealized here as a single point in spacetime, is the most difficult to represent visually since it should have an infinitesimal size. The single dot of the present is the only thing in spacetime that exists on this view of spacetime, making it a much more limited and precise view of the present than the more general form of presentism previously represented.*

rigorous resistance to the block universe picture on the part of those in founda-tional physics is relatively new, so it is only fair that we give it time. However, we certainly don't see anything in the resistance yet that ought to make much of an impact on the block universe consensus. In any case, again, none of this is very surprising because relativity doesn't *entail* a block universe. Let us just imagine a future in which SD or some competitor develops into a completely empirically equivalent and explanatorily powerful alternative account of GR; at the end of the day, the question is which account/interpretation of GR can explain, predict, and unify the most phenomena. This book is an attempt to show that the block universe posit with the appropriate formal machinery isn't just mere metaphysics, but actual physics that explains, predicts, and unifies a great many things. Thus the entire book constitutes our argument for the block universe, not just this chapter.

In chapter 3, we will learn that GR is perhaps even more damning for presentism than SR, and more importantly we will see how viewing the LSU paradigm in a block universe as fundamental resolves many of the puzzles, problems, and paradoxes in GR and GR cosmology.

Foundational Physics for Chapter 2

In the main thread, we used a set of four events with exaggerated times to explain time dilation, length contraction, and the relativity of simultaneity. In this thread, we provide the computational details omitted from the main thread.

First, we show how the second postulate leads to time dilation. Suppose that a light pulse bounces back and forth between mirrors separated by a distance D (top half of Figure 2.12). Alice, at rest with respect to the mirrors, measures a time T for the light pulse to leave from the bottom mirror (Event 1), bounce off the top mirror, and return to the bottom mirror (Event 2). Thus, Alice says

$$c = \frac{2D}{T} \tag{2.2}$$

Now suppose Alice is moving to the right with respect to Bob at speed v. Bob measures a time t between Events 1 and 2 with the light pulse taking the path shown in the bottom half of Figure 2.12. Since everyone measures the same speed of light, Bob computes

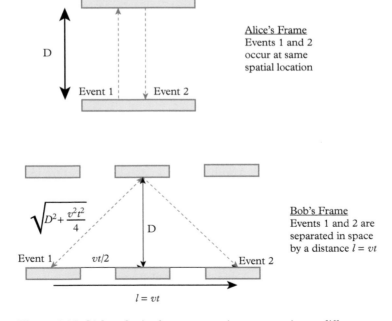

Alice's Frame
Events 1 and 2
occur at same
spatial location

Bob's Frame
Events 1 and 2 are
separated in space
by a distance $l = vt$

Figure 2.12 *Light reflecting between two mirrors as seen in two different frames.*

$$c = \frac{2\sqrt{D^2 + \frac{v^2 t^2}{4}}}{t} \tag{2.3}$$

Solving for $t(T)$ we find

$$t = \frac{T}{\sqrt{1 - \frac{v^2}{c^2}}} \tag{2.4}$$

T is called the "proper time" between Events 1 and 2 because those two events occur at the same location in space for Alice. For Bob, those two events occur at two different locations in space and his time t is longer than the proper time T. That observers measure the time between events occurring at two different locations in space to be longer than the proper time between those events is called "time dilation." Thus, Bob says Alice's clocks run slow. Of course since motion is relative, it must also be the case that Alice says Bob's clocks run slow compared to hers. This difference in the measurement of temporal duration between events leads to a different spatial distance between events as we now show.

Suppose Alice is crashing through two walls at rest with respect to Bob at Events 1 and 2. These walls are separated by a distance $\ell = vt$ according to Bob. Motion is relative, so Alice says Bob is moving to the left at speed v, that is, they must agree that their relative speed is v. Since Alice says the time T between Events 1 and 2 is shorter than the time t that Bob says passes between Events 1 and 2 by a factor of $\sqrt{1 - \frac{v^2}{c^2}}$, Alice must therefore also say her distance L between the walls of Events 1 and 2 is shorter than Bob's distance ℓ by the same factor, that is,

$$\ell = \frac{L}{\sqrt{1 - \frac{v^2}{c^2}}} \tag{2.5}$$

Bob measures what is called the "proper length" between the walls, since he is at rest with respect to the walls, so Alice measures a shorter length than the proper length. That observers in motion with respect to an object measure its length to be shorter than its proper length is called "length contraction." Thus, Alice says Bob's meter sticks are shorter than one meter. Of course since motion is relative, it must also be the case that Bob says Alice's meter sticks are shorter than one meter.

Here are the Lorentz transformations relating the boys' coordinates for the Events 1–4 of the main thread:

- Event 1: 20-year-old Joe and 20-year-old Sara kiss
- Event 2: 20-year-old Bob and 17.5-year-old Kim kiss
- Event 3: 22-year-old Bob and 20-year-old Alice kiss
- Event 4: 25.6-year-old Bob and 24.5-year-old Sara kiss

Joe is located at $x = 0$ (lower-case coordinates are the boys') and Bob is at $x = 1000$ km. When Joe passes Sara their clocks read $T = t = 0$ (upper-case coordinates are the girls') and Sara is located at $X = 0$, so $X = x = 0$ when $T = t = 0$. To find Kim's coordinates for when Bob passes her ($x = 1000$ km, $t = 0$), the Lorentz transformation of SR gives

$$T = \gamma \left(t - \frac{vx}{c^2} \right) = 1.25 \left(0 - \frac{0.6c(1000)}{c^2} \right) = -0.0025 \, \text{s} \qquad (2.6)$$

$$X = \gamma \, (x - vt) = 1.25 \, (1000 - 0.6c(0)) = 1250 \, \text{km} \qquad (2.7)$$

where $c = 300,000$ km/s and

$$\gamma = \frac{1}{\sqrt{1 - \frac{v^2}{c^2}}} = \frac{1}{\sqrt{1 - \frac{(0.6c)^2}{c^2}}} = 1.25 \qquad (2.8)$$

To get Alice's coordinates for Event 3 we have

$$T = \gamma \left(t - \frac{vx}{c^2} \right) = 1.25 \left(0.002 - \frac{0.6c(1000)}{c^2} \right) = 0 \qquad (2.9)$$

$$X = \gamma \, (x - vt) = 1.25 \, (1000 - 0.6c(0.002)) = 800 \, \text{km} \qquad (2.10)$$

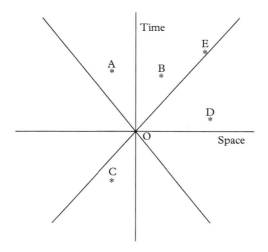

Figure 2.13 *Events A, B, and C are timelike related to the origin O. Event D is spacelike related to O and Event E is lightlike related to O.*

That is, per the girls, Bob was 450 km away from Alice at $T = -0.0025$ s and moving at $0.6c$. Since it took him 0.0025 s to get from Kim to Alice, that is, $\frac{450}{0.6c} = 0.0025$ s, Alice's clock reads $T = 0$ when Bob and Alice kiss. To get Sara's coordinates for Event 4 we have

$$T = \gamma \left(t - \frac{vx}{c^2} \right) = 1.25 \left(0.0056 - \frac{0.6c(1000)}{c^2} \right) = 0.0045 \text{ s} \qquad (2.11)$$

$$X = \gamma \left(x - vt \right) = 1.25 \left(1000 - 0.6c(0.0056) \right) = 0 \qquad (2.12)$$

That is, per the boys, Sara is 1000 km away from Bob at $t = 0$ and is moving at $0.6c$, so she arrives at Bob at $t = \frac{1000}{0.6c} = 0.0056$ s.

   Finally, we point out that there are three ways events can be related in spacetime (Figure 2.13) and Lorentz transformations conserve these relationships when transforming from one coordinate system to another.

3

Resolving Puzzles, Problems, and Paradoxes from General Relativity

We have observed the oldest and largest structures ever seen in the universe. These were the primordial seeds of modern-day structures such as galaxies, clusters of galaxies, and so on. Not only that, but they represent huge ripples in the fabric of space-time left from the creation period. If you're religious, it's like seeing God.

Nobel Laureate George Smoot [Croswell, 2001, p. 155]

3.1 Introduction

As the quotation from Smoot suggests, in the dynamical explanatory schema of cosmology the Big Bang and the early universe play a similar role to Leibniz's God or to Genesis in the Old Testament. As explained by Carroll in chapter 1, one inputs an initial state (initial conditions) and time evolves that state in the context of a space of states to obtain a physics solution per the Newtonian Schema. When a human agent (an experimentalist, for example) is responsible for establishing the initial conditions, the corresponding explanation of the physics solution per the Newtonian Schema Universe (NSU) is entirely without mystery, that is, there is no need to invoke "God-like" origins for the initial conditions. But, in Big Bang cosmology, conditions in the early universe (initial conditions) that are required to provide an NSU explanation of conditions in the universe today are inexplicable. Thus, this dynamical explanation per the mechanical universe is precisely the kind of thinking we are going to reject here. From the ant's-eye perspective we typically view Einstein's equations (in Hamiltonian form) as essentially dynamical, with the cosmic time frame pointing us to the initial conditions for existence: the beginning of the universe. Here we invite the reader to view Einstein's equations in that form as mere regularities of the ant's-eye perspective as an observer embedded in the block universe. We also invite the reader to view Einstein's equations from the Lagrangian Schema Universe (LSU) or God's-eye perspective as fundamental. We can't force the reader to accept this interpretation of GR but, as we shall see, it does resolve many outstanding mysteries of the dynamical universe.

Beyond the Dynamical Universe. Michael Silberstein, W.M. Stuckey and Timothy McDevitt, Oxford University Press (2018). © Michael Silberstein, W.M. Stuckey and Timothy McDevitt. DOI 10.1093/oso/9780198807087.001.0001

In this chapter, we will introduce general relativity (GR) and bring the adynamical global perspective or LSU to bear on general relativistic cosmology to answer Wilczek's question: "The account it gives—things are what they are because they were what they were—raises the question, Why were things that way and not any other?" [Wilczek, 2016, p. 37], as pertains to the flatness problem, the horizon problem, the origin of the universe, and the low entropy problem. Suffice it to say that the block universe interpretation of relativity only gets stronger in GR, as we will make clear in Philosophy of Physics for Chapter 3. We all know when it comes to the arrow of explanation or determination, such as causal chains or nomological[1] deductions, there are only three possibilities:

1. An ordered line ending in a terminus such as the so-called Theory of Everything discussed in chapter 1.
2. An ordered line that continues indefinitely (the "turtles all the way down" model), such as infinitely descending effective theories or infinitely nested sub-atomic particles.
3. Some sort of self-referential or closed loop.

Very few people have any interest in defending option 2. Most people would prefer option 1 for obvious reasons, but we saw in chapter 1 that this goal is probably illusory for the mechanical or dynamical model of explanation. Ultimately, we get stuck with trying to dynamically explain the initial conditions, for example, the Big Bang and early universe, and the values of the constants, not to mention the dynamical laws themselves, for example, why Einstein's equations and not some other? Even if we cooked up some more fundamental dynamical equation for quantum gravity, we can still ask, "why that equation and not some other?" Since dynamical laws and initial conditions are completely orthogonal in the mechanical paradigm, solving one mystery has no necessary bearing on the other, and of course we currently have no consensus explanation for either. So, we are asking the reader to consider option 3, an explanation based on empiricism and self-consistency where the most fundamental explanation (at least from a God's-eye perspective) isn't dynamical, but an adynamical global constraint over the block universe in accord with empirical facts.[2]

Hawking once defended an option-3-type view that he called the "no-boundary proposal." The idea is to model the universe as finite and as having a "beginning" and an "end" in a sense, but no boundaries. He makes the analogy with the Earth's North and South Poles, where the former is the "beginning" and the latter is the

---

[1] Having to do with laws or principles that logic doesn't require to be true, but are believed to be true nonetheless. Newton's law of gravity is one such example.
[2] A dynamical version of option 3 is not inconceivable. For example, one might attempt to circumnavigate the second law of thermodynamics and create an ouroboros-like spacetime whereby the universe ages then anti-ages in a repeated "circular" fashion. In that case, there would be no dynamical "beginning."

"end" in this analogy, the point being that on the Earth there are no edges or boundaries. That is, like the Earth, the universe is finite, and just as it doesn't make sense to ask what is north of the North Pole, it doesn't make sense to ask what came before the Big Bang, the temporal coordinate system of GR has no meaning beyond or outside of the spacetime manifold. Spacetime has a region that is naturally considered the furtherest point in the past (i.e., the Big Bang but without a singularity, just a point in spacetime) and one that is considered the furthest point in the future (the Big Crunch), but has no beginning in time as it were. Here is how Hawking describes the view:

> If space and imaginary time are indeed like the surface of the Earth, there wouldn't be any singularities in the imaginary time direction, at which the laws of physics would break down. And there wouldn't be any boundaries, to the imaginary time space-time, just as there aren't any boundaries to the surface of the Earth. This absence of boundaries means that the laws of physics would determine the state of the universe uniquely, in imaginary time. But if one knows the state of the universe in imaginary time, one can calculate the state of the universe in real time. One would still expect some sort of Big Bang singularity in real time. So real time would still have a beginning. But one wouldn't have to appeal to something outside the universe, to determine how the universe began. Instead, the way the universe started out at the Big Bang would be determined by the state of the universe in imaginary time. Thus, the universe would be a completely self-contained system. It would not be determined by anything outside the physical universe, that we observe.
>
> [Hawking, 1996]

Hawking has since abandoned this view primarily because his solution requires that the universe stops expanding infinitely and some anti-inflationary/anti-expansion process kick in that reverses the direction of time, otherwise the universe wouldn't be finite. However, it is now widely believed that the universe is ever-expanding and that the rate of that expansion is ever-accelerating.[3] Like Hawking, we will be defending a version of option 3 (adynamical in our case), but all similarity to his view ends there. The reason we demur at Hawking's particular model has nothing to do with accelerated expansion; it is his invocation of realism about the imaginary time of quantum mechanics. Hawking wanted to claim that somehow spacetime "emerges" from some quantum state of the universe in imaginary time. The view of quantum mechanics we defend (see chapters 4 and 5) does not allow for this kind of realism about imaginary time or quantum cosmology so conceived.

More specifically, in this book we are defending option 3 via "spatiotemporal ontological contextuality" in relativistic cosmology, quantum mechanics, quantum field theory, and quantum gravity. We will elaborate on contextuality in general and spatiotemporal ontological contextuality specifically in chapter 4, but their basis for option 3 will serve to resolve all puzzles, problems, and paradoxes

---

[3] We will actually take issue with this latter claim in chapter 6.

associated with GR in this chapter. On this view, there is nothing particularly mysterious or sacred about the initial conditions at the Big Bang or the cosmic time frame because conditions at any point in spacetime globally constrain conditions at the other points in spacetime. So, we could apply our adynamical global constraint (in this case Einstein's equations) to empirical information from any point in spacetime and compute conditions at the Big Bang or some other features of the "early" universe. Since dynamical laws for the NSU are just regularities or patterns from the ant's-eye perspective and the adynamical global constraint for the LSU is fundamental, and since we could pick any point in spacetime as our starting location for computational purposes (and these locations need not correspond to the locations of initial conditions in the Newtonian Schema),[4] the explanatory priority of initial conditions *a la* the Big Bang for dynamical laws disappears. And with that, the mystery of the origins of initial conditions also disappears. The mysteries of why those particular initial conditions at the Big Bang and why those dynamical laws, such as Einstein's equations, were both mysteries only because we took the dynamical/mechanical ant's-eye perspective as fundamental. In our view, while computationally we must still distinguish between initial computational input (such as initial conditions) and laws/constraints, ontologically no point in the manifold is more special than any other and dynamical laws are just patterns in the block universe, so we can say farwell to these mysteries. Conceived in this way, the universe really does explain itself in a self-consistent fashion, unless one insists on holding onto the mechanical paradigm and asking questions like "what 'caused' or 'created' the block universe as a whole?"

While it is difficult for us "ants," the trick is to let go of the mechanical paradigm; there is no dynamical or causal explanation for the block universe and there is no reason to think that it makes sense to talk about events "prior to" or "after" the events of the block universe. As an analogy, the block universe is more like a crossword puzzle than a finite automata such as Conway's Game of Life. We will emphasize this point again in Philosophy of Physics for Chapter 3, as well as drive home the fact that competitors to the block universe fair even worse in GR than they did in special relativity (SR). So again, we must ask readers to temporarily suspend their dynamical bias (the presumption of the fundamentality of dynamical explanation) in order to appreciate our "spatiotemporal contextuality writ large." This explanation in no way discounts the importance of research into resolving the singular nature of the Big Bang, which we will also address. We simply bring the God's-eye view to bear on the apparent special conditions associated with the early universe, its inexplicable origin, and the paradoxes associated with GR's closed timelike curves. For those who want a peak at the mathematics associated with the concepts of chapter 3, we have provided them in Foundational Physics for Chapter 3.

---

[4] More precisely, the initial value formulation of GR provides a solution to Einstein's equations for globally hyperbolic spacetimes (those that can be foliated into spatial hypersurfaces) using the spatial metric on a 3D spatial hypersurface and its generalized extrinsic curvature [Wald, 1984]. These are the GR Hamiltonian components of the initial state. In Foundational Physics for Chapter 1, the counterparts would be the initial position and initial velocity, respectively.

3.2 Abridged introduction to general relativity

Einstein found it necessary to correct Newtonian mechanics with special relativity (SR), as we explained in chapter 2. Later, he found it necessary to correct Newtonian gravity with general relativity (GR). It was necessary for two reasons. First, in Newtonian mechanics gravity was a mysterious force that acted instantaneously even at spacelike separation to connect massive bodies, and that account can't possibly work in the setting of Minkowski spacetime and given the two axioms of special relativity. Second, there were already observed heavenly anomalies to Newton's description of gravity.

As Weinberg shows throughout his GR textbook [Weinberg, 1972], GR follows from the Principle of Equivalence (or its alternative version the Principle of General Covariance, see below). The Principle of Equivalence states that the inertial mass of an object equals its gravitational mass, so that all objects near, say, the surface of Earth fall at the same rate. Per Newtonian mechanics and gravity, this follows simply from Newton's second law $F = ma$ and the gravitational force equation near the surface of Earth $F = \frac{GMm}{R_E^2}$ where M is the mass of Earth, R_E is the radius of Earth and m is the mass of the falling object. Setting these two equations equal to each other results in $a = \frac{GM}{R_E^2}$ for any mass m, as long as m is truly the same in both the second law and the gravitational force. Einstein took this thinking further realizing that this equivalence of inertial and gravitational mass means an observer falling from the roof of a house would effectively see no gravitational field, just as if he was in empty space far from any gravitational sources [Einstein, 1920]. Therefore, he reasoned that gravity isn't a force at all, rather it is the curvature of spacetime. According to Einstein, then, an object in free fall is not accelerating. Rather, an object is accelerating precisely when it isn't in free fall. Thus, an object at rest on the surface of Earth is not accelerating according to Newtonian physics, but it is accelerating according to GR. Conversely, an object falling at the surface of Earth is accelerating at $a = \frac{GM}{R_E^2}$ per Newtonian mechanics, but it is not accelerating per GR. This is the Principle of Equivalence.

The Principle of General Covariance says simply that any equation of physics that holds in a general gravitational field must hold in the absence of gravity by replacing the curved metric of GR with the flat metric of SR. And, the equation must be generally covariant, that is, it must be invariant under general coordinate transformations [Weinberg, 1972, pp. 91–3]. In summary, according to GR, spacetime isn't flat, as in SR, but it is curved and the timelike worldlines of objects in free fall are geodesics in this curved spacetime. This statement needs a little unpacking.

Geodesics are curves of extremal length. In a two-dimensional (2D) flat plane, a geodesic is the curve of absolute shortest length between two points, that is, the straight line connecting them. On the surface of a 2D sphere, geodesics are great circles, that is, a circle on the sphere that shares its center with that of the sphere (Figure 3.1). All lines of longitude are geodesics while the only line of latitude

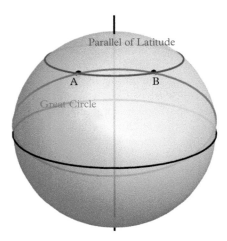

Figure 3.1 *Geodesics on the two-dimensional sphere share their centers with that of the sphere and are called "great circles." The center of the "Parallel of Latitude" shown is not at the center of the sphere, so it is not a geodesic.*

that is a geodesic is the equator (Figure 3.1). The short segment of the great circle connecting points A and B of Figure 3.2 is a path of absolute shortest length (the many paths shown nearby are all longer). The long segment of the great circle connecting points A and B of Figure 3.2 is a path of relative maximum length, that is, paths nearby it are all a little shorter. The relative acceleration of neighboring geodetic worldlines depends on the curvature of the spacetime in their vicinity, so that spacetime curvature replaces gravity as a force. For example, two straight lines in a 2D plane emanating from a point continue to diverge at a constant rate because the space they are in is flat. Conversely, the curvature of a 2D sphere means that two lines of longitude (geodesics) emanating south from the north pole diverge at an ever slower rate until at the equator they stop diverging and thereafter start to converge, eventually intersecting at the south pole. An object whose worldline is not a geodesic of the spacetime is under the influence of a (some) non-gravitational force(s). So, the geometry of spacetime is an important thing to know and in GR it is given by the "metric," which provides the infinitesimal length squared anywhere in spacetime. In order to obtain the metric for spacetime (hereafter simply "metric"), one must solve Einstein's equations of GR and these are a very complicated collection of the metric to include its first and second derivatives in the four coordinates of spacetime. In fact, Einstein's equations contain thousands of terms in the metric and its first and second derivatives when written in its most general form. In practice, the complexity is greatly reduced by finding solutions of Einstein's equations for highly symmetric spacetime metrics and the stress–energy tensor.

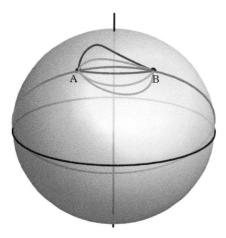

Figure 3.2 *Paths near a geodesic.*

For example, Schwarzschild obtained the first solution of Einstein's equations the year after GR was published by finding the spacetime metric in the vacuum surrounding a spherically symmetric, static mass. We will refer to the Schwarzschild solution later in this chapter and in subsequent chapters. We provide its precise form and black hole implications in Foundational Physics for Chapter 3.

3.3   Big Bang cosmology and its inexplicable initial conditions

Unlike the Schwarzschild solution in empty spacetime, Big Bang cosmology assumes a non-zero stress–energy tensor (SET) which is put into Einstein's equations. The SET describes the matter–energy momentum distribution in spacetime, so in order to provide the elements of the SET you have to know spatial and temporal distances for momentum, force, and energy. Of course, knowing spatial and temporal distances means you already know the metric. Therefore, you should view Einstein's equations as providing a self-consistency criterion or a "global constraint" between what you mean by spatial and temporal measurements and what you mean by momentum, force, and energy. Any combination of the metric and SET that solves Einstein's equations on the spacetime manifold M constitutes a solution of GR.

There are some important facts about GR that we need to point out before proceeding. First, to perform calculus on curved spaces (called "differentiable manifolds"), one assumes the manifold structure in small enough regions is essentially flat (like a football field on the surface of Earth), so that calculus can be done piecewise on the manifold just as it's done in flat geometry. The flat geometry results are connected between adjacent flat neighborhoods by, appropriately enough, "connections." This means SR obtains locally in GR, bringing with it the block universe ambiguity of a global now (more on this in Philosophy of Physics for Chapter 3). It also means there is no uniquely defined global conservation law. Instead we have SR's local conservation of momentum and energy.[5] Thus, the Einstein tensor and SET of Einstein's equations are said to be (locally) "divergence free," that is, roughly speaking, the amount of energy flowing into a location also flows out of that location or accumulates there. To turn that into a unique global conservation law would require a unique global foliation (carving) of spacetime into spatial surfaces and, as we said, a unique global foliation doesn't exist.[6] This will have important implications for our discussion of closed timelike curves below. Divergence-free sources are also part of the adynamical global constraint of RBW to be introduced in chapter 4.

To find the FLRW metric for Big Bang cosmology (so named for its cofounders, Friedmann, Lemaître, Robertson, and Walker), one makes assumptions that lead to a general metric form even simpler than that of Schwarzschild. In Big Bang cosmology, one assumes spacetime can be foliated into spatial hypersurfaces of homogeneity (same at every location) and isotropy (same in every direction) leading to a general metric form with just one unknown function a of one variable (time) which we denote $a(t)$. To understand what this function does, imagine a balloon with a coordinate grid painted on it. The distance between any two points on the balloon will change as it is inflated or deflated, but the coordinate distance won't change since it is painted on the balloon (Figure 3.3).

Figure 3.3 *A balloon with a coordinate grid painted on it. The distance between any two points on the balloon changes as the balloon is inflated or deflated, but the coordinate distance remains fixed. Thus, a so-called scaling factor $a(t)$ must be used to obtain actual distance from coordinate distance. "Co-moving observers" are objects like the galaxies depicted here that have fixed coordinates.*

---

[5] Momentum and energy combine into a single 4D vector in SR.

[6] There is also a problem of how to define energy density for the gravitational field in GR because a small region of spacetime (needed to define "density") is flat (by definition) and spacetime curvature is the source of gravity in GR [Wald, 1984, p. 84].

Thus, in order to get a distance between two points on the balloon using the fixed coordinate grid, we simply multiply the coordinate distance by a "scaling factor," that is, $a(t)$. The SET that corresponds to this so-called cosmological principle of spatial homogeneity and isotropy is that of a perfect fluid. The two most popular forms of a perfect fluid SET are those of pressureless dust and radiation with pressure, called "dust-filled" or "matter-dominated," and "radiation-filled" or "radiation-dominated" models, respectively. Einstein's equations then tell us the spatial hypersurfaces can be a 3D flat space (zero curvature), a 3D sphere (positive curvature), or a 3D hyperboloid (negative curvature) (Figure 3.4). In all three cases, the scaling factor is chosen to go to zero as time goes to zero, so all matter–energy is compressed to a point meaning the density is infinite.[7] This singular starting point is called the Big Bang, a name given to these models pejoratively by Fred Hoyle, a proponent of a competing theory without such a beginning called Steady State cosmology. Einstein's equations tell us the 3D sphere will expand from the Big Bang to a maximum size then recollapse to a Big Crunch. The 3D flat space will expand with "escape velocity" meaning the expansion rate goes to zero as time goes to infinity (asymptotic expansion rate equal to zero). And, the 3D hyperboloid will expand indefinitely with an asymptotic expansion rate greater than zero. These scenarios are depicted in Figure 3.5.

In order to obtain cosmological data from raw instrumental readings, one must assume some cosmological model. The name "concordance model" is used for the assumed cosmology model that produces the cosmological data affording the most robust fit. The current concordance model, ΛCDM, is the spatially flat, dust-filled, or matter-dominated model, also known as the Einstein–deSitter model, with a cosmological constant Λ added. "CDM" stands for "cold dark matter," which we will discuss in chapter 6. It is important to note that we cannot observationally prove the concordance model. In fact, there are theorems [Glymour, 1977; Malament, 1977; Manchak, 2009, 2011] which prove the "endemic under-determination of cosmological models in general relativity":

Zero Curvature   Positive Curvature   Negative Curvature

Figure 3.4 *Three shapes of space for FLRW cosmology.*

---

[7] The graphical version of GR called "Regge calculus" can help its continuous GR counterpart avoid this initial singularity, and perhaps that at the center of a Schwarzschild black hole, as we explain in Foundational Physics for Chapter 3.

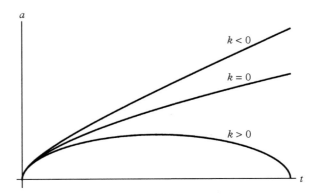

Figure 3.5 *The evolution of the scaling factor for each of the three possible shapes of space in FLRW cosmology assuming a(0) = 0. The 3D sphere will expand to a maximum size then recollapse. The 3D flat space and 3D hyperboloid will expand forever.*

Roughly, the theorems say that in almost every spacetime obeying general relativity, no observer, however long they live, could accumulate enough observations to exclude their being in another very different spacetime.

[Butterfield, 2014, p. 6]

So, one should understand the highly qualified nature of cosmological claims. With that caveat we proceed.

As we said at the beginning of the chapter, the solution of Einstein's equations for Big Bang cosmology can also be viewed as a block universe (Figure 3.6); more on this in Philosophy of Physics for Chapter 3. Once again, as Geroch says, from a God's-eye view "nothing ever moves therein; nothing happens; nothing changes" in this block universe view. While the solution depicted in Figure 3.6 was created in the time-evolved fashion of the Newtonian Schema, the solution itself, that is, the self-consistent metric and SET on the spacetime manifold M, is just a "static" entity. Viewing GR in this fashion we understand that its solutions are spatiotemporally global. While it is certainly the case that dynamical stories on M (Figure 3.6) can be told using the metric and SET, such dynamical stories are not fundamental according to our view. Instead, the spatiotemporally global solution is fundamental. If there are any locations on M, such as the Big Bang, that don't allow for dynamical explanation, then the existence and character of those locations are understood only adynamically, as required by the adynamical global constraint of the LSU, even if the solution was obtained using the Newtonian Schema. So, the situation *at any point* on M is contingent per the adynamical global constraint on the situation at points elsewhere on M. No point on M is the ultimate basis of explanation on our view, no point on M is ultimately any more or less special than any other. The Big Bang is unique in that it is a singular point, meaning the solution of GR breaks down there, but cosmologists assume quantum gravity will replace that singularity with a well-behaved event

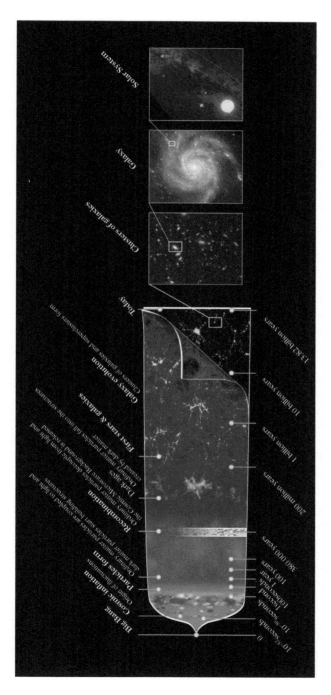

Figure 3.6 *Big Bang cosmology. Picture use authorized by the European Space Agency, copyright ESA–Christophe Carreau.*

such as a quantum fluctuation.[8] Having resolved this issue, there is no barrier left to interpreting GR in the LSU fashion.

Ultimately, a solution—a self-consistent metric and SET on the entirety of M—depends on two things: the adynamical global constraint (Einstein's equations) and information in accord with observations for *any* location on M. In this block universe perspective, one could still ask regarding spacetime, "why do we observe what we observe rather than something else?" This question replaces its counterpart in the mechanical universe, "why these initial conditions rather than some other?" In the LSU, conditions at any location on M are said to be *consistent with* conditions elsewhere on M. It is this spatiotemporal contextual consistency per the adynamical global constraint that ultimately *explains* the conditions at any particular point on M in relation to all other points on M. There is no explanatory priority of one location over another in the LSU. Accordingly, the only mystery would be the existence of M as a whole which is beyond empirical investigation and therefore beyond the purview of physics in this way of thinking. Thus, most would say that the principle of sufficient reason (PSR) cannot be satisfied on cosmological scales by empirical science in our "spatiotemporal ontological contextuality."

In contrast, per the NSU (the dynamical paradigm), conditions at any location on M are said to be *explained by* previous conditions on M *alone*, leaving initial conditions (potentially) unexplained and therefore (potentially) "mysterious." In many non-cosmological situations such as an experiment in a lab, a human agent is responsible for establishing the initial conditions, so the initial conditions are explicable and certainly not mysterious. But in FLRW cosmology per the NSU, the initial conditions are totally inexplicable and thus totally mysterious. The point is that the dynamical/mechanical model has no consensus explanation for the Big Bang, dynamical or otherwise, and the dynamical paradigm demands that they must eventually, in principle, come up with such an explanation.

Let us note that there are those who do not agree; they claim there is no need to explain the Big Bang or Past Hypothesis (the extremely low state of entropy at the Big Bang). In what follows, Callender canvasses several options one might take about this issue:

> Finally, let's return to a point made in passing about the status of the Past Hypothesis. Without some new physics that eliminates or explains the Past Hypothesis, or some satisfactory "third way," it seems we are left with a bald posit of special initial conditions. One can question whether there really is anything unsatisfactory about this (Sklar 1993; Callender 2004b). But perhaps we were wrong in the first place to think of the Past Hypothesis as a contingent boundary condition. The question "why these special initial conditions?" would be answered with "it's physically impossible

---

[8] Again, as we explained in Foundational Physics for Chapter 1, we are free to choose $a(0) = 0$ or otherwise in rendering a particular solution from the general solution of GR's differential equations. RBW cosmology is the Regge calculus version of Einstein–de Sitter (EdS) cosmology and, as we explain in Foundational Physics for Chapter 3, it does not possess an initial singularity, so the graphical (and more fundamental) version of GR may justify a non-zero choice for $a(0)$ in the continuous form of GR.

for them to be otherwise," which is always a conversation stopper. Indeed, Feynman (1965: 116) speaks this way when explaining the statistical version of the second law.

[Callender, 2016]

We think it is something of an evasion for someone who champions the NSU schema writ large to claim that the Big Bang need not be explained and we think most theoretical physicists, if not philosophers, would agree. Carroll also appears to adopt a similar attitude about the Big Bang itself except, since he is firmly ensconced in the dynamical paradigm, he still thinks the Big Bang is a mystery:

> So the Big Bang doesn't actually mark the beginning of our universe; it marks the end of our theoretical understanding. . . . The Big Bang itself is a mystery. We shouldn't think of it as 'the singularity at the beginning of time'; it's a label for a moment in time that we currently don't understand.
>
> [Carroll, 2016, p. 51]

> Even if the universe had a first moment in time, it's wrong to say that it 'comes from nothing'. . . . There is no state of being called 'nothing,' and before time began, there is no such thing as 'transforming.' What there is, simply, is a moment of time before which there were no other moments. The big bang is the moment prior to which there were no moments: no space, no time.
>
> [Carroll, 2016, p. 200]

Of course, saying there are no moments prior to the Big Bang is like saying there is no Earth north of the North Pole. The phrase "north of" doesn't even have meaning once you get to the North Pole. So, what Carroll is really saying is that the Big Bang is a mystery from the perspective of the dynamical/mechanical worldview. We agree completely with him that one shouldn't think of the Big Bang as "the singularity at the beginning of time." And, we agree with him even more when he says:

> [The Big Bang] is also, most likely, not real. The Big Bang is a prediction of general relativity; but singularities where the density is infinitely big are exactly where we expect general relativity to break down—they are outside the theory's domain of applicability.
>
> [Carroll, 2016, p. 51]

In our Relational Blockworld (RBW) concordance model, there is no initial singularity for the reasons given in Foundational Physics for Chapter 3. And, cosmological fine-tuning problems associated with cosmological initial conditions in the NSU are avoided in the LSU, as we will now see.

So, let us extend this block universe attitude to explain other mysteries in cosmology: the flatness, horizon, and low entropy problems. Planck 2015 data [Planck Collaboration, 2016] find that the universe is spatially flat to less than 5 parts in 1000, that is, $0.995 < \Omega_o < 1.005$ where $\Omega_o = 1$ means space is flat. The spatially flat model resides precisely between all possible positively curved and negatively curved models as shown in Figure 3.5, so it is a very unique situation. And, in order for space to be as close to flat as it is now it must have been much closer to flat earlier in time. For example, just one second after the Big Bang, it must have been the case that $0.99999999999995 < \Omega < 1.00000000000005$. Why does this extremely unique situation obtain? That is the flatness problem.

The second mystery is the homogeneity of the source of the "cosmic microwave background" (CMB) radiation. This is radiation assumed to have been "emitted" or freed from scattering when the universe was only 380,000 years old (this is called "recombination"). Prior to recombination the universe was so hot that electrons could not stay bound to protons to form electrically neutral atoms, so electromagnetic radiation was constantly being scattered by all the charged particles. Once the universe cooled to the point where hydrogen atoms could form, the background radiation was freed from scattering and cooled via the expansion of the universe to its present temperature of 2.725 K. As it turns out, the temperature of the CMB is the same in all directions to within 0.002 K. This implies that the temperature of matter and radiation at recombination must have been the same everywhere (homogeneous) to a remarkable degree. Yet, regions of CMB origin separated in the sky by just 1.5 arcsec were causally disconnected (beyond their causal horizons) at that time, so they could not have come into thermal equilibrium had they started at different temperatures. Why was the universe so homogeneous at recombination? Why would such an extraordinarily violent event like the Big Bang result in an extremely uniform temperature distribution? That is the horizon problem.

"Inflationary cosmology" was instituted to solve these problems. According to cosmic inflation, as the universe expanded and cooled it was possible that some fields failed to "condense" into lower temperature configurations. An analogous effect sometimes occurs in our atmosphere.[9] If the night air cools fast enough, water doesn't condense onto the grass (forming dew) immediately after the air temperature reaches the dew point. This causes the air to become "supersaturated" with water vapor. If such a "supersaturated" state (called a "false vacuum," equivalent to a negative energy density) occurred in some region of our universe, then that region would have undergone an exponential expansion instead of a slowing rate of expansion. Perhaps, as the story goes, we live inside one of those expanded regions, thereby explaining why space looks so flat and the CMB is so isotropic. But, inflation requires as much "fine-tuning" as it was intended to explain:

> From the very beginning, even as I was writing my first paper on inflation in 1982, I was concerned that the inflationary picture only works if you finely tune the constants that control the inflationary period. Andy Albrecht and I (and, independently, Andrei Linde) had just discovered the way of having an extended period of inflation end in a graceful exit to a universe filled with hot matter and radiation, the paradigm for all inflationary models since. But the exit came at a cost—fine-tuning. The whole

---

[9] This is only an analogy and not an explanation.

point of inflation was to get rid of fine-tuning—to explain features of the original big bang model that must be fine-tuned to match observations. The fact that we had to introduce one fine-tuning to remove another was worrisome. This problem has never been resolved.

[Paul Steinhardt in Horgan, 2014]

It can accommodate virtually any data thereby rendering it unfalisfiable:

Given the issues with inflation and the possibilities of bouncing cosmologies, one would expect a lively debate among scientists today focused on how to distinguish between these theories through observations. Still, there is a hitch: inflationary cosmology, as we currently understand it, cannot be evaluated using the scientific method. As we have discussed, the expected outcome of inflation can easily change if we vary the initial conditions, change the shape of the inflationary energy density curve, or simply note that it leads to eternal inflation and a multimess.[10] Individually and collectively, these features make inflation so flexible that no experiment can ever disprove it.

Some scientists accept that inflation is untestable but refuse to abandon it. They have proposed that, instead, science must change by discarding one of its defining properties: empirical testability. This notion has triggered a roller coaster of discussions about the nature of science and its possible redefinition, promoting the idea of some kind of nonempirical science.

[Ijjas et al., 2017, p. 39]

Ijjas and colleagues here are just echoing the kinds of concerns we raised in chapter 1. To be fair, Steinhardt has also pointed out that inflation is not without its virtues [Steinhardt, 2011], but whether or not these outweigh its disadvantages is absolutely irrelevant in the adynamical view. As with the mystery of the Big Bang, the problems inflation was constructed to solve are only problems in the dynamical perspective of the mechanical universe. In the adynamical perspective of the block universe no place on the spacetime manifold M is more explanatorily important or improbable than any other location, so LSU explanation doesn't "paint itself into a corner" as does the NSU explanation. The curvature of space and the homogeneity of the energy distribution in the early universe are no more mysterious than the situation elsewhere on M. The values of the metric and SET at *any* location on M are globally contingent on the metric and SET at all other points on M.

And, finally, we bring the block universe perspective to bear on the low entropy problem, that is, the special thermodynamic condition at the Big Bang used by some (the so-called Past Hypothesis) to ultimately explain or characterize[11]

---

[10] "Multimess" is the authors' pejorative replacement for the "multiverse" obtained from eternal inflation, since the multiverse contains "an infinite number of different possible outcomes, with no kind of patch [outcome], including one like our visible universe, being more probable than another."

[11] As Maudlin points out [2011], the second law of thermodynamics tacitly contains a notion of "forward time" in order to make sense of "increasing entropy over time." We will say more about this in chapters 7 and 8, and show that the Past Hypothesis doesn't explain the thermodynamic arrow of time or the Direction of time.

the thermodynamic arrow of time per the mechanical universe. Some people in turn claim that the thermodynamic arrow of time even explains the Direction of time. Simply put, the second law of thermodynamics says the universe evolves to higher entropy and it is this thermodynamic fact that is taken to characterize the Direction of time per the mechanical universe. The statistical mechanics believed to underwrite the second law says there are more configurations of a system with high entropy than there are for that same system with low entropy. If all configurations are equally likely, then the universe at present is more likely to be in a high entropy state than a low entropy state. However, the current configuration is, relatively speaking, a low entropy state. And, since the universe evolves to higher entropy, the universe must have started in an extremely low entropy state that, per statistical mechanics, is extremely unlikely. So, again, the dynamical perspective is faced with a "fine-tuning" problem. And, again, our block universe response is the same: the thermodynamic situation in the early universe is no more mysterious than the thermodynamic situation elsewhere on M because the early universe does not carry any more explanatory weight than any other region of spacetime. Thus, the flatness of space, homogeneity of temperature, and low entropy at their associated initial state (Big Bang) all follow from the adynamical global constraint (Einstein's equations) for the FLRW model given the input of current observations.

Again, one could ask, "why is there an M at all?" And, certainly one could engage in speculation concerning M with metric-SET configurations and/or adynamical global constraints that do not represent our experience. Such counterfactual speculation wouldn't lend itself to empiricism, by definition, but we wouldn't condemn it as an "unworthy academic exercise" either. The point is that while we speak of doing physics from a God's-eye view, given our contextuality and relationalism, there is no literal "view from nowhere" [Nagel, 1986] from which to ask "why does the entire relational block universe exist?" Such questions presuppose the dynamical perspective and the only answer one can give to such questions in our view will be in terms of counterfactual adynamical global constraints and alternative metric-SET configurations, that is, answers residing outside the purview of empirical science.

Again, these mysteries arise because the time-evolved bias of our ant's-eye view demands dynamical explanation and a dynamical story about the universe traced backward in time leads ultimately to conditions in the very early universe. Again, per Wilczek, "The account it gives—things are what they are because they were what they were—raises the question, Why were things that way and not any other?" [Wilczek, 2016, p. 37]. The key to avoiding this explanatory problem is to relegate dynamical explanation based on time-evolved stories to secondary (non-fundamental) status and accept that the more general block universe explanation based on a spatiotemporally global constraint is truly fundamental. This is adynamical explanation per the Lagrangian Schema Universe. In this more general adynamical explanation, Einstein's equations are understood as a global constraint, that is, a self-consistency criterion for the metric and SET on the spacetime manifold M. While time-evolved stories can certainly be told in

GR solutions, there well may be events in a GR solution that resist such dynamical explanation, for example, the origin of the universe or the question "why were things that way and not any other?" In those cases, we just have to accept that reality is best understood adynamically in spatiotemporally holistic fashion. As Wharton states:

> When examined critically, the Newtonian Schema Universe assumption is exactly the sort of anthropocentric argument that physicists usually shy away from. It's basically the assumption that the way we humans solve physics problems must be the way the universe actually operates.
>
> [Wharton, 2015, p. 177]

We now bring this block universe view of GR to bear on the paradoxes associated with closed timelike curves.

3.4   The paradoxes of closed timelike curves

Kip Thorne, Nobel Laureate and executive producer of the movie, *Interstellar*, has this to say:

> Closed timelike curve is the jargon for time travel. It means you go out, come back and meet yourself in the past.
>
> [Kip Thorne quoted in Carroll, 2013]

Having introduced GR and brought the God's-eye view to bear on general relativistic cosmology to resolve the flatness problem, the horizon problem, the low initial entropy problem, and the puzzle of an origin for the universe, we now bring this perspective to bear on the paradoxes associated with closed timelike curves (CTCs), that is, Kip Thorne's "time travel."

Recall that objects in free fall have timelike geodetic worldlines in the GR spacetime. Using a vacuum solution of Einstein's equations we can find its geodesics and imagine it harboring a small object where "small" means that it doesn't change the spacetime geometry appreciably. For example, this is done for GPS satellites in the Schwarzschild solution in order to correct for different clock rates between Earth-bound clocks and orbiting clocks [Stuckey, 1993; Ashby, 2002] (see Foundational Physics for Chapter 3). Indeed, as Ashby states, "If these relativistic effects were not corrected for, satellite clock errors building up in just one day would cause navigational errors of more than 11 km, quickly rendering the system useless" [Ashby, 2002, p. 43]. As it turns out, there are some GR spacetimes that possess timelike geodesics that loop back onto themselves: CTCs (Figure 3.7). For example, Gott found a well-known case using cosmic strings [Gott, 1991]. This can happen because GR employs curved spacetime manifolds and Einstein's equations are satisfied locally on that manifold; that is, you can imagine locally (small), flat regions of spacetime being connected to form a globally

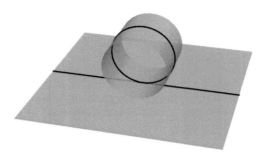

Figure 3.7 *A spacetime depiction of a closed timelike curve. If the local light cone structure permits it, a worldline can loop back onto itself.*

(large) curved spacetime (like flat football fields covering Earth). Thus, one can construct a spacetime satisfying Einstein's equations whereby time is forward-directed locally along a path while that path bends around globally into the past (picture a line of football fields joined end to end to circle Earth). This leads to two types of paradoxes [Lobo, 2008]: consistency paradoxes and causal loops. Consistency paradoxes are those like the "grandfather paradox," that is, someone goes into the past and kills his grandfather before he was born, thereby preventing his own birth. Here is how Arntzenius and Maudlin frame it:

> The standard worry about time travel is that it allows one to go back and kill one's younger self and thereby create paradox. More generally it allows for people or objects to travel back in time and to cause events in the past that are inconsistent with what in fact happened. . . . A stone-walling response to this worry is that by logic indeed inconsistent events cannot both happen. Thus in fact all such schemes to create paradox are logically bound to fail. So what's the worry?
>
> Well, one worry is the question as to why such schemes always fail. Doesn't the necessity of such failures put prima facie unusual and unexpected constraints on the actions of people, or objects, that have traveled in time? Don't we have good reason to believe that there are no such constraints (in our world) and thus that there is no time travel (in our world)?
>
> [Arntzenius and Maudlin, 2013]

Worries about the grandfather paradox apply to any kind of time travel where one can travel back into their own past. However, in the context of GR we only care about what people call "Thornian time machines" that Earman and colleagues describe as follows:

> This . . . kind of time machine was originally proposed by Kip Thorne and his collaborators (see Morris and Thorne 1988; Morris, Thorne, and Yurtsever 1988). These articles considered the possibility that, without violating the laws of general relativistic physics, an advanced civilization might manipulate concentrations of matter-energy

so as to produce [local] CTCs where none would have existed otherwise. In their example, the production of "wormholes" was used to generate the required spacetime structure. But this is only one of the ways in which a time machine might operate, and in what follows any device which affects the spacetime structure in such a way that CTCs result will be dubbed a *Thornian time machine*. We will only be concerned with this variety of time machine, leaving the Wellsian variety to science fiction writers.

[Earman et al., 2016]

Imagine, as GR permits in principle, we did instantiate a local CTC that allowed us to travel into the past and create some inconsistency (known as "bilking") with past or present events. What if anything would stop us from creating such an inconsistency? Logic alone? Some existing facts about physics? Something we must add to physics? Kutach lays out the possibilities in what follows:

The canonical answer to the grandfather paradox is laid out in Lewis (1976) where it is pointed out that the paradox trades on an equivocation between two different notions of possibility. The key idea can be summarized in two parts. First, the local fragment of the world that instantiates the time traveler's attempted attack on his grandfather is such that the time traveler is the same as non-time travelers with regard to causation, influence, action, and related notions. For example, in the scenario the potential assassin actually misses his target or loses his nerve, but hypothetically altering him to aim more accurately or to strengthen his resolve would by physical law very likely lead to his grandfather's quick demise. Second, there is no global state of affairs that both retains the continuous existence of the time traveler (as a normal human born in the usual way) around the temporal loop as well as the successful prevention of his birth. So it is locally possible for the time traveler to prevent his future birth, but it is not globally possible because that would be self-contradictory.

What appears to be a signature effect of CTC's is that some physical states that are possible according to the usual laws governing allowable local states cannot be extended via the usual laws of temporal evolution into a globally consistent solution. A restriction on the space of all physically possible fragments of history to those that can be extended by law into a globally consistent solution of a maximally large space-time is called a consistency constraint. Consistency constraints reveal themselves not only in cases of bilking but also in scenarios where a macroscopic object exists only along a CTC. For example, a book that is never manufactured but is passed along from a time traveler to her younger self will need to evolve around a CTC so that any weathering of the pages will vanish somehow when it persists all the way around its circular history.

[Kutach, 2013, pp. 306-307]

Suppose you agree with Arntzenius and Maudlin that appealing to logical consistency is 'stone-walling', and that adding something extra law to physics such as Stephen Hawking's "chronology protection conjecture" [Hawking, 1992] just to rule out such inconsistencies violates parsimony, is cheap and ugly, etc., is there an alternative? Fortunately yes, if one is willing to give up the dynamical way of thinking and treat adynamical global constraints as fundamental – not logical constraints, but physical adynamical global constraints. As Kutach notes, whether or not the particle collides with itself isn't determined by the initial conditions and Einstein's equations in the neighborhood of the wormhole, so the wormhole itself does not fix the relative likelihood of the two scenarios. That is, "the physical evolution is indeterministic, yet neither chancy nor wholly unconstrained by laws" [Kutach, 2013, p. 306]. In this case we don't need to add adynamical global constraints to Einstein's equations viewed in the dynamical/Newtonian fashion. Rather, our claim is that if you think of Einstein's equations themselves as global constraints in LSU fashion, then there is no worry about CTC's.

When we think about the grandfather paradox involving people, the focus is on what thwarts the free will of the time traveler, thus conflating a number of different questions such as the relationship between human action and physical laws. Fortunately physicists have figured out a way to instantiate the grandfather paradox physically, at least as a thought experiment, so we can analyze the consistency question without any extra baggage. Such consistency paradoxes associated with "Thornian time travel" (CTC's) are what we will discuss here from Novikov (Figure 3.8). In Novikov's paper, a wormhole connects regions A and B such that a ball going into region B at time T on trajectory $\alpha 1$ comes out of region A at an earlier time $T - \Delta T$ on trajectory $\alpha 2$ so that it strikes itself at point Z, scattering in two directions, $\alpha 2$ and $\alpha 3$. But, neither $\alpha 2$ nor $\alpha 3$ enters region B, so in fact the ball never enters region B to come out of region A to strike itself at point Z, which is a contradiction. The other paradox associated with CTCs per Lobo [2008] is the closed causal loop.

Closed causal loops provide for the existence of something that doesn't have an origin, at least not a dynamical or causal one. For example, in the movie *Back to*

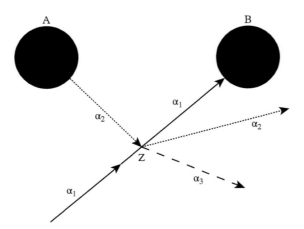

Figure 3.8 *A self-inconsistent trajectory. This figure mirrors that in Novikov, I.D.: Time machine and self-consistent evolution in problems with self-interaction.* Physical Review D 45, 1989–1994 (1992).

the Future, Marty McFly enjoys playing guitar in 1985. One of his favorite songs to play is Chuck Berry's hit "Johnny B. Goode," released in 1958. Marty time travels back to 1955 and plays the song at a dance where Chuck Berry's cousin, Marvin Berry, hears the song and calls Chuck to let him hear it. So, Chuck Berry got the song from Marty McFly in 1955, released it in 1958, Marty McFly hears it in 1984, time travels to 1955, and plays it for Chuck Berry to hear. Who actually wrote "Johnny B. Goode?" No one wrote it in this scenario. There is also the issue of how something without an origin ages as, again, pointed out by Kutach [2013, p. 306]: "a book that is never manufactured but is passed along from a time traveler to her younger self will need to evolve around a CTC so that any weathering of the pages will vanish somehow when it persists all the way around its circular history."

These are the paradoxes we will resolve via adynamical global constraint in the block universe. This in no way discounts the important research concerning technical issues associated with the existence of CTCs [Morris et al., 1988] or the possibility of building a time machine [Earman et al., 2016]. We are simply resolving the paradoxes of CTCs using adynamical explanation associated with the block universe perspective.

Consider the consistency paradox of Figure 3.8 as described earlier. Exactly how am I stopped from rolling the ball on trajectory $\alpha 1$? Is there a "physics police" who will arrest me before I can do so? Will I be struck by lightning? As Arntzenius and Maudlin point out, it is "cheap" to simply claim logical impossibility:

> If time travel entailed contradictions, then the issue would be settled. And indeed, most of the stories employing time travel in popular culture are logically incoherent: one cannot "change" the past to be different from what it was, since the past (like the present and the future) only occurs once. But if the only requirement demanded is logical coherence, then it seems all too easy. A clever author can devise a coherent time-travel scenario in which everything happens just once and in a consistent way. This is just too cheap: logical coherence is a very weak condition, and many things we take to be metaphysically impossible are logically coherent.
>
> [Arntzenius and Maudlin, 2013]

There must be something in the physics *proper* to rule out consistency paradoxes, and there is, but not the laws viewed dynamical (the Hamiltonian version of Einstein's equations in this case). Start by realizing that, again, the mystery obtains because these are questions based on the spatiotemporally local Newtonian Schema problem solving method that aligns so well with our dynamical experience. But, the NSU is not fundamental, however, despite our strong bias. Rather, *you have to adopt the Lagrangian Schema Universe (LSU) which says a GR solution isn't a solution unless it accounts for the entire spacetime manifold M.* Piecewise local solutions to Einstein's equations are worthless in and of themselves. When you imagined sitting in the spacetime of Figure 3.8 and rolling the ball on trajectory $\alpha 1$, you were making an assumption that clearly doesn't hold true. You were assuming, for all practical purposes, that the spacetime geometry of Figure 3.8 would not change due to your presence, just as the vacuum Schwarzschild spacetime geometry doesn't change due to the presence of a GPS satellite. So, as with the satellite in the Schwarzschild solution, you just ignored the SET associated with the ball. In the Schwarzschild solution that's OK because you can certainly create a SET for the satellite in orbit that is everywhere divergence-free and well-defined on M. Remember, the SET is a requirement of Einstein's equations even though it doesn't seem to affect the metric. If you want to introduce an object to M, you have to construct the corresponding SET element for Einstein's equations, no matter how seemingly innocuous that object is, geometrically speaking. Otherwise, you don't have a GR solution. And, if you imagine a situation that is ruled out by GR such as the paradoxical trajectory $\alpha 1$ of Figure 3.8, then it is ruled out by physics.

In this fashion, we have resolved the consistency paradox resulting from the spatiotemporally local ant's-eye view by adopting the spatiotemporally global God's-eye view of GR. We end up turning the original question on its NSU inquisitor by telling him that if he wants to know what will happen in Figure 3.8, he will have to construct a GR solution that embodies his premises as well as consequences. For example, Echeverria and colleagues did precisely that with similar cases [Echeverria et al., 1991] by simply varying the time delay created by the wormhole on the otherwise contradictory trajectory, thus creating a self-consistent solution for a trajectory starting along $\alpha 1$ in Figure 3.8. In that case, the answer to the question posed by our NSU inquisitor is that the result of rolling the ball on the trajectory $\alpha 1$ will be a self-consistent solution as shown by Echeverria and colleagues. As with the other puzzles and conundrums we are writing about, the paradoxes associated with CTCs arise because our time-evolved bias demands dynamical explanation. That is, our perceptions are formed in time-evolved fashion without knowledge of the future, so we are predisposed to think dynamically and, therefore, we want to understand reality as we experience it. In the context of CTCs, we can imagine paradox-generating initial conditions such as rolling the ball on trajectory $\alpha 1$ in Figure 3.8, and it would seem that these initial conditions can be instantiated without having to consider future outcomes, so we are genuinely mystified as to how nature will respond to that situation when our best dynamical laws say a violation of logic and dynamical determinism will result.

The short answer might be that "all laws of physics obey the laws of logic. Assuming a phenomenon must satisfy the laws of physics in order to occur, all logic violating phenomena are ruled out as being possible." But again, however, most of us feel that this answer is a dodge. After all, one can still ask, "why couldn't there be worlds where the logical principle of non-contradiction fails to obtain?" There is nothing to rule out the possibility of such worlds, not even logic, unless you just beg the question. The fact that we can't conceive of such worlds doesn't mean they are ruled out. Regarding the God's-eye view, we have Novikov's Principle of

Self-Consistency [Novikov, 1989] which states simply that local solutions of the laws of physics are only permissible if they lead to self-consistent global solutions. Carlini and colleagues [1995, p. 1] showed the following: "For the case of a 'hard-sphere' self-interaction potential we show that the only possible trajectories (for a particle with fixed initial and final positions and which traverses the wormhole once) minimizing the classical action are those which are globally self-consistent, and that the 'Principle of self-consistency' (originally introduced by Novikov) is thus a natural consequence of the 'Principle of minimal action.'" Einstein's equations are actually the Newtonian Schema version of a Lagrangian action (details are in Foundational Physics for Chapter 3). That Einstein's equations are a Newtonian Schema formulation of GR leads some to view GR dynamically instead of in block universe fashion which then fuels the CTC paradoxes. Following Carlini and colleagues, if you rather assume the LSU, Einstein's equations are understood as an adynamical global constraint and one must find a metric *and* stress–energy tensor (SET) on the entire spacetime manifold M in order to have a GR solution.

A similar kind of account can be given for closed causal loops. If you have a solution of Einstein's equations that permits a causal loop and there is no reason to believe you couldn't instantiate that solution, for example, it doesn't require an SET that we have no way of producing, then you would just have to accept the fact that not everything has a (causal or dynamical) origin. The belief that everything, like a song, must have an origin is clearly a bias based on our dynamical experience. In the case of Kutach's "weathering book," it must be possible that the book continues to age while becoming its younger version at some point along the CTC. If such a situation could be instantiated physically (we don't know how that could be, but we are not ruling it out),[12] then a divergence-free SET for the book could be constructed globally on M in accord with Einstein's equations and you would have a GR solution. We admit that there ought to be a more clever adynamical global constraint account that rules out anti-aging, but we couldn't generate one. If such anti-aging isn't ruled out by Einstein's equations viewed as adynamical global constraints, the problem is it isn't necessarily ruled out by the second law of thermodynamics given that this law is only well defined locally. As with the song, this "cyclically aging" book would have no causal or dynamical origin.

If you are bothered by "unusual and unexpected constraints on the actions of people, or objects, that have traveled in time," then once again, let go of the ant's-eye view and start thinking globally. Once you do, there is nothing spooky or unusual about adynamical global constraints at all. In chapter 4, we will see that the conundrums of quantum mechanics support our LSU block universe view even more strongly than the puzzles, problems, and paradoxes of GR.

---

[12] For example, the second law of thermodynamics wouldn't be violated in the case of a "reverse-weathering book" since the existence of a CTC kills the foliation into spatial hypersurfaces needed to define a universe and a universe is needed to account for the universal increase of entropy required by the second law of thermodynamics. Still, this doesn't supply the needed mechanism for "reverse weathering" and we don't know of any such mechanism.

Philosophy of Physics for Chapter 3

3.5   Introduction

We want to highlight two things in this philosophy of physics thread.[13] First, GR is no friend to the A-series of time in any of the guises we discussed in chapter 2, for example, dynamical presentism, despite what some people have claimed. Second, even among avid blockworlders, dynamical thinking in theoretical physics is deeply ingrained. We will focus in particular on the origins of the Big Bang itself and issues surrounding the thermodynamic arrow of time, such as the Past Hypothesis. As always, the lesson is that most of the explanatory mysteries remaining in physics are due to the mechanical/dynamical paradigm. In chapter 8, we will more directly and properly deal with the Direction of time question for which the thermodynamic arrow of time is so frequently invoked.

3.6   General relativity and the block universe

There are many solutions to GR which do not provide a global foliation like the cosmic time frame does. Many people think this makes GR even less hospitable to presentism or a preferred frame than SR alone. Often the argument runs as follows: there is nothing in Einstein's equations that guarantees a preferred frame or a global foliation and therefore its existence is contingent because it depends on initial conditions, which are themselves contingent. This is allegedly Gödel's argument for the unreality of "time" based on his own solution to GR in which the entire universe instantiates a CTC:

> In arguing from the mere possibility of the Gödel universe, in which time disappears (the failure of a global-foliation), to the nonexistence of time in the actual world, Gödel was employing a mode of reasoning in which he had more confidence than most of his philosophical colleagues. In the case of the Gödel universe, he reasoned that since this possible world is governed by the same physical laws (Einstein's equations) that obtain in the actual world—differing from our world only in the large-scale distribution of matter and motion—it cannot be that whereas time fails to exist in that possible world, it is present in our own. To deny this, Gödel reasoned, would be to assert that 'whether or not an objective lapse of time exists (i.e., whether or not a time in the ordinary sense exists) depends on the particular way in which matter and motion are arranged in this world.' . . . But it is provable that time fails to exist in the Gödel universe. It cannot, therefore, exist in our own."
>
> [Yourgrau, 2005, pp. 130–1]

Whatever you think of Gödel's argument in general, and there is much to query about it, having read the main thread to chapter 3 it should be obvious that in our

---

[13] See [Hoefer and Smeenk, 2016] and [Beisbart, 2016], for recent introductions to the philosophy of astrophysics and cosmology, respectively.

view this argument gets zero traction. Why? Because from the LSU perspective, the only sense in which either the dynamical laws or the initial conditions (the Big Bang) are contingent is in the trivial logical sense. The fact that there is a cosmic time frame is determined by the adynamical constraint operating over the entire block universe. Unless the entire block universe is contingent, there is nothing physically contingent about the initial conditions at the Big Bang.

On the other end of the spectrum there are those who argue that the cosmic time frame provides a preferred foliation:

> The move from SR to GR thus seems to reinvigorate the naturalist presentist's enterprise. As already James Jeans (1936) recognized, with apparent relief, the FLRW spacetimes make "a real distinction between space and time," such that we have "every justification for reverting to our old intuitional belief that past, present, and future have real objective meanings, and are not mere hallucinations of individual minds—in brief that we are free to believe that time is real." (23; cited after Lockwood (2005, 116f)) Many presentists have followed Jeans in imbuing cosmological time with ontological significance. But this move is not without its shortcomings. Michael Berry (1989, 105) resists the inference from the fact that there is a uniquely most natural reference frame for FLRW spacetimes—the one at rest with respect to the local matter of the universe averaged over vast distances—to the conclusion that there is absolute space and time.
>
> [Wüthrich, 2012, p. 16]

However, as we pointed out in Philosophy of Physics for Chapter 2 and the main thread of chapter 3, the existence of a cosmic time frame is not a preferred frame in any sense that negates the block universe implications of relativity. Some have argued that the cosmic time frame of FLRW cosmology can be used to establish a unique foliation of spacetime into spatial hypersurfaces thereby producing an unambiguous notion of absolute space, that is, a preferred frame, in GR cosmology. In fact, it is a perfectly homogeneous and isotropic matter–energy distribution that is required by FLRW cosmology to establish unambiguous spatial hypersurfaces of constant proper time lapse. Once you understand that actual observers will reside on planets of different sizes inside galaxies of different sizes inside clusters of different sizes inside superclusters of different sizes separated by enormous voids, the spatial surfaces of homogeneity and isotropy required to establish the unambiguous, global cosmic time of FLRW cosmology are destroyed. That is because proper time lapse for observers near massive objects is less than proper time lapse for observers far from massive objects (see Eq. 3.14 in Foundational Physics for Chapter 3). There will be the additional complication of an observer's motion relative to the cosmic microwave background (CMB), which some might argue provides a physical basis for absolute space, that is, being at rest with respect to the CMB means you're at rest in absolute space. In our case, for example, the Milky Way galaxy is moving at 650 km/s through the CMB, so our proper time lapse will be reduced relative to any hypothetical observer at rest with respect to the CMB. Of course, inhomogeneous matter distribution also affects the isotropy of the CMB (see the CMB angular power spectrum calculation in Foundational

Physics for Chapter 6), so the hypothetical CMB rest frame cannot be established physically. In short, while there is a hypothetical physical basis for absolute space (establishing a preferred frame) in the idealized FLRW cosmology, reality falls well short of actually instantiating the idealized FLRW cosmology for this purpose. Thus, just as in Minkowski spacetime of SR, there is no physical basis for a preferred frame (absolute space) in our universe per GR.

Specifically, the claim seems to be that you can define a "most natural reference frame" by averaging matter density over the universe at large. But, what spatial hypersurface (which defines "the universe") do you use to do the averaging, Isn't that as arbitrary as the choice to use that average to define the "most natural reference frame?" Even if you got everyone to agree on which surface to use to compute the average,[14] who would actually occupy this frame? Us, with our 650 km/s peculiar velocity living on Earth near the Sun in the Milky Way of the Local Group in the Virgo Supercluster? Isn't that a "most natural reference frame" from our perspective?

If nothing else, by the end of the book we hope the reader will appreciate that the argument from relativity (both SR and GR) to the block universe is quite powerful and the burden of replacing or reinterpreting relativity in a way that supports the A-series and captures all the explanatory power of relativity is non-trivial. In the words of Wüthrich:

> As a general reminder to compatibilist and incompatibilist presentists alike, let me finish by stressing that the strictures of SR are quite strong; Lorentz symmetry is fantastically well confirmed in many disparate contexts and for many different phenomena. As a consequence of this high degree of experimental and observational confirmation, it would be rational to expect Lorentz symmetry to be part of the true fundamental theory—although there is admittedly more to be said here about possible high-energy corrections of exact Lorentz symmetry. Rather than as a theory which has been supplanted by GR, relativistic quantum field theory, and—ultimately—a quantum theory of gravity, we should regard SR as a 'second-order constraint' on these more fundamental theories, as I have explicated in Wüthrich (2010, section 3.4). Quite generally, presentists often underestimate the dialectical work that needs to be done to get around SR's ruling that simultaneity is relative.
>
> [Wüthrich, 2012, p. 23]

So while the defender of the A-series or presentism can take heart that relativity doesn't *entail* the block universe, they should appreciate that escaping that implication will require a great deal of not only metaphysics, but empirically equivalent alternative physics. For an excellent, detailed book-length argument to this conclusion see Vesselin Petkov's *Relativity and the Nature of Spacetime* [Petkov, 2005].

---

[14] Think in terms of the actual situation for each galaxy, not the idealized situation of FLRW cosmology which has no counterpart in reality, and you quickly realize how difficult it is in practice to choose such a hypersurface for this purpose.

3.7   The Big Bang and the Past Hypothesis in the dynamical universe

When it comes to explaining the origins of the Big Bang or anything else that depends on those initial conditions (the Past Hypothesis), one can see the dilemma faced here by the dynamical worldview: if the principle of sufficient reason (PSR) is true and all fundamental explanation is dynamical, then the Big Bang should be explicable dynamically. It is hard to see how the dynamical paradigm can escape either explanatory infinite regress or giving itself key features of the universe, such as the dynamical laws or the Big Bang, as de facto deus ex machina. Some physicists are clearly nervous now with the burden of having to explain everything within the mechanical paradigm. For example, "But we should be at peace with the possibility that, for some questions, the answer doesn't go deeper than 'that's what it is'" [Carroll, 2016, p. 47]?

Of course Carroll's sentiment here is open to interpretation. He might mean that the origins of the Big Bang, the values of the constants, and the fundamental dynamical laws are just brute facts (i.e., axiomatic), or he might mean we have to accept that explaining these things is beyond our ability. The former is certainly a reasonable possibility but it means rejecting the fundamentality of dynamical explanation and/or PSR, both of which would be big blows to the explanatory self-image of physics as it currently stands. The latter is also reasonable but it is of course just giving up. Let us look for more clues in what Carroll says. Here, for instance, is what Carroll says about the explanatory importance of the Big Bang in the dynamical paradigm:

> Everything interesting and complex about the current state of our universe can be traced directly to conditions near its beginning, the consequences of which we are living out every day.
>
> [Carroll, 2016, p. 54]

> The origin of time's arrow, therefore, is ekinological: it arises from a special condition in the far past. Nobody knows exactly why the early universe had such a low entropy. It's one of those features of our world that may have a deeper explanation we haven't yet found, or it may just be a true fact we need to learn to accept.
>
> [Carroll, 2016, p. 58]

> The reason why there is a noticeable distinction between past and future isn't because of the nature of time; it's because we live in the aftermath of an extremely influential event (time-symmetry of laws).
>
> [Carroll, 2016, p. 55]

> The appearance of complexity isn't just compatible with increasing entropy it relies on it. . . . The only reason complex structures form at all is because the universe

is undergoing gradual evolution from very low entropy to very high entropy. "Disorder" is growing, and that's precisely what permits complexity to appear and endure for a long time.

[Carroll, 2016, p. 235]

In the preceding passages Carroll is referring to what Albert and others call the "Past Hypothesis." Carroll makes it clear that the Past Hypothesis does a great deal of explanatory work in the dynamical paradigm. Carroll here invokes the Past Hypothesis to explain many things including the thermodynamic arrow of time which he says in turn explains the Direction of time; a topic we will deal with at length in chapters 7 and 8. He also invokes the second law of thermodynamics to explain (at least in part) the fact that there are local pockets of the universe that have increasing complexity such as the existence of life and ecosystems. Here is an analogy he makes himself elsewhere. When you pour cold cream into a mug of hot coffee, we all know that per the second law of thermodynamics it will soon form a lukewarm tan-colored mixture with a high degree of entropy relative to the state it started in. However, around the mid-point of this process complexity will increase in the form of tendrils of milk and coffee, new structures appear, something like eddies or turbulence. He compares the mid-point of the coffee-cream process to the mid-point of the universe we happen to inhabit now. His claim therefore is that increasing entropy partially fuels increasing complexity. The claim that if the universe were not cooling off from the Big Bang then there wouldn't be life in the universe is certainly open to debate. However, here we just want to note that, as we will make clear in chapters 7 and 8, Carroll's claim presupposes a Direction of time; it doesn't fix one.

These quotations from Carroll illustrate the essential importance of the initial conditions at the Big Bang for the dynamical paradigm and they also illustrate that there is no consensus explanation for it. Again, even for ardent block universe proponents like Carroll, the initial conditions (the Past Hypothesis that explains the second law of thermodynamics on his view) and the dynamical laws are doing all the explanatory work. Carroll seems to buy into the following dilemma, either the Big Bang has a dynamical explanation or it is a brute fact. But as we suggested in the main thread of this chapter, if one is willing to jettison the dynamical paradigm in favor of the adynamical global constraint as fundamental (something the block universe makes possible), then there is an alternative.

Here is what Carroll says about explaining the fundamental physical laws themselves:

Physicists sometimes fantasize about discovering that the laws of physics are somehow unique—that these are the only ones there possibly could have been. That's probably an unrealistic pipe dream. Perhaps there could be a universe that had no regularities at all, one where there would be nothing we would recognize as a 'law of physics.'

[Carroll, 2016, p. 203]

As you can see from the previous passages, Carroll is prepared to acknowledge that both the fundamental dynamical laws and the Big Bang might have to be taken as brute facts if they cannot be explained dynamically. But to many of us this seems like an extremely large explanatory gap for the dynamical/mechanical worldview, especially in light of the promissory notes physicists keep issuing about their explanatory scope and ambitions. How is the invocation of such brute facts much better than invoking any other fundamental unexplained explainer that does *all* the explanatory work, such as God is alleged to do by Spinoza? We assume that unlike Spinoza and the God case, nobody wants to claim that the Big Bang and the fundamental laws are self-causing. If Carroll is right and both the initial conditions at the Big Bang and the fundamental dynamical laws are brute facts, then, again, theoretical physics must radically downgrade its explanatory ambitions, which we admit were perhaps over zealous to begin with, explanation does have its limits. Of course, as we saw in chapter 1, one can invoke the multiverse of eternal inflation and string theory to do all the explanatory work. But if so, the case for the multiverse had better be independently compelling: an entailment of some physical fact/theory we believe is highly likely in its own right and not itself a brute fact or a mere explanatory interpretative convenience. Otherwise, the multiverse will explain anything and everything, which violates the very idea of explanation as telling us why one thing is true and not other logical or physical possibilities. The good news is that with the LSU these explanatory mysteries from cosmology that were all generated by the dynamical paradigm just go away. So we are not stuck with either inexplicable brute facts or the multiverse.

We want to be clear that we are not picking on Carroll here. On the contrary, we are using him as a foil out of respect for his clarity on these issues. He represents such dynamical views beautifully. We also chose Carroll because he is a block-worlder who cares deeply about issues in foundations. We are hoping therefore that he is at least open to the possibility that the block universe viewed adynamically can do more than just fuel metaphysical discussions about the nature of time, it can solve problems in foundational physics as well.

Foundational Physics for Chapter 3

In the main thread, we introduced general relativity (GR), Big Bang cosmology, and closed timelike curves conceptually to show how the NSU leads to the puzzle of the origin of the universe, the flatness problem, the horizon problem, the low entropy problem, and paradoxes associated with closed timelike curves. The puzzle, problems, and paradoxes were then shown to be non-existent in the corresponding LSU. In this foundational physics thread, we will provide mathematical details on the spacetime metric, geodesics, Einstein's equations, the Schwarzschild solution, black holes, mathematical versus physical singularities, the event horizon, FLRW cosmology, the stress–energy tensor, the cosmological constant, and the Lagrangian formulation of GR. We conclude this thread with quantitative details on how Regge calculus cosmology avoids the initial singularity of Big Bang cosmology. It might be possible to extend this to the singularity at the center of black holes.

In GR, the geometry of spacetime is an important thing to know and it is specified by the "metric," which gives the infinitesimal length squared anywhere in spacetime. For example, the metric on the 2D plane in Cartesian coordinates is

$$ds^2 = dx^2 + dy^2,\tag{3.1}$$

or in matrix form,

$$g_{\alpha\beta} = \begin{pmatrix} 1 & 0 \\ 0 & 1 \end{pmatrix}\tag{3.2}$$

while the metric on the 2D sphere of radius r in spherical coordinates is

$$ds^2 = r^2 d\theta^2 + r^2 sin^2\theta\, d\phi^2,\tag{3.3}$$

or in matrix form,

$$g_{\alpha\beta} = \begin{pmatrix} r^2 & 0 \\ 0 & r^2 sin^2\theta \end{pmatrix}\tag{3.4}$$

These can be used to compute the curvature of the plane or sphere, respectively, and thereby provide equations which can be solved for their geodesics. In order to obtain the metric for spacetime (hereafter simply "metric"), one must solve Einstein's equations of GR:

$$G_{\alpha\beta} = \frac{8\pi G}{c^4} T_{\alpha\beta}\tag{3.5}$$

where $G_{\alpha\beta}$ is called the Einstein tensor and is a very complicated collection of $g_{\alpha\beta}$ to include its first and second derivatives in the four coordinates of spacetime.

In fact, the left-hand side of Einstein's equations has thousands of $g_{\alpha\beta}$ terms when written in its most general form. Here is a quick overview to give you an idea of how complicated it is (you don't need to follow in detailed fashion to get the idea):

$$G_{\alpha\beta} = R_{\alpha\beta} - \frac{1}{2}g_{\alpha\beta}R = \frac{8\pi G}{c^4}T_{\alpha\beta} \qquad (3.6)$$

where

$$R = g^{\alpha\beta}R_{\alpha\beta} = g^{11}R_{11} + g^{12}R_{12} + g^{13}R_{13} + \ldots + g^{42}R_{42} + g^{43}R_{43} + g^{44}R_{44} \qquad (3.7)$$

Whenever you see a repeated upper and lower index you have an implied summation over all four coordinate values (spacetime is 4D), so in general R (called the scalar curvature) contains 16 terms (10 independent terms since the inverse of the metric and the Ricci tensor are symmetric, that is, $g^{\alpha\beta} = g^{\beta\alpha}$ and $R_{\alpha\beta} = R_{\beta\alpha}$). Continuing, we have the Ricci tensor $R_{\alpha\beta}$ constructed from the Riemann curvature tensor $R^{\delta}_{\alpha\lambda\beta}$ in these four terms:

$$R_{\alpha\beta} = R^{\lambda}_{\alpha\lambda\beta} \qquad (3.8)$$

The Riemann curvature tensor is obtained from Christoffel symbols $\Gamma^{\mu}_{\alpha\beta}$ as follows:

$$R^{\lambda}_{\alpha\tau\beta} = \frac{\partial \Gamma^{\lambda}_{\beta\alpha}}{\partial x^{\tau}} - \frac{\partial \Gamma^{\lambda}_{\tau\alpha}}{\partial x^{\beta}} + \Gamma^{\lambda}_{\tau\sigma}\Gamma^{\sigma}_{\beta\alpha} - \Gamma^{\lambda}_{\beta\sigma}\Gamma^{\sigma}_{\tau\alpha} \qquad (3.9)$$

which, because of the implied sums, contains 10 terms. And finally, each Christoffel symbol is constructed from the metric, its inverse, and derivatives of the metric per

$$\Gamma^{\mu}_{\alpha\beta} = \frac{g^{\mu\sigma}}{2}\left(\frac{\partial g_{\sigma\alpha}}{\partial x^{\beta}} + \frac{\partial g_{\sigma\beta}}{\partial x^{\alpha}} - \frac{\partial g_{\alpha\beta}}{\partial x^{\sigma}}\right) \qquad (3.10)$$

an expression that contains 12 terms. That means Riemann has 1,176 terms in the metric, its inverse, and derivatives of the metric, so Ricci has 4,704 such terms and R has 75,264 such terms. Not all the terms are independent, due to symmetries, but there are thousands of terms in the metric, its inverse, and its derivatives in the most general form of the Einstein tensor. In practice, the complexity is greatly reduced by finding solutions of Einstein's equations for highly symmetric spacetime metrics and stress–energy tensor $T_{\alpha\beta}$.

For example, Schwarzschild obtained the first solution of Einstein's equations the year after GR was published by finding the spacetime metric in the vacuum surrounding a spherically symmetric, static mass M. Einstein's equations in vacuum reduce to $R_{\alpha\beta} = 0$ and if one starts with the most general spherically symmetric, static form of the metric

$$ds^2 = -f(r)dt^2 + h(r)dr^2 + r^2 d\theta^2 + r^2 \sin^2\theta \, d\phi^2 \qquad (3.11)$$

and solves $R_{\alpha\beta} = 0$ for $f(r)$ and $h(r)$, one obtains Schwarzschild's solution

$$ds^2 = -c^2 \left(1 - \frac{2GM}{c^2 r}\right) dt^2 + \left(1 - \frac{2GM}{c^2 r}\right)^{-1} dr^2 + r^2 d\theta^2 + r^2 \sin^2\theta \, d\phi^2 \qquad (3.12)$$

We will refer to this solution in subsequent chapters. The first thing to notice is that there are two values of r where $g_{\alpha\beta}$ is singular: $g_{\alpha\beta} = \infty$. Those two places are $r = 0$ where $g_{00} = \infty$ and $r = \frac{2GM}{c^2}$, the so-called Schwarzschild radius denoted R_S, where $g_{11} = \infty$. Keep in mind that this solution is only valid outside the mass M, so in order to be valid as $r \to 0$, the mass M must be compressed to infinite density. Thus, $r = 0$ is generally considered to be a "physical singularity," that is, a place where the theory of GR breaks down. One might also believe $r = R_S$ is a physical singularity since, for example, R_S for M the mass of Earth is about 1 cm. Squeezing the entire Earth into a ball 1 cm in diameter seems a bit extreme, but as it turns out R_S is a "mathematical singularity," not a physical singularity. That means the singularity disappears with an appropriate choice of coordinates, for example, the "Kruskal–Szekeres" coordinates named after Martin Kruskal and George Szekeres who introduced them. In Kruskal–Szekeres coordinates it is easy to see that timelike and light-like worldlines may pass smoothly into $r < R_S$, but they cannot pass out of this region. Since not even light can escape $r < R_S$, Wheeler called this region a "black hole" and R_S is called an "event horizon," that is, no event inside R_S can be seen outside of R_S. GR says that all the mass M is compressed to infinite density at the center of the black hole, but again the consensus is that this simply means GR breaks down in that regime and must be replaced. Most researchers believe this is precisely where a theory of quantum gravity is needed, just as with the Big Bang singularity of FLRW cosmology described in the main thread. [More on that in the following.]

Here we point out that time passes more slowly for an observer in the stronger gravitational field closer to M than for the observer in the weaker gravitational field farther from M. Along the worldline of a timelike observer, ds^2 is a measure of their proper time lapse, that is, $ds^2 = -c^2 d\tau^2$. For a stationary observer we have $dr = d\phi = d\theta = 0$, so

$$-c^2 d\tau^2 = -c^2 \left(1 - \frac{2GM}{c^2 r}\right) dt^2 \qquad (3.13)$$

giving

$$\Delta\tau = \sqrt{\left(1 - \frac{2GM}{c^2 r}\right)} \, \Delta t \qquad (3.14)$$

where Δt is the coordinate time lapse, which we see is the proper time lapse for our stationary observers very far ($r \to \infty$) from M. Thus, proper time lapse for a stationary observer near M is smaller than the proper time lapse for a stationary observer farther from M for the same Δt. Plugging $r = \frac{2GM}{c^2}$ into Eq. 3.14 we see $\Delta \tau = 0$; thus one often hears "time stands still at the event horizon of a black hole."

For a satellite in a planar, geodetic (free fall) Earth orbit at $r = r_o$ and $\theta = 90°$, ϕ is the only time-dependent coordinate and Eq. 3.12 gives

$$-c^2 d\tau^2 = -c^2 \left(1 - \frac{2GM}{c^2 r_o}\right) dt^2 + r_o^2 d\phi^2 \qquad (3.15)$$

In order to produce a direct relationship between $\Delta \tau$ and Δt for the satellite as we did for the Earth-based observer (Eq. 3.14 with $r = R_E$, the radius of Earth, and M the mass of Earth), we need ϕ as a function of τ or t that we can plug into Eq. 3.15. We can get that from our geodesic equations, since our satellite is in free fall around Earth,

$$\frac{d^2 x^\alpha}{d\tau^2} + \Gamma^\alpha_{\beta\delta} \frac{dx^\beta}{d\tau} \frac{dx^\delta}{d\tau} = 0 \qquad (3.16)$$

where again there is an implied sum over repeated indices. The $\alpha = 1$ equation where $x^0 = t$, $x^1 = r$, $x^2 = \theta$, and $x^3 = \phi$ is

$$\frac{d^2 r}{d\tau^2} + \Gamma^1_{00} \frac{dt}{d\tau} \frac{dt}{d\tau} + 2\Gamma^1_{30} \frac{d\phi}{d\tau} \frac{dt}{d\tau} + \Gamma^1_{33} \frac{d\phi}{d\tau} \frac{d\phi}{d\tau} = 0 \qquad (3.17)$$

having used $dr = d\theta = 0$ to simplify the equation a bit. The only non-zero Christoffel symbols relevant to this $\alpha = 1$ geodesic equation are

$$\Gamma^1_{33} = -\left(1 - \frac{2GM}{c^2 r_o}\right) r_o \qquad (3.18)$$

and

$$\Gamma^1_{00} = \frac{GM}{r_o^2} \left(1 - \frac{2GM}{c^2 r_o}\right) \qquad (3.19)$$

using Eq. 3.10 for our Christoffel symbols. Eq. 3.17 then gives

$$r_o \left(1 - \frac{2GM}{c^2 r_o}\right) \left(\frac{d\phi}{d\tau}\right)^2 = \frac{GM}{r_o^2} \left(1 - \frac{2GM}{c^2 r_o}\right) \left(\frac{dt}{d\tau}\right)^2 \qquad (3.20)$$

We can reject the $r_o = \frac{2GM}{c^2}$ solution because that lies inside Earth and the metric is only valid outside M. That means

$$\left(\frac{d\phi}{dt}\right)^2 = \frac{GM}{r_o^3} \qquad (3.21)$$

which looks very Newtonian: $\frac{v^2}{r} = \frac{GM}{r^2}$ where $v = \omega r$ and $\omega = \frac{d\phi}{dt}$. However, t in the Newtonian expression is the time measured by all observers whereas t in our GR expression is only time measured by stationary observers at infinity. Now Eq. 3.15 gives

$$\Delta\tau_S = \int \sqrt{\left(1 - \frac{2GM}{c^2 r_o}\right)dt^2 - \frac{r_o^2}{c^2}d\phi^2} = \sqrt{\left(1 - \frac{3GM}{c^2 r_o}\right)}\Delta t \qquad (3.22)$$

for the proper time lapse of the satellite relative to coordinate time lapse. Notice that the observer on Earth will measure a shorter proper time lapse for a given Δt than the satellite clock as long as $r_o > 1.5R_E$. If it wasn't for the third term under the square root in Eq. 3.22, that is, the "$\frac{v^2}{c^2}$" term, the orbiting clock would run faster than its Earth-based counterpart for any r_o. So, as is sometimes mentioned in popular presentations of SR, there is an "SR time dilation" effect for orbiting objects. This "SR" effect dominates at $r_o < 1.5R_E$ so that the orbiting clock actually runs slower than the Earth-based clock at very low altitude orbits. Of course, GPS satellites have a much larger orbital radius than $r_o = 1.5R_E$, so they run faster than Earth-based clocks.

In order to obtain the metric for Big Bang cosmology, we need to include the right hand side of Einstein's equations with the stress–energy tensor (SET) that has the following general form

$$T_{\alpha\beta} = \begin{pmatrix} \text{energy density} & \text{momentum flux} \\ \text{momentum flux} & \text{stress or pressure} \end{pmatrix} \qquad (3.23)$$

Notice that in order to provide the elements of the SET you have to know spatial and temporal distances for momentum, force, and energy. Of course, knowing spatial and temporal distances means you already know the metric. Therefore, you should view Einstein's equations as providing a self-consistency criterion or a "global constraint" between what you mean by spatial and temporal measurements and what you mean by momentum, force, and energy. Any combination of $g_{\alpha\beta}$ and $T_{\alpha\beta}$ that solves Einstein's equations on the spacetime manifold M constitutes a solution of GR.

There are some important facts about GR that we need to point out before proceeding. First, to perform calculus on curved spaces (called "differentiable manifolds"), one assumes the manifold structure in small enough regions is essentially flat, so that calculus can be done piecewise on the manifold just as it is done in flat space. The flat space results are connected between adjacent flat neighborhoods by, appropriately enough, "connections." This means SR obtains locally in GR, bringing with it the block universe ambiguity of a global now. It also means there is no uniquely defined global conservation law. Instead we have SR's local conservation of momentum and energy expressed as

$$\nabla^\alpha G_{\alpha\beta} = \nabla^\alpha T_{\alpha\beta} = 0 \qquad (3.24)$$

Thus, the Einstein tensor and SET are said to be "divergence free." Divergence-free sources are also part of the adynamical global constraint of our Relational Blockworld (RBW) approach to be introduced in chapter 4.

Now we are ready to describe the FLRW metric for Big Bang cosmology (so named for its cofounders, Friedmann, Lemaître, Robertson, and Walker). Here, one makes assumptions that lead to a general metric form even simpler than that of Schwarzschild's two unknown functions of one variable each, $f(r)$ and $h(r)$. One assumes that spacetime can be foliated into spatial hypersurfaces of homogeneity (same at every location) and isotropy (same in every direction) leading to a general metric form with just one unknown function of one variable $a(t)$

$$ds^2 = -c^2 dt^2 + a^2(t)\,(\text{spatial part}) \tag{3.25}$$

The spatial part can be a 3D flat space (zero curvature), 3D sphere (positive curvature), or 3D hyperboloid (negative curvature). The SET is that of a perfect fluid with either a pressureless dust with energy density ρ (called the "dust-filled" model)

$$ T_{\alpha\beta} = \begin{pmatrix} \rho & 0 & 0 & 0 \\ 0 & 0 & 0 & 0 \\ 0 & 0 & 0 & 0 \\ 0 & 0 & 0 & 0 \end{pmatrix} \tag{3.26}$$

or radiation with isotropic pressure p (called the "radiation-filled" model)

$$ T_{\alpha\beta} = \begin{pmatrix} \rho & 0 & 0 & 0 \\ 0 & p & 0 & 0 \\ 0 & 0 & p & 0 \\ 0 & 0 & 0 & p \end{pmatrix} \tag{3.27}$$

Einstein's equations are then solved for $a(t)$. In the flat case, $\frac{da}{dt}(t \to \infty) \to 0$, that is, the universe is expanding at "escape velocity" and will expand forever at an ever slower rate that approaches zero as t approaches infinity. The hyperboloid will also expand forever, but with an asymptotic growth rate larger than zero, that is, $\frac{da}{dt}(t \to \infty) > 0$. The spherical universe will grow to a maximum size then shrink to zero. In all three scenarios, tracing the evolution of $a(t)$ backward in time we find $a(0) = 0$, the so-called "Big Bang," that is, the origin of the universe.[15] The concordance model ΛCDM is the flat, dust-filled model, also known as the Einstein–deSitter model, with a cosmological constant Λ added.

[15] As we explained in Foundational Physics for Chapter 1, $a(0) = 0$ is one of two initial conditions chosen freely for the general solution of the second-order differential equation resulting from Einstein's equations. The graphical version of GR called "Regge calculus" can help its continuous GR counterpart avoid this initial singularity by justifying a non-zero choice for $a(0)$, as we will explain later.

$$R_{\alpha\beta} - \frac{1}{2}g_{\alpha\beta}R + \Lambda g_{\alpha\beta} = \frac{8\pi G}{c^4}T_{\alpha\beta} \qquad (3.28)$$

The left-hand side of this equation is actually the most general divergence-free form, so some cosmologists believe there is no reason *a priori* to ignore the possibility of Λ. We will have more to say about the concordance model in chapter 6 when we discuss "dark energy" (Λ). We should point out that Einstein's equations are actually the Newtonian Schema version of the following action:

$$S = \int \left[\frac{c^4}{16\pi G}R + \mathcal{L}_M \right] \sqrt{-g}\, d^4x \qquad (3.29)$$

where R is the scalar curvature, $\mathcal{L}_M$ is the matter–energy Lagrangian density, and $\sqrt{-g}\, d^4x$ is the invariant spacetime volume element. That Einstein's equations are a Newtonian Schema formulation of GR leads some to view GR dynamically instead of in block universe fashion which then fuels the mystery of the Big Bang and CTC paradoxes.

As for the singular nature of the Big Bang, most physicists believe a theory of quantum gravity will resolve that issue (as with the $r = 0$ singularity of the Schwarzschild solution). As Smeenk writes:

> Talk of the 'Big Bang' can be misleading as it suggests that there is a 'singular region' within spacetime. It would be inconsistent to treat singular points as points in spacetime. . . . Singularities are perhaps best regarded as global properties of the spacetime reflected in features such as the existence of incomplete geodesics. In any case as Torreti (2000) remarks, modern cosmology evades Kant's first antinomy through mathematical subtlety; there is no 'first instant' or beginning in time, yet the universe does not have an infinite past.
>
> [Smeenk, 2014, p. 207]

In the case of RBW, as we will see in chapter 6, the graphical version of GR called "Regge calculus" [Regge, 1961] is assumed to be fundamental to GR, not the converse per conventional thinking. The Regge calculus solution for $a(t)$ in the Einstein-deSitter (EdS) model reproduces that of GR's EdS solution when the spatial links are small. Spatial links are "small" when the "Newtonian" graphical velocity v between spatially adjacent nodes on the Regge graph is small compared to c, that is, $\left(\frac{v}{c}\right)^2 \ll 1$. In that case, the dynamics between adjacent spatial nodes is just Newtonian and the evolution of $a(t)$ in Regge EdS cosmology is equal to that in GR's EdS cosmology. When $v \approx 2c$ Regge EdS cosmology encounters the "stop point problem" [Lewis, 1982; De Felice and Fabri, 2000; Khavari and Dyer, 2009], that is, the backward time evolution of $a(t)$ halts, so $a(t)$ has a minimum. Of course, this is not a real problem for Regge EdS cosmology if one is simply using it to model GR's EdS cosmology, since one can regularly check v in the computational algorithm and refine the size of the lattice to ensure v remains small. However, in RBW the graphical approach is fundamental, so lattice refinements are not mere mathematical adjustments, but would constitute new "mean" configurations of matter. Of course, such refinements are certainly required in earlier cosmological eras, but one would expect there exists a smallest spatial scale (associated with a smallest nodal mass) so that eventually (evolving backward in time) $v \approx 2c$ could not be avoided and the minimum $a(t)$ would be reached. Thus, Regge EdS cosmology, which is fundamental to its continuous GR counterpart, helps its continuous GR counterpart avoid an initial singularity by justifying a non-zero choice for $a(0)$.

Early models of stellar collapse used spheres of collapsing EdS dust surrounded by Schwarzschild vacuum [Stuckey, 1994], so perhaps Regge calculus will avoid the $r = 0$ singularity of black holes as it avoids the initial singularity of FLRW cosmology. Anyway, since RBW explains the SCP Union2.1 supernova type Ia data without Λ [Stuckey et al., 2012a,b, 2016b], Regge EdS cosmology is our proposed concordance model.

4

Relational Blockworld and Quantum Mechanics

In the customary view, things are discussed as a function of time in very great detail. For example, you have the field at this moment, a different equation gives you the field at a later moment and so on; a method, which we shall call the Hamiltonian method. We have, instead [the action] a thing that describes the character of the path throughout all of space and time . . . . From the overall space-time point of view of the least action principle, the field disappears as nothing but bookkeeping variables insisted on by the Hamiltonian method.

Nobel Laureate Richard Feynman [Brown, 2005b, p. xv]

4.1   Introduction

The overall goal of this chapter is to show how the conundrums of quantum mechanics (QM) follow from the dynamical perspective of the Newtonian Schema Universe (NSU) and how they can all be resolved using our Relational Blockworld (RBW) version of the Lagrangian Schema Universe (LSU). In chapter 3, the puzzle of the Big Bang, the flatness problem, the horizon problem, the low entropy problem, and the paradoxes of closed timelike curves in general relativity (GR) were resolved using a "global contextuality,"; that is, a self-consistent/adynamical explanation in the block universe. Let us remind ourselves how that game was played.

Taking the block universe view as fundamental, all that is needed to explain a 4D pattern in the block universe is to provide an adynamical rule that leads computationally to that 4D pattern. The least action principle is an example of such a 4D computational method and Wharton calls this formalism the "Lagrangian Schema." In the case of GR, Einstein's equations provided the adynamical rule of self-consistency between the metric for spatiotemporal measurements and the stress–energy tensor for momentum, force, and energy. Even though the reader may have found it somewhat mentally jarring to think of GR in terms of adynamical global constraints, there was much the reader could still retain of the mechanical "worldview" because GR is a theory of classical physics. That is, the reader could still imagine that particles (or excitations of a field) had worldlines, and such particles had definite intrinsic properties independently of interactions

Beyond the Dynamical Universe. Michael Silberstein, W.M. Stuckey and Timothy McDevitt, Oxford University Press (2018). © Michael Silberstein, W.M. Stuckey and Timothy McDevitt. DOI 10.1093/oso/9780198807087.001.0001

with other particles and independently of other contextual features in their physical environment. Thus, the reader could still imagine a particle or field as given by classical physics. In RBW, every assumption of the dynamical/mechanical paradigm will be challenged.

First, we will interpret QM in an LSU fashion, relegating NSU accounts to "bookkeeping." Hence the RBW interpretation of QM is a realist, psi-epistemic account. Second, the fundamental ontology of RBW includes neither particles, waves nor fields as those are traditionally conceived. Indeed, in our ontology there are no fundamental QM entities with individual autonomy and intrinsic properties. In RBW, the so-called classical realm does not "emerge from" some exclusively QM realm and there is no "quantum" behavior without a "classical" context (and vice versa, actually); there are no non-contextual properties on our view. We will argue that the outcomes observed in quantum experiments are the result of "spatiotemporal ontological contextuality," a 4D contextuality that includes the experimental setup from initiation to termination. Thus, our fundamental ontology is 4D spacetimesource elements which we will describe detail, but for now note that they include properties that would normally be considered classical and properties that would normally be considered quantum. The distribution of spacetimesource elements throughout the block universe of classical physics is governed by an adynamical global constraint per the Lagrangian Schema. But understand that spacetimesource elements do not reside *in* spacetime but they are *of* spacetime (i.e., RBW is a background-independent theory). Third, in RBW, the quantum is characterized by unmediated "direct action," that is, direct connection between source and sink with no intervening worldlines, as alluded to by Feynman in the opening quotation of this chapter.

We begin in section 4.2 with a conceptual introduction to QM. The purpose of this section is to introduce the reader to textbook QM and to problems and mysterious phenomena that are considered essentially QM, such as entanglement. In this section, we want to drive home for the reader how far QM weirdness forces even purportedly dynamical interpretations to deviate from the mechanical/dynamical paradigm. RBW is thus an interpretation of QM borne out of the attitude that it is time to give up the ghost of Newton completely. In section 4.3, we provide an introduction to the mysterious phenomena said to define QM, namely, quantum superposition, entanglement, and quantum nonlocality. Again, we want to drive home for the reader how far QM weirdness forces even purportedly dynamical interpretations to deviate from the mechanical/dynamical paradigm. In section 4.4, we give three simple experimental examples of "quantum nonlocality," that is, the Zeilinger, Hardy, and Mermin experiments, and show how their conundrums result from dynamical explanation per the NSU. We then resolve those conundrums using an adynamical global constraint for QM (the Feynman path integral) and RBW's graphical spacetimesource element. In section 4.5, we canvass how RBW resolves all the key interpretational issues of QM and quantum information theory. In Philosophy of Physics for Chapter 4, we introduce RBW as a form of ontological contextual emergence, of which spatiotemporal ontological contextuality is a subset, and we show specifically that RBW is a realist, psi-epistemic account of QM. In Philosophy of Physics for Chapter 4, we distinguish RBW ontology from other views such as ontological structural realism [Ladyman, 2007] and instrumentalism, such as Quantum Bayesianism [Stuckey et al., 2016d]. In Foundational Physics for Chapter 4, we introduce Schrödinger's equation, explicitly compute the probability amplitudes for the Mermin experiment using the Feynman path integral, formally introduce the adynamical global constraint (AGC)[1] of RBW, and apply the AGC to the twin-slit experiment.

We acknowledge that RBW's form of adynamical explanation in the block universe is a high price to pay for a mere interpretation of QM, so we point out now that the payoff for this radical departure from the NSU is that the AGC and ontology of RBW also provide:

1. An interpretation of quantum field theory that resolves its unique technical and interpretive issues (chapter 5).

2. A new approach to unification and quantum gravity with applications to dark energy and dark matter (chapter 6).

3. A basis for our neutral monist explanation for time and change as experienced (chapters 7 and 8).

While we certainly do not expect anyone to discard their favorite QM interpretation based on the contents of this chapter alone, we believe the larger explanatory benefits and potential of RBW make it a viable interpretation of QM.

4.2   Abridged introduction to quantum mechanics

There are many ways to introduce QM, indeed there are several books devoted to that,[2] so our conceptual introduction will be extremely brief by comparison. Nonetheless, it will suffice as an introduction for our purposes.

When we introduced special relativity (chapter 2) and general relativity (chapter 3), the formalism dealt with objects, substances, or fields in spacetime. By contrast, the standard textbook wave-mechanics formalism of quantum mechanics (QM) deals not only with actual measurement outcomes in spacetime but also the wave function represented in 3N-dimensional "configuration space" or "Hilbert space" where N is the number of quantum systems. Specifically, the formalism of QM is used to compute the "wave function" or "probability amplitude" for possible measurement outcomes. The Hamiltonian formalism for QM is the Schrödinger equation (see Foundational Physics for Chapter 4) which is

---

[1] AGC will refer specifically to RBW's adynamical global constraint.
[2] For an introduction to the foundations of quantum mechanics we highly recommend Lewis, 2016.

a differential equation in the wave function $\psi(x, t)$. Once you have an outcome, both the configuration x_o, that is, specific spatial locations of the experimental outcomes, and time t_o of the outcomes are fixed, so the wave function $\psi(x, t)$ of configuration space [Gao, 2017] becomes a probability amplitude $\psi(x_o, t_o)$ in spacetime, that is, a probability amplitude for a specific outcome in spacetime. In other words, the time evolution of the wave function in configuration space before it becomes a probability amplitude in spacetime is governed by the Schrödinger equation. When there are different ways to instantiate a particular measurement outcome, such as {source → slit 1 → detection event} or {source → slit 2 → detection event} in the twin-slit experiment (Figure 4.7), the probability amplitudes for each of the options are added ("superposed") before squaring to produce the probability. This is where QM differs from statistical mechanics, for example, in that it is the probabilities proper that combine in statistical mechanics. Since probabilities are always positive, the different options in statistical mechanics can only combine constructively (accumulate), whereas the probability amplitudes of QM can combine constructively or destructively (cancel each other) in a process called "quantum interference via quantum superposition." This is a bit counterintuitive because QM says it is possible to reduce the probability of a particular outcome while increasing the number of ways that outcome could occur. This adding of probability amplitudes then squaring the result to produce a probability is called the "Born rule." One can also use Maxwell's equations of electromagnetism to find the electric field $\vec{E}$ throughout spacetime and produce a probability per unit volume for photons in analogy with the Born rule. That is, the energy density of the classical field $\vec{E}$ goes as $\vec{E}^2$ and $\vec{E}$ is also produced by adding all contributions by vector addition (superposition), so it is also subject to interference via superposition (classical). Since the energy per photon is given by hf where h is Planck's constant and f is the frequency of the photon, the number of photons per unit volume (which when normalized over some volume gives the probability per unit volume) is proportional to $\frac{\vec{E}^2}{hf}$ in complete analogy with the Born rule [Serway and Jewett, 2014, pp. 1267–68].

When the possible measurement outcomes are denumerable (discrete) their configuration space is represented in vector form and called "Hilbert space." In this case, the basis vectors correspond to the possible outcomes of a particular measurement. These basis vectors are the eigenvectors of a "measurement operator" (matrix) and the possible measurement outcomes are the eigenvalues of those eigenvectors. The wave function is now the "state vector" in Hilbert space and its projection onto any particular basis vector squared gives the probability of measuring that particular eigenvalue. When two different operators commute, that is, AB = BA, they share a basis in Hilbert space and it is possible to measure A without changing B and viceversa. If A and B don't commute, then they don't share a Hilbert space basis and measuring A then B then A again can produce two different values for measurement A, so it is not possible to know simultaneously (by measurement) the value of A and the value of B for that quantum system. That measurement can be order-dependent in QM is quite unlike classical mechanics where, for example, we routinely assume we can know the position of an object and its momentum. In QM, the operators for position and momentum do not commute, so you can't measure where an object is without changing what you measured previously for its momentum and viceversa. This "noncommutivity" of measurement operators is a characteristic that distinguishes quantum mechanics from classical mechanics.

The amazing thing about the various formalisms of QM is that they have such a high degree of predictive accuracy yet there is absolutely no consensus as to a physical model of quantum phenomena. There are several different such competing models and we call them "interpretations." An interpretation must at least be empirically equivalent to textbook QM, should tell us what the theory means physically, and should resolve the various mysteries, such as the measurement problem and the nature of QM entanglement. We can't begin to canvass the many interpretations that exist, but we can define and classify some of them by looking at the axioms of "textbook QM":

(a) Quantum mechanics is universal in its application (e.g., it applies to measuring devices).

(b) The state of a system S is represented by a vector or wave function (in $3N$-dimensional Hilbert or configuration space) and is time-evolved in accord with a deterministic linear dynamical law (i.e., Schrödinger's equation).

(c) A magnitude Q (e.g., position) is associated with a Hermitian operator O and Q has a determinate value when the state is an eigenstate of O. That is, a system has a (determinate) property if and only if the quantum state is an eigenfunction of the Hermitian operator associated with that property. Key point: this condition fails for superposition states, so such states are presumably meaningless or indefinite.

(d) On making a measurement of Q on S the probability of obtaining a particular result is given by Born's rule as described earlier.

Sometimes (d) is characterized in terms of a measurement "collapsing the wave function" whose observable and thus definite outcomes are distributed according to the Born rule. All this begs the following interpretative physical questions:

1. What if anything physical is a wave function? What is it "a wave of?" People often retort "it is a wave of probabilities," but that is obviously a category mistake.

2. If the wave function is real and something physical, and it lives in configuration space, does that mean configuration space is real and it is only an illusion that we live in three spatial dimensions?

3. If the wave function really collapses, what causes the wave function to collapse? How and exactly when does it happen?

4. Why do QM systems evolve deterministically in configuration space until measured and then yield stochastic outcomes?

These points will be characterized more formally in Foundational Physics for Chapter 4 as needed.

Different interpretations will provide different answers to these questions. In other words, each interpretation must resolve the question of measurement in QM. Indeed, more specifically, we can now define the infamous "measurement problem" in QM as follows: Axioms (a)–(c) seem to contradict axiom (d). Given axioms (a)–(c), when we go to measure some quantum state of a system then we ought now to have an entangled state of the quantum system plus the (presumably classical) measuring device. This is because quantum mechanical systems are described by wave-like mathematical objects (or vectors) of which sums (superpositions) can be formed. The time evolution of the Schrödinger equation preserves such sums. So when observers or measuring devices interact with QM systems that should simply create an entangled state of said QM system plus the observer, and so on ad infinitum. Yet axiom (d) tells us, and experience confirms, that when a measurement is performed we always observe a definite outcome. We do not experience the world as exhibiting superposition or entanglement, or so we think. The Schrödinger dynamics (b) are wrong about what happens when we make measurements, but right otherwise with stunning accuracy. The Born rule (d) is stunningly accurate only when we make measurements.

We can classify some of the main interpretations of QM with respect to how, if at all, they alter the axioms of textbook QM. Denying axiom (a) would mean finding some physical states of affairs, perhaps related to spatial or temporal length scales, mass, gravity, etc., where QM breaks down in some limit and classical or quasi-classical mechanics with definite values takes over. These are called macrorealist accounts of QM [Knee et al., 2016]. So far such attempts have failed, QM seems to apply to the so-called macroscopic as well as microscopic. Altering axiom (b) means rewriting the Schrödinger equation, by adding a nonlinear term for example, such that collapse of the wave function happens spontaneously under the right microscopic physical conditions. We will discuss such a view below. Denying axiom (c) means exploiting certain loopholes in no-go theorems (see shortly) and positing that even though the formalism indicates that a QM property is not in an eigenstate and is therefore in a superposition state, this property really is definite. Such theories are typically no-collapse theories. The property in question is often position (hidden variable theories), but can be other properties as well (modal theories). We will discuss such a view shortly. Finally, axiom (d) as such, the Born rule, modulo any baggage about collapse talk, is purely phenomenological and can't be denied—it's what we observe. However, some different interpretations do provide different explanations as to why the Born rule is true (see Lewis, 2016 for more details).

Perhaps the most well-known interpretation at the moment is the Everett or Many-Worlds interpretation and it actually accepts (a)–(d), but (d) isn't interpreted in terms of collapse. The Many-Worlds interpretation is a no-collapse account which says that, for example, when a z-spin measurement is performed on a QM state that is not a z-spin eigenstate (it is in a superposition of spin states according to axiom (c) above), it results in both possible results, spin-up and spin-down, each existing on its own branch of the universal wave function of the universe, which has indenumerably infinite branches and forever evolves deterministically in configuration space. There was no pre-existing definite property of the QM system that determined a unique outcome of the measurement because there was no unique outcome. The Many-Worlds interpretation says only the universal wave function is real; since it is fundamental it tends to remain undefined. The wave function lives in configuration space so there is a debate about whether one should accept that there are several extra spatial dimensions or if there is a way to avoid this ontological commitment. So why does the universe appear to behave classically and what causes the splitting branches according to this interpretation? The answer is "environmental decoherence," an allegedly interpretation independent fact about QM systems, which says, keeping it non-technical for now, that when QM systems interact even slightly with the environment, they very quickly start behaving classically for all practical purposes:

> The lesson is that if you want to observe interference effects, you need to keep the system in question isolated from outside influences, since even an interaction with a single particle can destroy the effect. This is difficult for microscopic systems, but effectively impossible for macroscopic ones: such systems will inevitably interact with photons, air molecules and other aspects of their environment in complicated and untraceable ways. Decoherence for macroscopic systems is rapid, very complete, and highly irreversible. This means that if the state of a macroscopic system comes to have two components, those components will not exhibit any appreciable interference effects. And since interference is just a name for interaction between two wave components, this means that for all practical purposes the two components do not interact. They can be regarded as separate branches of physical reality. Even though Everett's theory has no collapse process, decoherence results in effective collapse giving the appearance of collapse from within a branch.
>
> [Lewis, 2016, p. 109]

Keep in mind that decoherence doesn't solve the measurement problem. If it did, there wouldn't be branches representing every possible measurement outcome in Everett. We will provide our own unique account of decoherence below, which is to say that we think it is a phenomenon open to interpretation, even though it is an observed behavior of QM systems that everyone must prima facie accept.

We will also provide our own RBW resolution to the measurement problem shortly, but for now notice that the problem is a direct result of treating QM dynamically, on a par with Newtonian mechanics. The assumption that the Schrödinger equation (the Hamiltonian or NSU version of QM), like classical mechanics, is really a dynamical equation of motion that physically time evolves the wave function (whatever that is) of a QM system, living in configuration space (which we have no observable reason to believe exists), is the primary culprit in the measurement problem. RBW will deny this offending assumption.

There are many other interpretations that have different ways of dealing with the measurement problem, but the other, obviously related, huge question in QM is how different interpretations explain the observed phenomena of quantum superposition, entanglement and quantum nonlocality. This will be the subject of the next section and will also help motivate our own approach.

4.3 Quantum superposition, entanglement, and quantum nonlocality

While the expression "quantum nonlocality" has become a catch-all term for QM deviations from classical physical intuition, not all quantum weirdness involves anything like nonlocality in the sense of violating the spirit or letter of special relativity (SR). The most basic deviation from classical physics in QM is known as "quantum superposition," as we alluded to earlier. Take the twin-slit experiment as instantiated by Zeilinger (Figure 4.1), for example. Most people are acquainted with the basic weirdness—when a "momentum measurement" is made the quantum system in question (e.g., electrons) behaves in a wave-like fashion (though the interference pattern is made up of discrete points and can be built up one electron at time from the "source") and when a "position measurement" is made the quantum system behaves in a particle-like fashion (Figure 4.2). This is sometimes called "wave/particle duality," but that name is merely descriptive; it explains nothing. Textbook QM says that an electron with a definite momentum is in a superposition of various position states (and vice versa). Whether the electron has definite momentum or definite position is determined by the experimental context, that is, the measurement being made (Figure 4.1). Notice that no interpretation of QM can escape this behavioral contextuality, the type of measurement (i.e., position versus momentum in this case) being made makes all the difference to the outcomes and the behavior of the "system." We should also note that the physical interpretation of superposition states is up for grabs; different interpretations of QM have very different accounts of the ontology for a superposition state [Lewis, 2016].

So far we are only discussing the behavior of a single quantum system. When one quantum system interacts with another one in the right way, what obtains is "quantum entanglement." With regard to entangled states, no measurable property of particle 1 alone, and no measurable property of particle 2 alone, has any definite value. The smallest system to which any state can be assigned a definite value of any measurable property is the two-particle system or entangled state. To get slightly more technical, a set of entangled quantum systems compose a system whose quantum state is represented quantum mechanically by a tensor-product state-vector which does not factorize into a vector in the Hilbert space of each individual system. Entangled states then are not Cartesian products of their components and thereby appear to violate, among other things, classical probability theory, though some interpretations might dispute this. Here is where quantum nonlocality as such comes into the picture: a measurement on one particle may "affect" the state of another particle (measurement outcomes) even if the measurements are spacelike separated (chapter 2).

In other words, if there were some sort of causal connection or transference of information between spacelike separted measurements of entangled particles (that explains why when one particle property is measured the other entangled particle instantaneously registers an almost perfectly correlated measurement outcome), then the process in question would have to be faster-than-light (FTL) by definition of "spacelike separated." This at least violates the spirit of SR. And this is one of the reasons we became suspicious of many NSU explanations for so-called quantum nonlocality, such as Bohmian mechanics and spontaneous collapse interpretations, such as Ghirardi–Rimini–Weber (GRW), because they really do seem to demand such nonlocal FTL connections [Lewis, 2016, chapter 5]. Since such connections are spacelike (outside the lightcone), they must be FTL, which means they violate Einstein causality which is always timelike—threading the lightcone. Thus, the ordering of such detection events in Minkowski spacetime will be reference frame-dependent which violates the idea that causes always precede their effects. In the standard textbook terminology of wave function collapse, in SR it is ambiguous as to which measurement at which detector "caused" the wave function to collapse. As Lewis states, "[i]f the theory in question says that the subsequent behavior of an object depends on what is going on 'right now' at some distant location, then the theory is prima facie incoherent, because 'right now' cannot be objectively defined for spatially separated points" [Lewis, 2016, §5.2]. This is the relativity of simultaneity, as we explained in chapter 2. This leads some to conclude that a preferred reference frame providing absolute simultaneity to Minkowski spacetime must be added to make SR complete. As Lewis says correctly, while such an addition to spacetime isn't ruled out by SR, this preferred reference frame would be undetectable and do nothing other than solve one problem for such interpretations of QM. Thus, many of us feel such an addition is completely ad hoc. From our perspective, this is another example of how far dynamical accounts of QM have to go to preserve their dynamical bias, especially in a relativistic setting. More on this shortly.

The motivations for these dynamical interpretations are certainly laudable and are namely to provide an explanation for QM behavior (entanglement and nonlocality) that gets around Bell's theorem and to provide a resolution to the measurement problem. We will deal with Bell's theorem first as it and the Kochen–Specker theorem are the two most famous "no-go theorems" in QM. It is well known that Einstein and many of his colleagues were concerned about the "completeness" of QM because the theory seemed to entail that in order to make the right predictions about the behavior of entangled states at spacelike separation one must accept that, prior to measurement, not every property, such as spin values, of entangled particles can be definite. In 1935, the famous EPR paper (Einstein, Podolsky, and Rosen) argued for the incompleteness of QM and the need to supplement it with "hidden variables." As discussed earlier, "hidden" means adding definite properties not given by the textbook theory. EPR argued the quantum system must have pre-existing definite properties that correspond to any measurement whatsoever that someone might perform on the system in question. This classical assumption is sometimes known as "realism."

Given two explicit assumptions, what Bell's theorem [Bell, 1964] shows is no assignment of (in this case) spin properties to the two particles can reproduce the measurement results we observe that are indeed correctly predicted by QM. Hence this result is known as a no-go argument against any hidden variables interpretation of QM. [Many subsequent versions of Bell's proof now exist. See for example Hawthorne and Silberstein, 1995. In the following, we are adhering to Lewis' presentation as inspired by Mermin, 1981.]

However, both the explicit (classical) assumptions Bell made can be questioned and they are as follows:

- Independence assumption: Measurements performed on a quantum system in the future cannot affect the current state of its physical properties. For example, the spin properties of electrons now are completely independent of the measurements to be performed on them in the future.

- Locality assumption: Measurements performed on one particle at spacelike separation from another cannot influence the state of that other particle, regardless of what QM seems to imply. For example, the spin properties of entangled electrons at spacelike separation cannot affect one another.

So the idea here is straightforward: the assumptions are that the properties of particles "now" cannot be affected by future measurements or by events at spacelike separation "now" (though this claim becomes problematic in spacetime for the reasons discussed). This is exactly what the lightcone structure of SR would seem to suggest. The point, however, is that one can cook up a hidden variable account of QM if one is willing to reject one or either of these assumptions. Bohmian mechanics and spontaneous collapse theories both reject the locality assumption, hence their conflict with SR. As we will see shortly, there is a further question as to whether one can cook up a *local* hidden variable theory.

The Kochen–Specker theorem [Held, 2013], another no-go theorem against hidden variables, is formally non-trivial but in a nutshell it shows that it can't be the case that every property for which QM makes predictions can simultaneously be ascribed a definite value. Any attempt to do so results in ascribing contradictory physical properties to the system and doesn't recover the predictions of QM.

The Kochen–Specker theorem also relies on the independence assumption. While there are ways around this no-go result as well, the general consensus regarding the implication of the Kochen–Specker theorem is as follows:

> Independence is essentially an assumption that the physical properties in question are non-contextual. As noted in the previous section, choosing what measurement to perform on a system is tantamount to choosing the physical environment of the system. So if physical properties depend on the physical context of a system as well as its intrinsic state, then they can depend on the measurement performed on the system. Taken at face value, the lesson of the Kochen–Specker theorem seems to be that if there are any physical properties responsible for the measurement results predicted by quantum mechanics, they must be contextual properties.
>
> [Lewis, 2016, p. 44]

What is important for the purposes of motivating our interpretation of QM is to see that even dynamically motivated interpretations such as Bohmian mechanics end up straining the dynamical paradigm to the breaking point. And this is true regardless of whether they preserve or reject locality in their explanations of EPR correlations. We think this shows that it is time to simply let go of NSU thinking when it comes to the quantum.

Lewis notes the following:

> At first glance, though, the kinds of causal [dynamical] explanation given by the main interpretations of quantum mechanics look just like those given in classical physics. All three interpretations, Bohm, GRW and Everett explain measurement outcomes (at least in part) in terms of the evolution of the wave function according to the Schrödinger equation. The Schrödinger equation has the form of a wave equation, so the explanation looks analogous to those given in classical electromagnetism in terms of the propagation of waves in a field. Explanations in Bohm's theory also add the motion of particles according to the Bohmian dynamical equation; but again, causal explanation in terms of the motion of particles is entirely familiar from classical mechanics. So superficially, at least, the causal explanations given by the major interpretations of quantum mechanics seems to fit the 'classical ideal' very well. But as we shall see, the causal explanations offered by each interpretation involve significant non-classical causal features that are unfamiliar from the perspective of classical physics.
>
> [Lewis, 2016, p. 120]

So, upon further examination these particles, waves, and fields (whatever they are) violate their classical counterparts in various ways. Think about what we have already learned: realism to one degree or another is threatened in QM on almost any interpretation and it seems unlikely that any interpretation of QM can escape *some sort* of contextuality. If one wants to escape the type of contextuality implied by the no-go theorems then locality and/or independence is threatened, which means the behavior of QM systems, even given definite properties prior to measurement, is going to be in partially determined by events at spacelike separation or in the future. So for example, while it is true that Bohmian particles have definite position values prior to measurement, those values are given in part by other events at spacelike separation. All of this violates the spirit if not the letter of dynamical and causal explanation as described in chapter 1. The only non-instrumentalist interpretation to preserve locality and independence, the Many-Worlds interpretation, forces us to accept that "measurements" have no unique outcomes. This is quite a blow to the dynamical perspective where, if only indeterministically, the past state of a system plus a dynamical law gives rise to a unique outcome.

Looking at specific interpretations in detail only serves to drive this point home. For example, both the nonlocal Bohmian mechanics and GRW spontaneous collapse accounts exploit the space allowed by the no-go theorems and take the property of position as basic and as having a definite value. Bohmian mechanics posits particles with definite position as basic (i.e., the position of the particle is the hidden variable). In the case of Bohmian mechanics we have both waves that "guide" or "push" the particles around and the particles themselves, where the law or force governing the wave function and the particles with position is deterministic [Lewis, 2016, §3.4]. This is how Bohmian mechanics avoids the measurement problem, the wave function always obeys the Schrödinger equation and never undergoes a collapse; that is, it is only the particle positions, not the wave function that produces measurement outcomes. In Bohmian mechanics the behavior of the particle depends on the positions of all the particles in the system at the time in question (i.e., the other particles it is entangled with), even at spacelike separation. Furthermore, at least as Bohm and Hiley tell the story, Bohm's interpretation also violates independence.

> What all this means is that quantum properties cannot be said to belong to the observed system alone and, more generally, that such properties have no meaning apart from the total context which is relevant in any particular situation. (In this case, this includes the overall experimental arrangement so that we can say that measurement is context dependent.) The above is indeed a consequence of the fact that quantum processes are irreducibly participatory . . . .
>
> [Bohm and Hiley, 1993, p. 108]

So somehow, relativity aside, the wave function instantaneously and dynamically conveys to all the relevant particles what the other entangled particles are doing, as well as all other relevant environmental contexts anywhere in the universe. This sounds more like the occult "Force" from the film *Star Wars* than any physically believable force or law of nature. In the 1713 edition of the *Principia*, regarding his "occult" force of gravity that acts instantaneously at a distance, Newton famously said, "I feign no hypothesis" as to what it is or how it works. We would say that this applies here as well, regardless of whether one thinks of it as a force or a law as some do.

GRW, on the other hand, cashes out particle talk in terms of the behavior of waves—as with Everett only the wave really exists. In GRW and other spontaneous collapse theories the law dictating the behavior of the waves is truly indeterministic, hence the "spontaneous" collapse of the wave function, and the collapse of the wave function causes each "particle" (a point-like wave packet) to acquire a determinate location. So, a measurement on one particle instantaneously causes itself and any particle it is entangled with to acquire a determinate position, again, even at spacelike separation [Lewis, 2016, chapter 5].

Echoing the point made earlier, as Lewis says, at the very least, Bohm and GRW are both nonlocal *and* irreducibly "holistic" (what was earlier called contextual):

> So the fact that entanglement holds between a specific pair of particles means that non-local theories still need emergent properties. Bohm's theory and the GRW theory are non-local, but non-locality cannot recover Humean supervenience. In Bohm's theory, the entangled state cannot be fully captured by an ascription of non-relational properties to the two particles involved, or to the non-relational properties of the two regions of space occupied by their respective wave packets. In the GRW theory, the entangled state cannot be fully captured by an ascription of non-relational properties to the two regions of space occupied by the particle 1 and particle 2 wave packets. In each case, specifying the properties of the parts doesn't suffice to specify all the properties of the system: we also need to specify irreducible properties of the whole.
>
> [Lewis, 2016, p. 197]

Attempts to construct local hidden variable theories only creates more tension with the dynamical paradigm. One well-known attempt is called "flashy GRW" and we needn't bother you with the details, just the consequences:

> There is no perspective-neutral [frame independent] causal story. Second, since flashes are discrete space-time events, causation at the level of the flashes is intermittent rather than continuous. A flash here and now can cause a flash over there later without any intermediate causal links. In fact, the causal elements in flashy GRW are very sparse. A macroscopic object is a galaxy of flashes, but for a microscopic system, it is very unlikely that there will be any flash at all within a humanly reasonable time-frame. So flashy GRW is rather nihilistic when it comes to micro-ontology, for most microscopic systems there is nothing there at all! The causal account of the behavior of such systems is packed into the flashes that occur in the macroscopic objects involved in the preparation and detection of this system, with the intermediate system itself not existing as a concrete causal process at all.
>
> [Lewis, 2016, §5.4]

So not only does flashy GRW not provide a deterministic causal/dynamical process, but rather one with stochastic jumps, it is stuck with discontinuous processes (what we earlier called immediate or direct action), all of which is supposedly produced by some dynamical equation of motion acting on an unobservable wave of we-know-not-what. Admittedly we are biased, but it seems to us if one's dynamical paradigm leads one into such a picture, again, maybe it is time to seriously consider giving up the ghost of Newton and adopting the God's-eye view and thinking about such correlations in terms of an adynamical global constraint.

As we noted, another way to attempt a local hidden variable theory is to reject the independence assumption. There is a class of such interpretations known as retrocausal interpretations of QM. The Kochen–Specker theorem can be avoided by giving up the assumption that the properties of a QM system now are statistically independent of any measurements that will be made in the future. The claim then is that future events can causally/dynamically influence earlier events, hence the name "retrocausal" [Price, 1996]. Presumably then, if we are willing to except retrocausation or "backward causation," then there is no formal reason we can't have a hidden variable theory in which every determinable property of a QM system always has a determinate value. But again, the price of these properties with definite values before, during, and after measurement is that the values of those properties are not in any way intrinsic but rather are determined in this case by contextual features *in the future*, namely, future measurement settings. Such properties are perhaps always definite but not intrinsic. This seems to us to at least violate the spirit of a hidden variable theory. And secondly, with regard to motivating our own interpretation, it is necessary to ask what retrocausation means in a block universe (which tends to be the picture in such accounts, since they require determinate future states of affairs), in which, from a God's-eye point of view, all events past, present, and future are just there. Why think dynamical explanation is fundamental in such a case? As we will see in the remainder of this chapter, Price simply concedes that all such talk of causation or dynamical explanation is strictly ant's-eye and "perspectival" in a block universe. So even when such interpretations are motivated by the block universe, the proponents still tend to try to cash things out dynamically or causally. Again, we say just give up the ghost of Newton and fully commit to the God's-eye view.

4.4 Experimental instantiations of quantum weirdness

The Zeilinger, Hardy, and Mermin experiments described below all involve particles that are entangled which, as explained earlier, means that the result of a measurement on one particle is inextricably related to the result of a measurement on its entangled partner. In an even bigger blow to the mechanical paradigm, particles can become entangled even though they have never interacted; the mere threat of measurement[3] can cause entanglement [Aharonov et al., 2016].

---

[3] It is also interesting to note that measurements can be made without interaction with the measured particle [Elitzur and Vaidman, 1993].

Entanglement makes QM difficult to understand,[4] at least in any mechanistic manner, as we explained earlier. In this section, we briefly review challenges to the dynamical/mechanical paradigm posed by the Zeilinger, Hardy, and Mermin experiments. We will, in each case, resolve the "quantum mystery" using our brand of adynamical explanation in the block universe (called Relational Blockworld (RBW)) and explain why dynamical explanation in the mechanical universe caused it to be a mystery in the first place.

The Zeilinger experiment is shown in Figure 4.1. The mystery of this experimental outcome in Figure 4.2 does not require any familiarity with the formalism of QM. A laser beamed into a crystal produces a pair of entangled photons. Again, "entangled" simply means that a measurement outcome on one of the photons (at detector D1 here) is inextricably related to the outcome of a measurement on the other photon (at detector D2 here), as we will soon see. One of the photons proceeds down through the double slit to detector D2 while its entangled partner travels to the right through a lens to detector D1. The experimentalist can decide whether to locate D1 at a distance of one focal length (f, the "momentum measurement") or two focal lengths (2f, the "position measurement") behind the lens

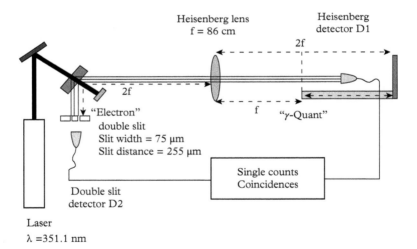

Figure 4.1 *Zeilinger's delayed choice experiment. Anton Zeilinger, "Why the quantum? 'It' from 'bit'? A participatory universe? Three far-reaching challenges from John Archibald Wheeler and their relation to experiment," in* Science and Ultimate Reality: Quantum Theory, Cosmology and Complexity, *John D. Barrow, Paul C.W. Davies and Charles L. Harper, Jr. (eds.), (Cambridge Univ Press, Cambridge, 2004), pp 201–20. Reproduced by permission of Cambridge University Press.*

---

[4] For example, Amir Aczel has a book titled *Entanglement: The Greatest Mystery in Physics* [Aczel, 2002].

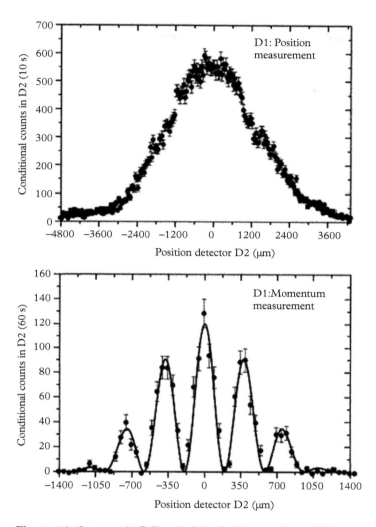

Figure 4.2 *Outcomes in Zeilinger's delayed choice experiment. Ibid. Reproduced by permission of Cambridge University Press.*

(Figure 4.1). The photon detected at D2 then produces two different corresponding patterns, that is, the upper pattern (labeled "D1: Position measurement") and the lower pattern (labeled "D1: Momentum measurement") of Figure 4.2. That the photon at D2 acts like a particle (upper outcome) or a wave (lower outcome), depending on how its entangled partner, is measured at D1 is interesting, but that is not the conundrum. Notice that the distance from the photon source to D2 is much shorter than the distance from the photon source to D1. The two photons are emitted at the same time traveling at the same speed, so the first photon to reach a detector is at D2. But, the photon's behavior at D2 is supposedly determined by the experimentalist's choice of where to locate D1. How does the photon at D2 know where D1 will be placed? Doesn't the experimentalist have a real (delayed) choice that can be made after the photon has been detected at D2 and before the other photon has been detected at D1? If causal influences must proceed from past to future and the outcomes at D1 and D2 are causally related then we have to conclude that the photon at D2 determines the experimentalist's choice of where to put D1, thus our conundrum. Apparently, QM doesn't care about "choice," delayed or otherwise; it simply predicts the correlation in the outcomes without regard for which outcome occurs first.

The reason we have this confusion over the interpretation of QM is that we are seeking to understand the quantum state of affairs in spacetime while the Hamiltonian method of QM is one of time-evolved states in configuration space. This fact concerned Einstein who wrote, "I cannot seriously believe in it [QM] because the theory cannot be reconciled with the idea that physics should represent a reality in time and space, free from spooky actions at a distance" [Born et al., 1971, p. 158]. In other words, QM without interpretation simply provides the probabilities of various experimental outcomes in various experimental configurations without regard for a subsequent explanation of the phenomena in spacetime. If we think of QM explanation along the lines of

$$\text{Why} = \{\text{What, Where, How}\}$$

we have the NSU explanation:

$$\text{NSU} = \{\text{initial state, configuration space, Schrödinger equation}\}$$

and we have the LSU explanation:

$$\text{LSU} = \{\text{fundamental ontological element, spacetime, adynamical global constraint}\}$$

We can get an adynamical explanation in the block universe (LSU) for QM using the Lagrangian method for QM called the Feynman path integral, since it resides in spacetime.

The Feynman path integral assigns a probability amplitude to the spacetime region containing a specific experimental arrangement to include a specific experimental outcome, so it is what we are calling an adynamical global constraint for QM.[5] We should be clear that "global" doesn't mean "the entire spacetime

---

[5] This is not the RBW adynamical global constraint for QM. The RBW adynamical global constraint for QM comes from the non-relativistic limit of the graphical form of quantum field theory called lattice gauge theory, as we will introduce in Foundational Physics for Chapter 4.

manifold," it means the region in space occupied by the experimental equipment from the beginning to the end of a particular experimental trial. Accordingly, the probability that the Feynman path integral assigns to a particular experimental arrangement to include a specific experimental outcome refers to the frequency of occurrence of these "experimental regions in spacetime" per the God's-eye view (Figure 4.4). The Feynman path integral is an adynamical global constraint in spacetime, so all we need to complete our LSU for QM is a fundamental ontological element.

In the adynamical 4D view of RBW, the quantum exchange of energy is accomplished in unmediated/direct fashion by amalgams of space, time, and sources we call spacetimesource elements."[6] Recall what Rovelli said in chapter 1, that we need to erase the dualism of spacetime + matter, and that is exactly what we are doing here. The properties of a spacetimesource element are determined by its classical physics context. For example, the spatial length of the various pieces of the spacetimesource element in Figure 4.3 are determined by the distance from Source to slits, slits to D2, Source to lens, and lens to D1. Thus, quantum physics per RBW depends on the concepts of classical physics, because of what we will call "spatiotemporal ontological contextuality" (see the philosophy of physics thread for this chapter). Spacetimesource elements constitute our fundamental ontology and, since they don't represent 3D entities moving in space as a function of time, they are best viewed as being *of* spacetime, not *in* spacetime (i.e.,

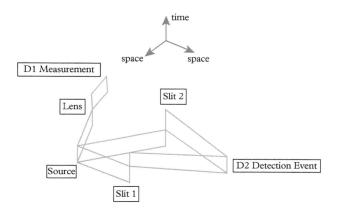

Figure 4.3 *The spacetimesource element for a particular trial in Zeilinger's delayed choice experiment. All the orange rectangles constitute a single spacetimesource element for the single pair of entangled outcomes at D1 and D2.*

---

[6] The word "source" is used here as in quantum field theory to mean the emission or absorption of quanta of energy. When we want to designate source to mean the emission of quanta we will use "Source."

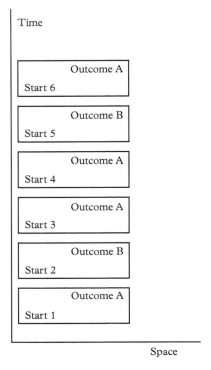

Figure 4.4 *A spacetime depiction of six trials of an experiment with two possible outcomes A and B. The Feynman path integral has assigned the probability $\frac{2}{3}$ to the outcome A and the probability $\frac{1}{3}$ to the outcome B.*

this is a background-independent theory), although we will talk about their distribution throughout the spacetime of classical physics per an adynamical global constraint. See, for example, the spacetimesource element for a trial in Zeilinger's experiment shown in Figure 4.3. In RBW, the fundamental ontological entity, the graphical spacetimesource element with direct action, is motivated by our broader search for new fundamental physics, such as, quantum gravity, unification, dark matter, and dark energy (chapter 6).

With regards to unmediated/direct action, we think of it in this way. Suppose there exists a "quantum entity" with some set of causal properties traversing the space between the Source and detector to mediate a quantum exchange. According to environmental decoherence, this quantum entity cannot interact with its environment or it will cease to behave quantum mechanically, for example, it will act as a particle instead of a wave and cease to contribute to interference patterns. Now if this quantum entity doesn't interact in any way with anything in the universe (it is said to be "screened off"), then it is not exchanging bosons of any sort with any object in the universe of classical physics. Thus, it doesn't affect the spacetime geometry along any hypothetical worldline (except at the

Source and detector) according to Einstein's equations (or it would be interacting gravitationally, i.e., exchanging gravitons). Accordingly, there can't be any stress–energy tensor associated with the worldline of this quantum entity any place except at the Source and detector. So, practically speaking, this posited "screened-off quantum entity" is equivalent to direct action. Thus, according to RBW, QM is simply providing a probability amplitude for the distribution of 4D fundamental ontological elements through the region of spacetime occupied by the experimental equipment and process from initiation to termination, to include a particular outcome. That is, in RBW there are no "QM systems" such as waves, particles, or fields that exist independently of spatiotemporal contexts.

Recall that in Zeilinger's experiment the photon at D2 acts like a particle (upper outcome, Figure 4.2) or a wave (lower outcome, Figure 4.2) depending on how its entangled partner is measured at D1. The distance from the photon source to D2 is much shorter than the distance from the photon source to D1 and the two photons are emitted at the same time traveling at the same speed, so the first photon to reach a detector is at D2. Does that mean the photon's behavior at D2 is determined by the experimentalist's delayed choice of where to locate D1? That would entail effects preceding causes.[7] The only other causal option is that the experimentalist's choice was caused by the photon at D2 and most people don't like the idea that our "freely made" decisions can be the result of a single photon striking a distant detector.

Our conclusion is that we are letting our desire for a spatiotemporally local, forward-time-evolved (NSU) explanation lead us to this conundrum. If instead we ignore our anthropocentric bias and allow for the possibility that objective reality is fundamentally the LSU, then we can accept that there are some LSU explanations that don't allow for NSU counterparts, at least NSU counterparts without serious baggage, such as that discussed in the last section. The LSU explanation here is the probability for the distribution of fundamental ontological entities in the experimental setup and procedure through the block universe according to the Feynman path integral (the adynamical global constraint).

So, in summary, why do we see these mysterious outcomes in Zeilinger's delayed choice experiment? *The spatiotemporal distribution of spacetimesource elements (Figure 4.3), which account for the outcomes in the spacetime configuration of Zeilinger's experiment, follows from the adynamical global constraint.* The outcomes are only mysterious when one tries to formulate a dynamical counterpart to this adynamical explanation. Again, we have to accept that in such cases adynamical explanation in the block universe (LSU) is fundamental and this is just one of those cases that doesn't possess an (obvious) corresponding dynamical explanation in the dynamical universe (NSU). The light ray example allowed for a perfectly reasonable NSU counterpart to the LSU. Indeed, classical physics usually[8] allows for a reasonable NSU counterpart to the LSU explanation and, as

---

[7] More on this notion of "retrocausality" in Philosophy of Physics for Chapter 4.
[8] Counterexamples from GR were presented in chapter 3.

creatures with time-evolved perception, we prefer the NSU, so we are spoiled into believing a reasonable NSU explanation exists for *any* physics solution. But, as with Zeilinger's experiment, there are phenomena that produce all sorts of puzzles, problems, paradoxes, and conundrums when viewed in the NSU. QM is a particularly egregious violator in this regard, as the Hardy and Mermin experiments also show.

The Hardy and Mermin experiments challenge counterfactual definiteness, also known as realism or the faithful measurement assumption; the Mermin experiment does so by violating Bell's inequality. We use RBW to resolve the conundrums of this entanglement/quantum nonlocality for the Hardy and Mermin experiments immediately after their introduction.

The Hardy experiment is shown in Figure 4.5 (from Mermin, 1994). As with the Zeilinger experiment, one doesn't have to be familiar with the formalism of QM to appreciate this mysterious experimental outcome. Each detector has two settings (1 or 2) and two possible outcomes (R or G). Here are the results:

1. In experimental trials (called "runs") where the detector settings are not the same—left detector set at 1 and right detector set at 2 ("12" for short) or the converse "21" – one never sees both detector 1 and detector 2 outcomes of G ("GG" for short). So, 12GG and 21GG never happen.

2. One occasionally sees 22GG.

3. One never sees 11RR.

What is so mysterious about that? Well, let's try to explain the situation for the particles in a 22GG run using typical mechanical thinking. The particles don't "know" how they will be measured in any given run, in fact the detector settings can be chosen an instant before the particles arrive at their respective detectors. And, assuming the measurements are spacelike separated, no information about the setting and outcome at one detector can reach the other detector before the particle is measured there; again, the measurements are said to be "local." Thus, the particles don't "know" how they will be measured when they are emitted and they can't transmit that information to each other during the run. So, it seems that the possible outcomes with respect to each possible detector setting must be coordinated between the particles at emission to prevent the proscribed outcomes 12GG, 21GG, and 11RR. Mermin calls the coordination

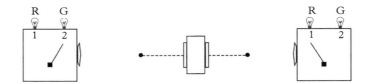

Figure 4.5 *The Hardy experiment.*

of outcomes "instruction sets" [Mermin, 1994] and the fact that the particles have definite properties corresponding to possible measurements whether they are carried out or not is called "realism" or "counterfactual definiteness." The combination of these assumptions is often called "local realism." Both assumptions seem reasonable. Let us start with locality, that is, no FTL transference of information.

Locality is reasonable because the temporal order of spacelike separated events is frame-dependent according to SR (chapter 2). That is, an FTL signal from event A to event B as seen by Bob could be seen going from B to A for Alice who is in motion with respect to Bob. Since motion is relative, the temporal order of spacelike separated events A and B is ambiguous. Thus, SR would seem to rule out FTL information transfer.[9] Realism is a reasonable assumption because the particles don't "know" how they will be measured until the last moment; they can't communicate between detection events, and they have to avoid proscribed outcomes. As Mermin states, "[i]t cannot be proved that there is no other way [to account for the observed results], but I challenge the reader to suggest any" [Mermin, 1981, p. 942]. So, let's continue with the analysis.

Since the result is 22GG, we know the particles were both 1X2G (X = R or G). We know particle 1 wasn't 1G2G, because had the detectors been set to 12 we would have reached a 12GG outcome and that never happens. Likewise, particle 2 couldn't be 1G2G or a setting of 21 would have produced a 21GG outcome which never happens. Thus, both particles must have been 1R2G in order to give the 22GG outcome and rule out the possibility of a 12GG or a 21GG outcome. But both particles being 1R2G can't be right either because then a setting of 11 would have produced a 11RR outcome which never happens. So, what are the counterfactuals, that is, unmeasured properties of the particles, in a run giving a 22GG outcome? We have exhausted all the logical possibilities and none of them work! If the particles don't have any unmeasured properties (no instruction sets), how do they "know" how to behave during a measurement so as to avoid proscribed correlations? There are only conjectural answers to that question in the foundations community; as we saw earlier each interpretation has its own explanation for the apparent lack of counterfactual definiteness. We will shortly provide an RBW explanation after introducing the QM violation of Bell's inequality per the Mermin experiment.[10]

The Mermin device (Figure 4.6) is a bit more complicated than that in the Hardy experiment (Figure 4.5) in that the Mermin device has three settings (1, 2, or 3) instead of just two. Again, there are just two possible outcomes (R or G) for each setting. The experimental results here are as follows:

---

[9] This is different than timelike separated events, that is, events that could be connected by a slower-than-light signal. All observers agree on the temporal ordering of timelike related events regardless of their relative motion.

[10] A quantitative explanation using the Feynman path integral is provided in Foundational Physics for Chapter 4.

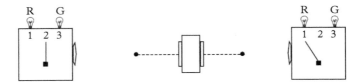

Figure 4.6 *The Mermin device.*

1. The outcomes of the two detectors are always the same when the settings are the same: half the time both detectors register R and half the time they both register G.
2. Overall, the experimental outcomes agree half the time, regardless of the settings.

As with the Hardy experiment, the particles must avoid the proscribed outcomes of result 1, that is, 11RG, 11GR, 22RG, 22GR, 33RG, 33GR. Further, they don't "know" how they will be measured and the experimental outcomes in any given run are spacelike separated. Thus, again, by strictly mechanical thinking, it seems we need instruction sets to avoid the proscribed outcomes of result 1. Looking at all runs in which each particle is 1G2G3R, for example, we have an overall $\frac{5}{9}$ agreement. That is, there are nine possible setting combinations (11, 12, 13, 21, 22, 23, 31, 32, 33) and there is agreement in the outcomes for five of those settings, 11GG, 22GG, 33RR, 12GG, and 21GG. In fact, there is $\frac{5}{9}$ agreement for any instruction set with two R(G) and one G(R). The only other possible instruction set would be all R (1R2R3R) or all G (1G2G3G) and in those cases you have $\frac{9}{9}$ agreement. Therefore, "the only apparent way to account for experimental result 1," that is, instruction sets, violates experimental result 2, that is, one must have overall agreement in only $\frac{1}{2}$ of the runs. That instruction sets entail agreement in more than $\frac{5}{9}$ of all runs is known as a "Bell inequality" and, as we have seen, QM can violate a Bell inequality. So, in this case, the violation of Bell's inequality seems to rule out counterfactual definiteness at least according to naive mechanistic thinking.

Recall that in the Hardy experiment we tried to explain the situation for the particles in a 22GG run. Given the proscribed, spacelike separated outcomes and the fact that the particles don't "know" how they will be measured seemed to imply that the particles' outcomes must be coordinated for all possible measurements, that is, we need Mermin instruction sets or counterfactual definiteness. We then attempted to account for the experimental outcomes and found that, having exhausted all the logical possibilities (all possible instruction sets), none of them worked. If the particles don't have any unmeasured properties (no instruction sets), how do they "know" how to behave during a measurement so as to avoid proscribed correlations?

Again, the mystery arises because we are looking for a NSU explanation, that is, a spatiotemporally local, time-evolved story of the ant's-eye view. In such a story, the state of each of the particles is time-evolved instant by instant from emission to detection, so we are simply asking for an instant-by-instant account of the particles in this experiment, as we tried unsuccessfully to provide earlier. But, if reality is fundamentally an LSU whose observed patterns are the result of fundamental 4D ontological elements distributed throughout the block universe per an adynamical global constraint, then *the observed patterns in the Hardy and Mermin experiments result from the spacetimesource elements distributed through spacetime per the Feynman path integral of QM*. We can't observe the particles between emission and detection. QM tells us and we find experimentally that additional detectors introduced between the Source and detectors already in place will change the final outcomes (contextuality!). Thus, we're forbidden from acquiring such intermediate information without changing the experimental configuration, that is, the block universe pattern that we are trying to explain. Consequently, the adynamical global constraint (Feynman path integral) doesn't have to say anything about counterfactual definiteness, it only has to account for the actual spatiotemporal experimental configuration and outcomes. Indeed, from our particular information-theoretic perspective, the absence of counterfactual definiteness simply means that there is no more information available in the spacetime region of the experiment than what is actually observed. Therefore, in the block universe view, we are not concerned with counterfactual outcomes, that is, block universe patterns that aren't "there." Strictly speaking, there is no hidden information to depict. Thus, without instruction sets, we just have to accept that our fundamental LSU in this case doesn't harbor a less fundamental NSU counterpart. In other words, while our account certainly violates independence (though not locality), unlike certain retrocausal interpretations, we have no compulsion to try and save "realism." The bottom line is that the seemingly nonlocal correlations that interpretations attempt to explain dynamically in NSU fashion, we explain as the result of a more fundamental underlying adynamical global constraint on the distribution of spacetimesource elements. And again, for a more formal RBW treatment of these experiments see Foundational Physics for Chapter 4.

4.5 RBW on the interpretational issues of quantum mechanics

In this section, we summarize how RBW deals with the key questions that any interpretation of QM must face. In the last section, we already outlined our account of entanglement and so-called nonlocality so we will not dwell on them here, but more will be said in the philosophy of physics thread. We start with the measurement problem.

If one constructs the differential equation (Schrödinger equation) corresponding to the Feynman path integral, the time-dependent foliation of spacetime gives the wave function $\psi(x, t)$ in concert with our time-evolved perceptions and the fact that we don't know when the outcome is going to occur. As we stated earlier, once one has an outcome, both the configuration x_o, that is the, specific spatial locations of the experimental outcomes, and time t_o of the outcomes are fixed, so the wave function $\psi(x, t)$ of configuration space [Gao, 2017] becomes a probability amplitude $\psi(x_o, t_o)$ in spacetime, that is, a probability amplitude for a specific outcome in spacetime. Again, the evolution of the wave function in configuration space before it becomes a probability amplitude in spacetime is governed by the Schrödinger equation. However, the abrupt change from wave function in configuration space to probability amplitude in spacetime is not governed by the Schrödinger equation. In fact, if the Schrödinger equation is universally valid, it would simply say that the process of measurement should entangle the measurement device with the particle being measured, leaving them both to evolve according to the Schrödinger equation in a more complex configuration space. However, modulo the Many-Worlds interpretation, we don't *seem* to experience such entangled existence in configuration space, which would contain all possible experimental outcomes. Instead, we experience a single experimental outcome in spacetime. As we learned, this contradiction between theory and experience is called the "measurement problem." However, the time-evolved story in configuration space isn't an issue with the path integral formalism because we compute $\psi(x_o, t_o)$ directly; that is, in asking about a specific outcome we must specify the future boundary conditions that already contain definite and unique outcomes. Thus, the measurement problem is a non-starter for us. When a QM interpretation assumes the wave function is an epistemological tool rather than an ontological entity, that interpretation is called "psi-epistemic." In RBW, the wave function in configuration space isn't even used, so RBW is trivially psi-epistemic (more on that and the purely epistemic status of the Schrödinger equation in Philosophy of Physics for Chapter 4).

Second, our fundamental ontological entity, the spacetimesource element, provides a new ontology for quantum interference, entanglement, and the mystery of wave–particle duality—it's not particles or waves or both, but neither. Our use of the spacetimesource element is another idea reflected in the Feynman quotation at the beginning of this chapter: "the field disappears as nothing but bookkeeping variables insisted on by the Hamiltonian method" [Feynman in Brown, 2005b, p. xv]. The properties of a spacetimesource element are obtained from its classical spacetime context, so RBW is a form of "spatiotemporal ontological contextuality" (more on that in Philosophy of Physics for Chapter 4). As we explain in Foundational Physics for Chapter 4, we modify lattice gauge theory (the graphical counterpart to quantum field theory) to model the direct (quantum) exchange of energy between sources (i.e., without worldlines), which we term "unmediated exchange." This modification to lattice gauge theory is a graphical form of "direct action theory." As we showed earlier, the adynamical global constraint behind the distribution of spacetimesource elements through the block universe corresponding to the correlated detector clicks in the Hardy and Mermin experiments is just the Feynman path integral. This psi-epistemic "unmediated exchange" avoids dilemmas associated with counterfactual definiteness because environmental aspects that are not germane to the experimental context between the Source emission event and the detection event(s) are not represented in a spacetimesource element or the adynamical global constraint. As discussed earlier, the adynamical global constraint explains spacelike separated, correlated outcomes that violate Bell's inequality (per entanglement) as 4D patterns in the block universe; this conception of entanglement will come into focus in the philosophy of physics thread for this chapter. This LSU explanation doesn't require an additional time-evolved story per the Schrödinger equation, nor does it require a retro-time-evolved story as in most retrocausal accounts. The adynamical global constraint explanation of the 4D pattern is the ultimate explanation, as illustrated by the Zeilinger, Hardy, and Mermin experiments.

Third, since spacetimesource elements represent psi-epistemic unmediated exchanges (i.e., they do not contain worldlines of counterfactual definiteness), there is no "screened off quantum entity" that must decohere to behave classically. So per RBW, environmental/dynamical decoherence is really just a dynamical take on what is in fact ontological spatiotemoral contextuality; it is a theory to explain the spatiotemporal extent of a spacetimesource element in its classical context. In the twin-slit experiment, for example, a detection event that contributes to an interference pattern results from a single spacetimesource element that spans the Source emission event, both slits, and a detector detection event (Figure 4.7). Conversely, a detection event that contributes to a particle-like (non-interference) pattern results from a spacetimesource element that does not connect the (original) Source emission event to the detection event. Instead, the spacetimesource element containing that detection event is connected to some location between the detection event and the (original) Source emission event. Environmental/dynamical decoherence explains the two different classical contexts for the two different spatiotemporal extents of these two different spacetimesource elements.

Fourth, a standard characterization of quantum versus classical behavior is the algebraic non-commutivity of observables, that is, matrices, as explained in section 4.2. As per RBW, the non-commutivity of observables means the spacetimesource elements for the different classical contexts represented by those combined observables are different. If the observables commute then the different classical contexts represented by those combined observables have the same spacetimesource elements. For example, consider the successive spin measurements in Figure 4.9 where we first have the Stern–Gerlach (SG) magnets oriented in the y direction (spin y measurement) with the y up outcome directed as an input to SG magnets oriented in the x direction (spin x measurement). Finally, we direct the x up outcome to the input of another spin x measurement. We find that 100% of the particles leaving this experimental arrangement have x spin up. Now we switch the first two spin measurements as shown in Figure 4.10 and we find

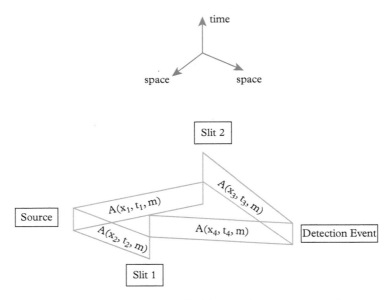

Figure 4.7 *Twin-slit interference. The boxes are the components of the spacetimesource element depicting mass m loss at the Source emission event and mass m gain at the Detection Event contributing to an interference pattern at the detector.*

that 50% of the particles leaving this experimental arrangement have x spin down and 50% have x spin up. Thus, the spacetimesource element for Figure 4.9 is distinct from that of the Figure 4.10. If the spin x and spin y operators commuted, the outcomes of these two arrangements and their spacetimesource elements would be the same. If instead we recombine the outputs of the spin y measurement as shown in Figure 4.11 before directing the spin y measurement output to the second spin x input, we find again that 100% of the particles leaving this experimental arrangement have x spin up. In this configuration the spacetimesource elements do not contain information about the intermediate y-oriented SG magnets, they are irrelevant, so the spacetimesource elements for Figure 4.9 and Figure 4.11 are the same. Thus, obtaining spin y information in the middle of a pair of spin x measurements disturbs the final spin x result. This is what we meant in our discussion of the Hardy experiment when we wrote, "We can't observe the particles between emission and detection, QM tells us and we find experimentally that additional detectors introduced between the Source and detectors already in place will change the final outcomes. Thus, we're forbidden from acquiring such intermediate information without changing the experimental configuration, that is, the block universe pattern that we're trying to explain (contextuality!)." Niels Bohr summed it up nicely when he wrote, "[t]he essential wholeness of a proper quantum phenomenon finds indeed logical expression in the circumstance that any attempt at its well-defined subdivision would require a change in the experimental arrangement incompatible with the appearance of the phenomenon itself" [Bohr, 1954, p. 72]. The algebraic counterpart is that the x and y spin operators don't commute.

Fifth, since there isn't any "thing" moving through the detector to cause the sequential clicks (a trajectory) per RBW, a probabilistic assessment is all that is available to RBW to make the quantum versus classical distinction. For example, the set of first detection events of each worldline in a set of particle trajectories is the first spacetimesource element in a sequence in the particle detector, as will be explained in Philosophy of Physics for Chapter 5. That first set of outcomes is highly probabilistic, as with the first event on the alpha particle trajectory in the cloud chamber per Mott [1929]. Subsequent detection events, however, fall along the classical trajectory with high probability, as shown by Mott, and are used computationally to assign particle masses and charges in particle physics detector events. That means sequential spin x measurements produce a classical trajectory, since the outcome of the second and subsequent clicks is given with a probability of 100%. If we follow a spin x measurement with a spin y measurement, the probability for either y outcome is 50%, so we have quantum behavior. This is contrary to standard thinking whereby spin is purely a quantum property. Indeed, if we immerse the spin measurement equipment in a cloud chamber to create particle trajectories through the SG magnet, we would expect to find what is typically called classical behavior/pattern (Figure 4.8). According to RBW, this is precisely in accord with the characterization of particles per Colosi and Rovelli [2009], that is, they are not entities with intrinsic properties

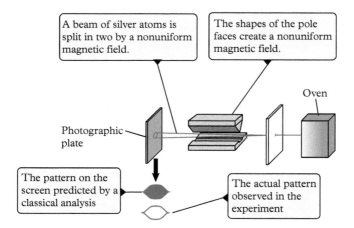

Figure 4.8 *A Stern–Gerlach spin measurement. Figure 42-16, p. 1315 of* Physics for Scientists and Engineers with Modern Physics, *9th edn, by Raymond A. Serway and John W. Jewett, Jr. Reproduced by permission of Brooks/Cole.*

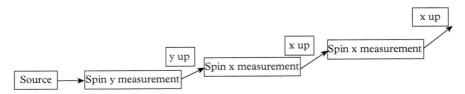

Figure 4.9 *A sequence of spin measurements. The probability of an x up outcome for this experimental arrangement is 100%. In other words, you will never see an x down outcome in this experiment because the only input to the second x measurement is the immediately preceding x measurement's up outcome direction.*

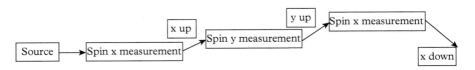

Figure 4.10 *A sequence of spin measurements. The probability of an x up outcome for this experimental arrangement is 50%. In other words, this experiment will produce half x up and half x down outcomes.*

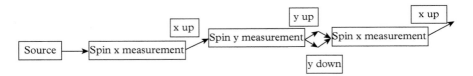

Figure 4.11 *A sequence of spin measurements. The probability of an x up outcome for this experimental arrangement is 100% because while there is a y-oriented SG magnet between x measurements, the two possible outcomes from the y-oriented SG magnet are combined and sent to the second x-oriented SG magnets with no way to know the y outcome. Therefore, you have effectively sent all preceding x up outcomes directly into another x measurement, that is, no intermediate y measurement is being made, so you will never see an x down outcome in this experiment.*

as defined by n-particle Fock space (more on that in chapter 5). Rather, they are defined relationally/contextually by their experimental context. So, given two different detectors (immersed in a cloud chamber or not), we can get two different "particles," that is, two different classical trajectories defined by relational properties, even though we have the same Source.

So again, the "emergence" of classical behavior from quantum reality is not the right way to think about RBW. The quantum phenomenon resides necessarily in the classical phenomenon. Any given worldtube (representing a diachronic entity of classical phenomenon) can be decomposed spatiotemporally with spacetime-source elements. True, but the properties of the spacetimesource elements are those of the classical context. So, classical behavior isn't emerging from quantum reality. There is no dynamic way to think about spacetimesource elements.

Suppose one sets up an experiment, turns on the Source (more on that next), and starts collecting data. Then one decides to change something about the experimental setup (length of an interferometer arm, say). It doesn't make sense to say "I changed the spacetimesource elements" because the entire experimental arrangement and procedure must be represented "at once" in a spacetime depiction and that is where the spacetimesource elements reside. As Geroch says, there's no changing that.

The most difficult aspect of the RBW ontology is in trying to model sources. What is emitting the mass or energy at the Source? What is receiving the mass or energy at the sink? It is fine to use direct action and rid oneself of "mediating quantum entities," but what are classical entities in RBW? One can't get rid of *those* in a classically contextual approach! And you know classical 3D objects can be broken into smaller and smaller pieces (thereby breaking a classical/diachronic entity into smaller and smaller classical/diachronic entities). Where is the end of that process? Since everything in RBW is based on a classical context, aren't the smallest classical entities the fundamental ontological entities of RBW?

The first step in addressing this concern requires us to accept that we use the NSU approach as a matter of pragmatism. That is, we use Newtonian physics rather than quantum and/or relativistic physics for everyday mechanical engineering problems because it works in those contexts. We *know* it's not the "right" physics, but who cares? It works and the "right" physics doesn't gain us anything. Likewise, we can use the label "electron Source" for an object even after it is removed from the context which led to that label. According to RBW, what it does in another context is the subject of the new context, but the label of "electron Source" and what that means in terms of supplying power, making sure the filament is intact, etc., is required, pragmatically speaking, for everyday attempts to conduct experiments. Yes, we are really saying that what makes something a source of electrons is determined by certain spatiotemporal contextual features. So-called electron guns don't fire context-free electron "bullets" with intrinsic properties. Of course, if every "thing" had to be defined for every conceivable context, no experiments would be performed because there would be no repeatability. We have to be able to describe experiments in terms of how we actually set them up and conduct them so others can replicate them. So, in RBW this is all good practice as a matter of experimental pragmatism.

With this in mind, we can establish what it means to be an electron Source context by context. Now, this process is far more involved and complex (but not complicated) than can be imagined because there are many aspects of the electron Source that aren't even germane to our experiments, for example, that the housing is painted grey. So, we're not going to ask, "what does it mean contextually to say the housing is painted grey?" This would involve a discussion of what "grey paint" means contextually, what "housing" means contextually, etc. However, as in experiments involving "the quantum," when we are surprised by the behavior of a collection of classical entities, we need to remember that the concordant physics was developed under the assumption of the NSU model for perfectly pragmatic reasons, but according to RBW, the whole enterprise of physics is one giant program based on self-referentiality, contextuality, and self-consistency. For RBW, the most fundamental ontological reality is not entity-based, the deepest explanation is in no way "compositional" or "realization"-based. Ontological spatiotemporal contextuality is truly fundamental.

So that answers the question about how the so-called QM and the classical relate in RBW. Obviously, this explanation is not ontologically reductive, as spacetimesource elements contain features associated with many different spatial and temporal length scales, features that are all spatiotemporally co-determining. Equally obvious, in RBW there is no radical dualism or discontinuity between the QM and the classical. The smallest diachronic entities (i.e., classical, in our language) are those that can be modeled by worldlines (1D objects in spacetime), they are trans-temporally identifying a series of individual clicks like the particles of particle physics experiments as described in Foundational Physics for Chapter 5. Since RBW is based on classical spacetime contextuality and classical physics is based on diachronic entities, then are the smallest diachronic entities the fundamental ontological entities for RBW? No, because even the smallest diachronic entity is composed of multiple clicks. Thus, the smallest ontological decomposition is in terms of individual clicks and each click is part of a spacetimesource element as we make clear in chapter 5. So, clearly classical behavior does not emerge from quantum reality. The snake eats its own tail, where the head of the snake is the so-called classical and the tail of the snake is the so-called QM.

Sixth, consistent with this RBW distinction between quantum and classical ontology is the RBW distinction between quantum and classical statistics, as characterized by the Born rule whereby one adds individual amplitudes then squares to get the probability, rather than squaring each amplitude then adding as in classical probability, such as statistical mechanics. The manner by which classical trajectories are decomposed per spacetimesource elements requires cancellation of possibilities as in the path integral, whereby the spacetime path of extremal action (classical trajectory) is obtained by interference of non-extremal possibilities which contribute with equal weight [Shankar, 1994, p. 224–5]. Classical statistics doesn't provide for this so-called quantum interference. In fact, per RBW, the reason classical statistics works for classical objects is precisely because a classical object is a set of definite (high probability) quantum exchanges, as we have just explained. That is, classical statistics assumes a distribution of classical objects and each classical object obtains from quantum statistics having removed non-extremal possibilities (by definition of "classical object"). Therefore, classical statistics follows from quantum statistics as the classical ontology of trajectories follows from the quantum ontology of spacetimesource elements. As with all the other mysteries of QM, the Born rule is vexing if one assumes the fundamental ontology is a distribution of time-evolved 3D objects[11] with intrinsic properties that interact via forces in accord with those properties. Moving to the 4D perspective allows one to consider an entirely new fundamental ontology, one without 3D objects at the fundamental level based on an adynamical global constraint. Since the fundamental ontological entities aren't themselves classical 3D time-evolved objects, there is no reason to believe the statistics for their distribution must be classical. All that is required is that the statistics governing the distribution of non-classical fundamental entities leads necessarily to classical objects and their associated classical statistics. According to RBW, quantum statistics as characterized by the Born rule does precisely that. Indeed, per the Feynam path integral for QM the most probable path is the classical path, and per the transition amplitude for QFT the most probable field configuration is the classical field configuration (as will be shown in Foundational Physics for Chapter 5). Now we briefly cover the possible connection between RBW and information-theoretic accounts of QM [Stuckey et al., 2016d].

4.6 RBW and quantum information theory

There are many diverse information-theoretic accounts and we will not attempt to survey them all here. Let us then say that any account of the wave function that is purely epistemic, that is, the wave function is not an element of reality at all, and for whatever reason it also rejects the need to posit more fundamental elements of reality such as Quantum Bayesianism, is not a realist account of the quantum. Of course, one might try to place stronger constraints on a fully realist account, such as fundamental entities must be local beables in spacetime that make up or compose classical entities as in classical mechanics, or at least upon which classical entities supervene. Such fully realist accounts tend to insist that information cannot be a primitive, but must be carried by some physical entity. However, it must be noted that there are attempts at fully realist information-theoretic accounts in which information is fundamental, such as Lloyd's claim that the universe is a quantum automaton [Lloyd, 2006, p. 240] or Wheeler's "it-from-bit" [Folger, 2002]. There are also information-theoretic accounts of the quantum which in some sense seem to fall between purely epistemic non-realist accounts and fully realist interpretations. Take the following, for example:

> The quantum information-theoretic interpretation of quantum theory (QIT) was fully formulated by Bub and Pitowsky, but had its origins in the earlier work of Clifton, Bub, and Halvorson, Bub, and Pitowsky. This interpretation is formulated fundamentally in terms of *information-theoretic constraints on the possibility of correlations between events*. It is argued that a small number of such constraints pick out the Hilbert space as the fundamental space in which the theory is formulated, which

---

[11] Or 0D objects in the case of point-like particles.

in turn imposes conditions on the possibilities of correlations between events. These information-theoretic principles have physical motivation, as they represent various no-go theorems of quantum mechanics. The *interesting feature of QIT, though, comes in considering these kinematical principles to be the fundamental formulation of quantum theory*. What quantum theory tells us about the structure of reality, according to this way of thinking, is that it conforms to a non-Boolean underlying event space, out of which an effectively classical macroscopic world can emerge.

[Dunlap, 2015, p. 360; italics added]

In characterizing his theory, Bub has often invoked Einstein's famous distinction between principle and constructive theories [Bub and Pitowsky, 2010]. The former, such as SR with its relativity principle and light postulate, need not make reference to an underlying ontology and dynamics, but need only provide the aforementioned constraints. Staying with this analogy, quantum theory is really a theory about the possibility of representing and manipulating information. QM is then a principle theory about information-theoretic constraints which does not need an underlying ontology to ground it. Of course, one might accept this as a purely epistemic or methodological claim and still want to know what, if any, fundamental ontology underlies the quantum. Here is Bub's response:

The question: What is information in the physical sense (if it's not about the properties of physical stuff)? should be seen as like the question: What is a field in the physical sense (if it is not the vibration of a physical medium)? The answer is something like this: Quantum mechanics represents the discovery that there are new sorts of information sources and communication channels in nature (represented by quantum states), and the theory is about the properties of these information sources and communication channels.

[Bub, 2009, p. 554]

Bub's response still begs important questions. Are these "information sources and communication channels" fundamental ontological elements of the universe? And if not, is he just rejecting the idea that quantum theory needs an underlying metaphysical picture? It is one thing to make no commitment to an underlying ontology, another to claim one need not search for the underlying ontology, and yet again quite another to say there is no underlying ontology. As with Lorentz's ether interpretation of SR, those who insist on a fundamental underlying ontology for QM assume that such an account will be constructive and thus the fundamental explanation will be dynamical or causal as in classical mechanics. Even Einstein assumed that the principles of SR would be explained constructively someday [Howard, 2015]. However, it seems that Bub and others want to take seriously the possibility that the fundamental explanations or axioms underlying the quantum or represented by it are not dynamical laws or entities, but information-theoretic constraints. RBW suggests another possibility: that there is a theory fundamental to QM, a principle theory, and that underlying theory is a realist, psi-epistemic account in spacetime where the fundamental explanans is a physically robust adynamical global constraint from which dynamical laws and various information-theoretic constraints could be derived. Thus, RBW could ground certain information-theoretic accounts and make good on their promise to provide profound new insights about the nature of physical reality without instrumentalism, quietism, abandoning scientific realism, or being forced to make vague claims about the fundamentality of information simpliciter.

So, the fundamental entities in RBW are 4D spacetimesource elements and the fundamental explanation is the adynamical global constraint, just as in some information-theoretic accounts whereby the Schrödinger equation is not understood as time-evolving physical systems. Rather than say, as Bub does, that some algebraic non-Boolean underlying event space is fundamental and the classical macroscopic world "emerges" in part from information-theoretic constraints, RBW says very clearly that the wave function and Hilbert space are epistemic, and 4D spacetimesource elements are the fundamental ontological entities. Accordingly, algebraic approaches, dynamical laws, and dynamical explanation are not fundamental but are to be recovered from the adynamical global constraint. With those caveats, the adynamical, graphical, 4D approach of RBW is quite compatible with certain information-theoretic accounts of QM and might underwrite them.

To summarize and look forward to chapter 5 on QFT, as we point out in Foundational Physics for Chapter 4, the fundamental empirical data are the individual clicks in a QM experiment. In our approach, the distribution of these individual clicks is given probabilistically by QM in a classical context (the relative positioning of diffraction gratings, beam splitters, mirrors, detectors, etc.). The fundamental entities of the classical context are the time-evolved things of dynamic narration, the "diachronic entites." These are put in by hand throughout physics, but the composition of their worldtubes in terms of individual discrete spacetime events is illuminated via graph theory (again, see Foundational Physics for Chapter 4). Since the individual clicks are distributed in the worldtubes of diachronic entities, the rule for the construct of the spacetimesource element relating sources—the sink(s) of which is(are) the individual click(s)—must be in accord with the rule for the construct of the worldtubes of diachronic entities. Further, quantum probability theory for the distribution of the spacetimesource elements in their classical context must make correspondence with classical probability theory dealing with the diachronic entities of the classical context. We explained how our adynamical global constraint (AGC) and the Born rule together satisfy these constraints in Foundational Physics for Chapter 4 and section 4.5 of the main thread, respectively. What about the diachronic entities of particle physics via QFT?

As we explain in the main thread of chapter 5, in order to construct the worldlines of individual particles in a particle physics "event" one must spatially localize the "hits in the tracking chamber," associate those clicks with a particular track, then geometrically parameterize the tracks. This is the process by which worldlines are inferred from particle physics detector data and they constitute what we mean by the diachronic entities of particle physics. Since diachronic entities (particles) require trans-temporal identification, the notion of a particle resides in its conserved quantities. Indeed, the conserved quantities are the properties defining a particle and particle physics is driven by the search for symmetries of the QFT Lagrangian density. Therefore, per Noether's theorem, the essence of particles in a particle physics experiment must reside in symmetries of the Lagrangian. However, since the particles are diachronic entities constructed from individual clicks, the AGC and Born rule dealing with the distribution of individual hits in the tracking chamber of a particle physics experiment are constrained in QFT just as they are in QM. And consequently, the symmetries of the Lagrangian density of QFT particle physics are no more fundamental than those of the Lagrangians for everyday classical items involved in QM experiments (diffraction gratings, etc.) where we understand the conservation laws may be obtained equivalently from either the Newtonian Schema or the Lagrangian Schema.

In conclusion, RBW works in the context of the diachronic entities of particle physics per QFT just as it works in the context of the diachronic entities of classical physics per QM. Indeed, we would define the diachronic entities of particle physics to be classical objects (see section 4.5). Since classical physics doesn't seem to support the fundamentality of the Lagrangian Schema over the Newtonian Schema or vice versa, there is nothing in particle physics per RBW that provides for it either.

Philosophy of Physics for Chapter 4

4.7   Introduction

In this philosophy of physics thread, the goal is to articulate the ontological implications of RBW more fully than has been achieved thus far and articulate what makes its global and adynamical explanatory methodology unique. First, we will characterize ontological contextual emergence and then argue that the spatiotemporal contextuality of RBW is a subset of that. In the process it will become clear why RBW, which shares some features with related views like ontological structural realism (OSR), is nonetheless very different. Along the way, the coherence and potential of all such ontological contextuality will be defended. Finally, we will contextualize RBW as a unique interpretation of QM by comparing and contrasting it with retrocausal interpretations of QM with which it shares some history and features. RBW is a realist, psi-epistemic account that invokes adynamical global constraints as fundamental. Much of the first section is from a book manuscript in progress called *Emergence in Context* by Bishop, Silberstein, and Pexton, and a forthcoming paper by Silberstein [2017]. Much of the last two sections is from Stuckey et al., 2015.

4.8   Contextual emergence defined

Talking about emergence, or emergence versus reductionism, is always tricky because in philosophy and across the sciences there is an growing body of work on the topic with ever-growing and more heterogeneous definitions of the term "emergence" [Silberstein, 2012]. For example, historically (such as the work of C.D. Broad) in analytic philosophy there is the notion of an emergent property or phenomena as requiring a brute bridge law to explain. Such laws, say psycho–physical bridge laws, only obtain contingently in the actual world with nomological (natural law-like) necessity; they may fail to obtain in other microphysically identical worlds. So on this view it is a contingent fact about the actual world that under the right neurochemical, functional, or information-theoretic conditions that conscious experience will "pop" into existence. This "strong" or "radical" kind of emergence will be critically discussed at length in chapters 7 and 8. Suffice it to say, however, that most philosophers and physicists, being physicalists or ontological/methodological reductionists, find such a strong notion of emergence problematic, spooky, dis-unifying and non-explanatory. As we see throughout this book, and as noted by Butterfield and Isham [Butterfield and Isham, 1999, p. 112], when foundational physicists say things like time, space, or spacetime is "emergent" from a more fundamental (possibly discrete) theory than general relativity or quantum field theory—so-called quantum gravity—they mean time or what have you isn't basic but in some sense can be derived from that which is (chapter 6). Let us call this weak or conservative emergence. Thus, the foundational physicists' notion of emergence is much more akin to what most philosophers of science would call reductionism. As an analogy, we could say that the universe is like a finite automaton (e.g., Conway's Game of Life), wherein the fundamental building blocks are essentially physical; they are evolved in time by some dynamical rule or law, and everything else emerges conservatively/weakly for free from their dynamical evolution, such as gliders in the life game. The kind of ontological emergence defended herein is neither the radical variety nor the physicists' variety. Both those conceptions of emergence presuppose two things we wish to deny: 1) that explanation is fundamentally dynamical and 2) that reality is compositional or made up of distinct levels that are defined in terms of spatial and temporal length scales. The primary focus in this chapter is not the emergence of mind from matter but rather the nature of both the quantum and the classical, and the relationship between them. For us that relationship is one of ontological spatiotemporal contextuality, but first we must define contextual emergence. What will become clear is that in RBW the classical does not emerge from the quantum in either the weak or radical sense of emergence. Indeed, the nature of both the quantum and the classical, and their relationship, must be fully re-conceived.

In the first part of this section the necessary background will be given to define contextual emergence properly. It will then be defined and finally compared and contrasted with views that might appear similar. The widely shared metaphysical assumption in the West is that reality is like an axiomatic system such that, at the end of the day, a feature of reality such as a particle or law is either an axiomatic one (fundamental such as the most basic physical entities, whatever they are) or theorem-like (a logical consequence of what is fundamental). If axiomatic, then it falls in the fundamental category. If a logical consequence of what is fundamental, then it falls in the conservative or weak emergence category, for example, gliders in the Game of Life. So for example, if there were brute psycho–physical bridge laws of the sort posited by David Chalmers [Chalmers, 1996], they would be just another part of the axiomatic base. Given this picture of reality, it seems there is only room for weak emergence or strong emergence in the sense of brute bridge laws that we add to the axiomatic base that somehow operate over essentially irreducible new types or kinds.

This picture of reality is often further grounded in the overwhelming assumption across the sciences in the West that more and more will be explained by digging deeper and deeper down into shorter temporal scales and smaller length scales for entities, laws, and mechanisms that somehow locally explain the behavior of the (relatively more abundant) macroscopic phenomena. In fact, many scientists and philosophers would assert that to explain scientifically is to reduce in some compositional/mechanistic sense or via some sort of intertheoretic reduction. There is another alternative here, however: let's call it contextual emergence. Simply put for now, the idea behind contextual emergence is that new properties, entities, laws, etc., emerge out of multiscale contexts of various sorts, that nature is inherently contextual. It is the contextual nature of reality that grounds emergence and makes it possible. It is important to note that, while we reject any sort of compositional reductionism, RBW does provide a theoretical unification of physics as we spell out in detail in chapter 6.

Contextual emergence provides a framework to understand two things: 1) how novel properties are produced and 2) why those novel properties matter. Contexts modally constrain systems. These contextual constraints represent both the screening off and opening up of new areas of modal space. Moreover, these modalities are the result of constraints that are multi-scale. In all such cases what we take to be basic parts and their dynamics get constrained/determined/overridden by contextual features allegedly outside the system, often at different interacting scales, and then new and stable patterns arise. The relevant and determining contextual features will vary from case to case, such constraints may be more or less concrete, but they include physical, structural, topological, and dimensional constraints. But in all cases, however, the relevant constraints are global such as to trump local dynamics and mechanisms.

The properties and behaviors in a particular allegedly more "fundamental" domain (including its laws) at smaller length and time scales offer at best some necessary but not sufficient conditions for the emergence of said phenomena. As with Mackie's INUS conditions, the "underlying" or "reducing" domain is a necessary part of a sufficient condition for properties and behaviors in that or other domains [Brennan, 2011]. Therefore, phenomena at many different scales can count as contexts for phenomena at many different scales. The universe is not divided into autonomous, closed levels or scales. Such constraints are multilevel. Think again of the way macroscopic measuring devices constrain the behavior and outcomes of quantum behavior. Laws are constraints not "bosses" or "governors"—the universe is not like a computer. Some constraints are more universal than others, such as conservation laws and the symmetries behind them. Some global or systemic constraints at multiple scales trump what we think of as dynamical laws "governing" a system; for example, various cases of universality wherein global dimensional and topological constraints trump local dynamics [Silberstein, 2017]. The implication of all this is that in cases of ontological contextual emergence the emergent phenomena do not even nomologically supervene on phenomena at smaller length and time scales.

Contextuality, in the ontic sense, means a particular confluence of circumstances that produce a combination of constraints and stability conditions, stability conditions allow certain constraints to be "activated," and the constraints can be heterogeneous in nature. Again, they can be topological, dimensional, or structural constraints, but they all limit the modal space available to the system (reduce its degrees of freedom) and open up new possibility spaces closed off outside that context (add new degrees of freedom). This is a form of multidirectional determination, since any causal process is bounded by these constraints and the constraints can be topdown, bottom up, side to side (as it were). In that sense there is no causation at all without a contextual limiting of modal possibility. Such emergent properties or features maybe causal, dynamical, etc., but they always result from global or systemic constraints or contextual features that are often immune to local perturbations.

Contextuality in the epistemic sense means a particular way of dividing the system and environment up so as to allow, for example, deductibility in the case of intertheoretic reduction (and some form of system/environment distinction is always required, and deductibility always implicitly relies on concepts taken from the environment side of the line), and these can include stability conditions for the definition for variables to be the object of projectability.

Let us start with the good news. Whereas in the past, the idea of contextuality as ontologically fundamental would have met with general skepticism or hostility, there are now a number of metaphysical views that are at least in the ballpark of contextual emergence. We will briefly discuss three such views here for purposes of contrast. The first view is ontic structural realism (OSR). There are by now many varieties of OSR but let us focus on the following definition:

> OSR is the theory that relations are all that exist. In opposition to the standard view, which tends to be defended by what we could call particularist ontologies, OSR says that the world is structure all the way down. What we call particular things and their natures are just invariant patterns in that relational structure . . . . According to OSR, there are pervasive relations that make the world a connected and interdependent structure . . . Thus OSR is committed to an irreducible relational holism.
>
> [Briceno and Mumford, 2016, pp. 198–9]

At minimum, OSR so defined is a rejection of both "primitive thisness" and intrinsic properties. Most of the objections to OSR as defined stem from the fact that it ultimately refuses to ground relations in any relata (something fundamentally non-relational). Of course, there are various ways a proponent of OSR could try to define fundamental relations (e.g., modal structures [Briceno and Mumford, 2016]), but the basic worry persists regardless. One version of this objection goes like this: if abstract, perhaps even Platonic, modal structures are fundamental (such as those of fundamental physical theories), then how do we ever get a world of what appears to be concrete physical objects. That is, how can OSR save the appearances? Take the following:

> Thus even those ontologies that have some sympathy for relational holism are compelled to ground those relations in something fundamentally non-relational: OSR has correctly identified a problem. But OSR offers the wrong solution: a solution that empties the world of all its concreteness. Unless our world is a Platonic world of exclusively ante rem universals, OSR is wrong. Other holistic ontologies seem to do equal justice to the interdependent character of the world without abandoning the realm of the concrete. Monism, process metaphysics and dispositionalism are good examples. In all these metaphysics, there is at least one concrete bearer of the interdependent structure. There is the field, the one spacetime manifold, processes all the way down, or a choreography of powerful substances. All of these offer the requisite

interdependence and holism. Unlike OSR, none of them claims that pure relations alone can do the job [Briceno and Mumford, 2016, pp. 216–17].

Monism of various sorts will claim that the relations are grounded in the whole. For example, "existence monism" claims that the universe has no parts since only the whole exists and "priority monism" holds that the parts exist but the whole is prior to the parts such that the universe is an integrated whole [Schaffer, 2010]. "Dispositionalism" is the view that properties are intrinsic dispositions in the sense that objects, entities and systems have no intrinsic properties except for dispositional properties. Objects, etc., have intrinsic dispositions to interact with the intrinsic dispositions of other objects in a symmetric fashion such that, "[i]n Martin's example, the solution of water and salt is the joint product of the soluble salt and the solvent water, but these substances have many other dispositions, depending on the particular interactions they undergo" [Dorato, 2016, pp. 239–40]. What makes these dispositions intrinsic is that they could presumably exist in a possible world in which only that object exists, and indeed, such dispositions can exist unmanifested in this world.

So again, the good news from our perspective is that OSR, monism, and dispositionalism will all involve some sort of ontological contextual emergence (OCE) and therefore we can consider them as allies and we hope the reverse. But it is important to see that OCE doesn't entail any of these views and is therefore not saddled with the objections that go along with these views. Let us start with OSR. Modal structure isn't fundamental in OCE, properties are, and modal relations are defined by property compatibility/incompatibility relations. Modal structure is crucial for mapping the world epistemically and as an indicator of the ontologically importance of properties. It is one thing to say that relations (in some sense) need not ultimately be grounded in anything non-relational, but quite another to claim that relations as modal structures or some other abstracta are fundamental. OCE has no commitment to structure over concrete and other relata, it has no commitment to relations being defined in terms of modal structures or anything else Platonic. Furthermore, the emphasis on structure in OSR misses the point of contextualism: that those structural properties, when they are ontologically important, are often produced in the contexts provided by relata not other structures. To place structure above relata, or relata above relations, is a category error, since there is no non-contextual ordering of those things one can make. OSR fetishizes all sorts of formalisms, whether they ground modal relations in the real world or not. Indeed, in many OSR examples it is highly dubious that it is the structural properties whose compatibility with other properties produces the modal structure. The problem with OSR is not the acknowledgement of the importance of structure for some things or particular contexts, it is the inductive leap to say that the only thing that ultimately matters is structure.

It would be understandable to read OCE as just a kind of (w)holism in which, "in the limit, there may be only one thing, the universe, whose break-down into separable parts is no more than our conceptual imposition or construction upon this vast singular being" [Serway and Jewett, 2014, p. 321]. Yet, in the case of OCE, we must understand that it is relations and contexts all the way down, up, and side-to-side; there need be no universe above and beyond these (e.g., no wave function of the universe). That is, there need be no "view from nowhere." OCE claims that even the entities, relations, and laws of "fundamental" physics are determined contextually. The context includes features at larger time scales and length scales. Therefore, in principle, the arrow of determination and explanation in contextual emergence— in accord with our best scientific understandings when we examine them closely— is not exclusively "bottom up" but multi-directional.

So with regard to monism, OCE could perhaps fit with it but certainly doesn't necessitate it. After all, the argument is that contextuality is fundamental, and contextuality is a relation between things, situations, etc. So we have a flavor of monism in that everything is connected, the whole isn't just the set of autonomous parts, but nor is it just ultimately one thing either. The different contexts and their interactions determine the nature of the whole and make it a whole. The necessity of contextual structure within that one thing cannot be eliminated, so why is that monism as such? Indeed, given that different property manifestations require different contexts, the sheer number of modally salient properties in OCE requires a multitude of contexts. That is, from the perspective of OCE the idea that one could "suck out" all the contexts and interactions from the world and there would still be "the one" left over is highly dubious. We would want to see an empirical or scientific motivation for such a holism. And since science is an activity that can only start in the context of screening off a subset of the many things/manifestations from each other, that will take some doing (see Schaffer, 2010 for such an attempt). In short, saying that at minimum you need at least one context for properties, etc., to emerge isn't the same as claiming the world we live in could be a world produced by only one overarching context.

With regard to intrinsic dispositions, we vacillate between a more or less harsh response. What exactly is an intrinsic disposition? Is that just an unanalyzed and fundamental feature of reality? How exactly does that work? Does every object have an infinite number of such dispositions, and is it like a hidden program or "instruction set" that tells the object what to do when it encounters one of an infinite number of things it might interact with? Presumably not, otherwise one is stuck with some metaphysical equivalent of the frame problem here. But if not, however, then what does it even mean to claim that X and Y have an intrinsic disposition to yield Z upon interaction? What makes such dispositions intrinsic? If the answer is that they exist even unmanifested, why would one ever be justified in believing that when by hypothesis they only manifest upon interaction? How can there be a property that is both inherently intrinsic and inherently relational? If process-like ontologies also embrace intrinsic dispositions then they are in the same boat and must answer the same questions.

The less harsh reply is this: OCE is ecumenical with respect to intrinsic dispositions. Although the notion of intrinsic would have to be modified as it is only intrinsic when placed in a context (again, this may violate the very idea of intrinsic), and the context part isn't the usual emphasis of dispositionalists. But whether one says "system X has the disposition to manifest pre-existing property Y only in context Z," or "acquires property Y only in context Z," seems tangential to the question of the necessity of context Z for property Y to influence the world. After all, for OCE the headline is that contexts are necessary to produce the properties that do stuff. So again, for this discussion to be of great interest to us it would have to be more than a merely metaphysical debate (see Dorato, 2016 for such an attempt).

We think OSR, monism, and dispositionalism entail OCE, but not the reverse. So no matter how these metaphysical debates settle out, OCE is in business. But the spirit of OCE, at least for us, is to be skeptical of the very idea that in order to explain order and stability in the universe there must be some second-order, metaphysical, and undetectable metaphysical glue hiding behind the world of experience. It doesn't matter to us where the glue is put: in transcendent governing laws that are like program rules, in the properties of objects, in the wholeness of the universe, etc. For us, the contextual nature of reality removes any motivation for such glue. Unlike strong emergence, there is nothing spooky at all about the emergence of novel and stable new phenomena in a world in which what is fundamental is the scale-invariant interdependence and of interactions of various phenomena. Furthermore, as we make clearer in the next section, the ultimate stability and robustness creating principles in RBW aren't dynamical, they are adynamical global constraints that "tell the parts what to do" in an adynamical fashion.

One important criticism to consider here is that perhaps there are good, purely metaphysical considerations for believing that there must be intrinsic properties, or that dispositions must be intrinsic in some sense. If so then OCE can't be true. There are certainly people who make that argument [Choi and Fara, 2016]. However, there are also those who claim that there need be no intrinsic properties or dispositions, or at the very least, that they are not be found in fundamental physics where one would most expect them [Choi and Fara, 2016]. There are those who argue that, based on how they are portrayed by physics, relatively fundamental physical properties such as mass, charge, and spin are purely relational or purely extrinsic because these properties are defined in terms of how these particles behave in certain contexts. Others retort that this is just a methodological fact about physics with no metaphysical implication [Choi and Fara, 2016]. The truth is, this is in fact an ancient discussion in metaphysics with no consensus and no end in sight. To this day, a standard argument for panpsychism is the claim that fundamental physical entities must have intrinsic properties and properties such as mass, charge, and spin are not intrinsic, they are extrinsic dispositions; therefore their intrinsic properties must be mental or experiential in some way. There are of course just as many arguments to the contrary [Choi and Fara, 2016]. A lot hinges in these discussions on exactly how one defines "intrinsic" and "extrinsic."

In what follows, in outlining one such position on extrinsic properties Carruth has perfectly captured what OCE is claiming in this regard:

> The model proposed by Martin and Heil, then, does not encourage us to think of powers as isolated, but rather as participating in a network or web of potency/dispositionality.. . .
>
> The model here is not a chain, but a net . . . .
>
> (Heil, 2005, p. 350)
>
> Start with any disposition partner and you find a network—a Power Net.
>
> (Martin, 2008, p. 87)
>
> Every disposition is, in this way, a holistic web, but not just an amorphous spread of potency. (p. 6)
>
> In every power or disposition, an ineliminable reference to the infinity of potential partners is inherent. The powers that an object instantiates locate it within the intricate structure of this network, they define its connections, its potentiality for interaction. But, as Martin insists, this potency is not shapeless, raw or blurred round the edges; on the contrary, it is brought into sharp definition by the reciprocal partnerings which are possible for that object in virtue of the particular genuine powers it possesses, the network is infinitely intricate and complex, but equally it is perfectly defined and delineated. The network (and any particular power that participates in it) is disposed towards far more than it could ever manifest—whilst the potentialities which this intricate filigree of reciprocal partnerings for mutual manifestation are directed towards run to infinity, the number of mutual manifestations which actualise will always be much lower. This plenitude of potentiality, Martin claims, is "carried" by the relatively limited number of actual dispositions, and it is "natural that so little can carry so much." (p. 60)
>
> [Carruth, 2008, p. 36]

Carruth also gives an excellent response to the "regress" problem which states, unsurprisingly, that "if the identity of powers is fixed by their relations to other powers, and these powers themselves only have their identity fixed in the same way, then either there is a vicious regress or else the fixing of the identities of powers relies upon circularity: a power F relies for its identity upon some other power G, whose very identity relies on the identity of F itself!" [Carruth, 2008, p. 38]. Carruth argues that this problem is resolved by Bird, as follows:

> Bird employs the resources of graph theory to explain how this is possible. In graph theory, the identity and distinctness of a vertex can be given as purely supervenient on the overall structure of that graph. There are some restrictions here: the graph must be asymmetric, in order that "such a graph would have no way of swapping vertices while leaving structure unchanged" (p. 528). This adhered to, for any vertex

which occurs within the structure of the graph, "the structure determines the identity of the vertices—the structure itself distinguishes each vertex from every other vertex i.e., the identity of vertices supervenes on the set of instantiations of the edge relation" (ibid.). The nature, and thus the identity, of each particular vertex which is occurent within the structure of the graph is given and determined by this structure. There can be no threat to these vertices of regress or circularity with regards to their identities, these are well grounded, albeit extrinsically, in the structure of the graph. Further levels of complexity can be built in so that structures can be generated to accommodate an infinity of vertices.

Bird argues that we can adopt this model for understanding the way in which the identity of a particular power is given extrinsically, through the determination of its nature by way of the structural properties of the dispositional network. A particular power, then, is like a vertex.

[Carruth, 2008, pp. 37–9]

It is unsurprising that both OSR [Ladyman and Ross, 2007] and OCE [Stuckey et al., 2005; Silberstein et al., 2007] have also invoked graph theory as way to think about such extrinsic dispositions. Of course, certainly nothing in this graph theory analogy entails that these properties or all properties must be fully relational. However, as Carruth writes, "I argue that if parity of reasoning is to be maintained, and we adopt the model Bird suggests in order to defend against the 'regress argument'—that the natures of powers are given extrinsically in a manner analogous to the fixing of the identity of vertices in graphs—then this analogy must be maintained" [Carruth, 2008, p. 43]. The point here is that while there is an ongoing debate in metaphysics about the possibility that all properties are fully relational, there is certainly no consensus on the matter, and that is because there is no knock-down no-go argument for the conclusion that dispositions must be intrinsic in some way.

Strong/radical emergence also stresses context or new conditions but OCE makes so-called radical emergence not spooky because new entities, properties, and laws emerging relative to various contexts is the fundamental nature of reality from the start and not some miraculous occurrence for chemistry, life, and mind. If there really was some autonomous fundamental physical entity such as quantum fields or quantum loops, if the microphysical really were causally closed, if physicalism were true, or ontological reductionism true, etc., then ontological emergence would be about "spooky," metaphysically brute and disunifying laws or causes that allegedly "explain" why chemistry, life, and mind just "pop" into existence under certain conditions.

We think it this very conception of strong emergence that generates such hostility even among proponents of views such as those above that entail OCE. More specifically, it is important to note again that there are other views out there now that share much with contextual emergence. Most prominently what comes to mind are the OSR of Ladyman and Ross in *Everything Must Go* [Ladyman and Ross, 2007] (hereafter, ETMG) and "the scale-free universe" described in

Thalos' recent book [Thalos, 2013]. What is noteworthy here is that both parties are critical of emergence talk. We will circle back to these views momentarily, but less us begin with stating some ontological implications of contextual emergence:

1. Physics is only fundamental in the sense that it applies everywhere in the universe. Physics provides the most universal constraints.

2. Physics constrains the special sciences but does not determine them—logically, metaphysically, nomologically, or otherwise.

3. Indeed, the physical facts of our universe are neither necessary nor sufficient for the special science facts of our universe.

4. Supervenience (global or otherwise) and thus minimal physicalism is false in that two worlds could have identical physics and yet diverge with respect to special science facts.

5. Contrary to the Game of Life analogy used by Daniel Dennett, the universe is merely just "real patterns" all the way down in that there are no individuals at any scale in the universe; that is to say, no entities with primitive "thisness" and intrinsic properties. Physical facts and special science facts are symmetric in this regard. So in keeping with the analogy, it is not just gliders and such that are merely just real patterns, so are the fundamental building blocks or "cells."

6. The only reason we use causation talk for special science real patterns and law talk for real patterns in physics is because some of the latter apply everywhere. So lawtalk indicates no special metaphysical or nomological glue at bottom; it is still just patterns.

7. There is no interest and context-free compositional reduction or intertheoretic reduction of special science facts to physical facts to be had.

8. So all ontology is "scale relative" or exhibits "relative onticity" as Atmanspacher and Kronz [1999] state. This includes both physics and the special sciences. Therefore the universe is not divided into autonomous levels with essentially different properties.

To reiterate, it seems clear to us that Ladyman and Ross, and Thalos do and would agree with most if not all of points 1 to 8, yet they reject talk of emergence even in this world where most everything is contextually emergent given those points. We think their rejection is driven by the aforementioned dilemma that emergence talk is typically either invoking weak or strong emergence, which are both very different than contextual emergence.

Ladyman and Ross make it clear in ETMG that for them, the use of the word "emergence" is never helpful, even when it designates a position with which they agree. For instance, for years various philosophers have been using the term emergence to describe how features of quantum mechanics, such as entanglement, strongly suggest we should reject the idea of particles as things with intrinsic properties [Silberstein, 2012]. This is certainly a point that Ladyman and Ross make in ETMG, yet there is little or no recognition of such philosophers in the book. One exception here is Andreas Hüttemann and here is what they write about him: "Hüttemann (2004, p. 52) [referring to quantum entanglement] is pleased to talk of 'emergence' whereas we never are" [Ladyman and Ross, 2007, p. 57].

Clearly we share some of Ladyman and Ross's concerns about needlessly wielding the term "emergence." There are the concerns about the ambiguity of the meaning of the word and the historical baggage it carries. For instance, they worry that emergence means vital forces, extra substances, or something mystical beyond scientific explanation (i.e., strong emergence). There is a particular worry that emergence implies downward causation and a violation of their primacy of physics (PPC) principle [Ladyman and Ross, 2007, pp. 44–5]. They state that rejecting the latter is unscientific [Ladyman and Ross, 2007, p. 45]. On page 57 they write, "when someone pronounces for downward causation they are in opposition to science" [Ladyman and Ross, 2007, p. 57]. What exactly they mean by "downward causation" is a tricky question and requires more exploration. The key is in their definition of emergence. "This doctrine warrants its name because it holds that 'higher' levels of organization 'emerge' indeterminably out of 'lower' level ones and then causally feedback 'downward'" [Ladyman and Ross, 2007, p. 56]. Clearly they want to reject the levels picture of reality (as in point 8 above) and they want to reject the mysterious configurational forces of the British emergentists.

As should be clear from points 1 to 8 we share Ladyman and Ross's rejection of "levels physicalism" and their rejection of emergence construed as meaning beyond scientific explanation, or invoking spooky new forces. Nevertheless, it turns out that many genuine scientific explanations are not strictly reductive, as we think they acknowledge in ETMG. We also share their skepticism of downward causation as construed by researchers like Kim and Papineau [Silberstein, 2012]. However, both Silberstein and Bishop have argued at length elsewhere [Bishop, 2005, 2012; Silberstein, 2006, 2012, 2013; Chemero and Silberstein, 2008; Silberstein and Chemero, 2011a] that processes at larger length and temporal scales do constrain and determine "lower-level" processes. Scales interact and the constraint or determination of the behavior of any given scale or system is almost always multi-directional. Whether or not one calls this "causation" depends on one's account of causation. Sometimes people call it reciprocal causation, sometimes just constraints, and sometimes global or systemic determination. In all cases, however, we call it OCE. We are not alone in making such points. There are by now several accounts of emergence that explicitly avoids all the offending elements Ladyman and Ross reject [Silberstein, 2012; Bishop, 2005, 2010a,b]. We hope to make it clear that the OSR of Ladyman and Ross is in real need of an account of emergence that is neither the weak nor strong variety.

We think that there is a fundamental tension in ETMG, given their asymmetry claim regarding physics and the special sciences: physics is often invoked in the special sciences but the reverse is never the case. For example, we did not need to change physics to explain biology and we do not appeal to biological processes to explain phenomena from physics. The tension is that this principle seems to imply the causal closure of the physical (CoP), which Ladyman and Ross explicitly reject. If it is a universal exceptionless truth that the special sciences are never needed to explain (relatively) physical phenomena, then presumably in the final analysis when it comes to explaining the behavior of the brain/body, we should not need to invoke anything mental or social such as conscious intentional states. In ETMG, they also explicitly affirm Papineau's argument for CoP from fundamental forces: "Some physical forces were found. None of the non-physical ones were" [Ladyman and Ross, 2007, p. 42]. We, however, would argue that there are cases where the behavior of physical and biological systems depend on certain contextual features that belong in the domain of one or another special science.

We see a similar story unfolding in Thalos' book *Without Hierarchy* [Thalos, 2013]. She argues that emergence, as typically conceived, posits that the universe is divided into a hierarchy of relatively autonomous, and discrete levels (often defined in terms of spatial and temporal length scales), each with their own intrinsic and essential properties. She also adds that such emergence is typically conceived as an expression of non-reductive physicalism that embraces minimal supervenience physicalism such that all the levels supervene on and emergence from a basic fundamental physical level. She does not use the language explicitly but it is clear that this brand of emergence accepts causal closure of the physical (CoP). Her criticism of this brand of emergence is as follows:

> Emergentists in today's intellectual climate, some of whom aim to defend autonomous sciences, proclaim that the unity of science does not lie in the reducibility of the various sciences to Physics, but rather in metaphysical relations of the entities and properties recognized by the "special sciences" to those recognized by Physics. Thus emergentism in contemporary hands still commits the error of acknowledging a Master Science, but it does it as a matter of metaphysics: it comes in the form of a master ontology of independent entities.
>
> [Thalos, 2013, p. 21]

She is especially keen to critique this account of emergence because she espouses a view close to contextual emergence in many ways that she calls "scale free." In a scale-free universe, the there is causal and other "activity" at every scale and all scales interact; there is no special or privileged scale at which to view activity in the universe [Thalos, 2013]. In other words, either the universe has a single fundamental scale or it has none, and she maintains the latter. It is clear that she means to reject CoP, minimal supervenience physicalism and the levels or hierarchy view, all consistent with our points 1 to 8 earlier. She does not want her view to be conflated with the kind of emergence she is rejecting:

> An alert reader might suggest that my view—by the inelegant name of the scale-free universe proposal—might be a very strong, perhaps even radical, form of emergentism, but a variation of it all the same. After all, both are apparently in pursuit of an

articulation of the (admittedly vague) credo that the whole is on some sense greater than, or transcends, the sum of the parts.

[Thalos, 2013, p. 33]

In addition to making it clear that she rejects the three aforementioned tenants of "emergence," she goes on to state:

Further, there is no 'emerging' on my view. True, there might have been eras in the life of the universe where there were fewer scales—or indeed more—at which there was real action than there are today. But it's not as if the action at higher scales keeps emerging from below in a sustaining way, as the emergentists imagine.

[Thalos, 2013, p. 33]

Our take on her argument is as follows: the essence of emergentism historically is a story about how essentially different or new higher-level phenomena such as life or mind can emerge from some absolutely more fundamental and essentially different underlying substrate. She wants to make two points about this. First, this conception of reality (which reifies essences and levels) is an empirically unjustified barrier to scientific and ontological unity; it only makes things harder. We agree. Second, in her scale-free account, nothing emerges in the odious strong sense described above and, therefore, her account is free of the many problems, mysteries, and inconsistencies associated with a non-reductive physicalist account of emergence that maintains, for instance, both minimal physical supervenience and "downward causation." In her own words:

I want to be very clear about my message: emergentism, as articulated by the conjunction of all of 1–4, is simply inconsistent, and therefore untenable, even unintelligible. And positively the only way out, while clinging to the sentiments regarding activity at every scale, is to dispense with the notion of levels altogether—and so with the very core idea of emergentism.

[Thalos, 2013, p. 43]

Why does it matter that the relevant [emergent] feature be "new?" What work is the concept doing in the analysis? Well, the emergentist has to have some means of identifying the features of the world that are in some sense "emergent." (Note that I by contrast don't have to identify "novelty:" I simply say there is activity at every scale—it's not confined simply to the micro. I refer to size scales, whereas they wish to refer to something else, something more subtle, something more metaphysical. Emergentists don't care about activity so much as they care about novelty at each scale or level.

[Thalos, 2013, p. 38]

Thalos' book suggests that one either embraces some sort of (w)holism as she does or one embraces emergence. This is no doubt right for strong emergence, but not for contextual emergence. One of our important messages is that it is not emergence versus holism. Even a wholist view must give an account of the transitions leading to new states and observables, and this is what contextual emergence seeks to do. Rather, it is precisely in part the scale-free nature of reality in Thalos' terminology (we might call it interdependence and interpenetration of scales) that makes contextual emergence possible. For us, emergence and wholism in her sense are two sides of the same coin.

One could state the problem in this manner: in spite of their claim that it is real patterns all the way down, Ladyman and Ross privilege fundamental physics, as this is the phenomenon that exclusively motivates them to champion OSR. And therefore they get into trouble with the special sciences in which individuals are not a secondary derived category. Thalos wants to abandon all sense of privileged scales or "levels" and Ladyman and Ross still privilege fundamental physics. On the other hand, Thalos does not need to justify individual oriented special sciences. Contextual emergence differs from both sides as it emphasizes that contextual emergence is the norm at various scales, across scales, and, most importantly of all, in mixed-scale interactions. For example, whether there is a privileged scale or not is itself an interest relative and context-dependent question. For a particular explanandum, there may well be a privileged scale at which to view the question. Even to define scales one needs to make reference to the real physical characteristics of systems and these pick out privileged scales. For example, the strong force has a scale associated with it and so does the gravitational force, but for nuclear binding the latter is largely irrelevant. Again, many current scales, etc., were not all there at the Big Bang; some of them come into existence through dynamic interaction and we ought to be able to explain this kind of emergence. This is one reason contextual emergence is often both epistemic and ontic. Indeed, what counts as a part or whole in any given case is contextually dependent and therefore dependent on the theoretical representation of the system and environment in question. This isn't anti-realism but an acknowledgment that, given contextual emergence, there will be many different equally useful ways to carve up the world into parts, wholes, systems, and sub-systems.

4.9   RBW and ontological spatiotemporal contextuality: A subset of ontological contextual emergence

While OCE in general does invoke global constraints of various sorts, it still tends to assume that dynamical explanation is fundamental in some sense. This is almost built into the concept of "emergence" wherein we envision essentially new phenomena such as life or mind "emerging from" more fundamental matter. RBW, of course, rejects that dynamical conception of emergence; it goes even further and claims multiscale adynamical global constraints in a block universe are fundamental, which is to say that QM systems are not more fundamental than the classical world, and ultimately "emergence" is conceived in terms of ontological spatiotemporal contextuality, as we saw in the main thread for chapter 4. Here we apply that idea to interpreting QM. Unlike almost all other accounts of the quantum, in RBW, spatiotemporal ontological contextuality is fundamental in that the classical world is not composed of or realized by autonomous quantum entities with intrinsic properties moving through the experimental apparatus with traceable worldlines as with particles or fields. In our account, the presumably classical experimental setup (or the many analogs of that in a natural setting) cannot be reduced away and plays an absolutely essential role in explaining so-called quantum outcomes. This bold abandonment of part/whole or smaller scale/larger scale reductionism may strike many readers as beyond reasonable, but it is nothing new in the history of QM. Here, for example, is Peter Lewis summarizing Niels Bohr's view about ineliminable contextuality in QM:

> Quantum mechanics requires a radical revision of our attitude towards the problem of physical reality (Bohr, 1935, 697). The radical revision he has in mind seems to be a kind of contextualism: what properties a system has depends not only on its quantum state, but also on its physical environment. In particular, Bohr claims that the z-spin of an electron is only defined when the physical environment of the system is such that we could actually measure its z-spin. But the physical environment can never be such that we can actually measure z-spin and w-spin at the same time, since the measuring magnets need to be aligned along different directions in each case.
>
> [Lewis, 2016, p. 75]

Here is the view that Carsten Held ascribes to Bohr:

> In this version of ontological contextualism the property $v(f(Q))$, rather than depending on the presence of another property $v(Q)$, is dependent on the presence of a Q-measuring apparatus. This amounts to a holistic position: For some properties it only makes sense to speak of them as pertaining to the system, if that system is part of a certain system-apparatus whole.
>
> [Held, 2013]

And here, John Wheeler offers a similar view:

> **It from bit**. Otherwise put, every *it*—every particle, every field of force, even the space-time continuum itself—derives its function, its meaning, its very existence entirely—even if in some contexts indirectly—from the apparatus-elicited answers to yes-or-no questions, binary choices, **bits**. It from bit symbolizes the idea that every item of the physical world has at bottom—a very deep bottom, in most instances—an immaterial source and explanation; that which we call reality arises in the last analysis from the posing of yes–no questions and the registering of equipment-evoked responses; in short, that all things physical are information-theoretic in origin and that this is a **participatory universe**.
>
> [Wheeler, 1990, p. 310–11]

It is important to understand that these are not claims that laboratory measurements, experimental observers, or conscious beings, as such, are essential in QM. Everything we are saying about the contextual nature of quantum mechanics would be true in a world with none of the above. Rather, at least as we interpret them, these claims are about the essentiality of "classical" measuring devices or their analog in nature, that is, the necessity of stipulating the classical context. Nor are we making any claims about the fundamentality of information. And we are not claiming, as Bohr is sometimes accused of doing, that there must be some essentially classical phenomena outside the scope of quantum phenomena to force the quantum to behave (or appear to behave) classically. In keeping with the ontological contextualism of Bohr and Wheeler, we are claiming that the universe is not compositional or part/whole reducible, as assumed by most physicists. Rather, inter-scale contexts or relations are fundamental. It is worth noting that more and more foundationalists are coming around to this sort of thinking [Kupczynski, 2016], or at least they are questioning the "Lego-compositional-conception" of reality:

> Quantum entanglement and superposition raise further questions concerning physical composition. If the quantum state completely describes a system but never collapses then it seems likely that the only system with definite physical properties is the entire universe—a dramatic failure of the properties of the whole to be determined by those of its parts! Moreover the radical indistinguishability of quantum "particles" often associated with the (anti)-symmetrization of the quantum state of a set of (fermions) bosons of the same species threatens to undermine their claim to exist as individual parts of the fusion of that set into a whole.
>
> The search for ultimate building blocks of the physical world in elementary particle physics appears to have run its course: the quantum field theories of the Standard Model present us with no clear candidates for ultimate building blocks—neither particles nor fields. A system may have more than one incompatible partition into constituent parts. Even if a system does have a unique or natural partition into basic parts, to explain its behavior it may be necessary also to advert to parts of a different kind—parts formed from basic parts through (one or more) different composition relations, and/or parts that supervene on basic parts without being composed of them. The intricate webs of physical composition relations in theories of condensed matter physics stand as beautiful testimony to the creativity of the physicists who have woven them.
>
> [Healey and Uffink, 2013, p. 20]

Whether it is quantum entanglement itself, a particular interpretation of QM such as Everett offers, the indistinguishability of QM particles, the inability to provide a coherent particle-like or field-like ontology for quantum field theory, or the mereological-intuition defying weirdness of condensed matter theory where given the right context wholes behave like parts and parts behave like wholes, more and more theorists are willing to consider the possibility of explanatory and ontological contextuality. Again, as we noted in the main thread for this chapter, even many of the relatively dynamical/mechanical interpretations of quantum mechanics are stuck with some sort of contextuality.

And again, as we noted in the main thread for this chapter, irrespective of differing interpretations, textbook QM claims the ubiquity of environmental decoherence phenomena for complex systems on appropriately long time scales means that one may, for all practical purposes, FAPP treat the quantum mechanical system as classical. That is, the interaction of a QM system with its environment entangles the two and distributes the QM coherence over so many degrees of freedom as to render it unobservable. An intuitive picture of the interaction between system and environment can be provided by the analogy with a measurement interaction: the environment is monitoring the system, it is spontaneously performing a measurement (rather, it is letting the system undergo an interaction as in a measurement) of the preferred states, but it isn't actually an irreversible measurement of course. The point is environmental decoherence alone, however you explain or characterize it, bespeaks of contextuality.

So again, keep in mind our contextuality is spatiotemporal (4D). For us the fundamental explanatory role goes to what we call an adynamical global constraint applied to the spatiotemporal distribution of our fundamental ontological entity, the spacetimesource element. Thus, it is the adynamical global constraint that "binds" the spatiotemporal contextual features in any given case. RBW shows how to make such contextualist intuitions very precise without instrumentalism or obscurantism. RBW agrees with OSR which says "we might as well dispense with things and assume that the world is made of structures, or nets of relations" [Kuhlmann, 2013, p. 46]. In contrast, however, the properties of our fundamental, relational spacetimesource elements are obtained from their classical context, that is they are contextually from relata, so RBW can also be characterized as a spatiotemporal form of "ontological contextuality" (Figure 4.12). There is no question that this view of reality requires a kind of circularity, recursion, and self-referentiality that most people find hard to swallow. But we don't see any a priori reason that reality couldn't be like this. It helps to remember that while, formally speaking, one could decompose the worldtubes of classical objects into spacetimesource elements, in so doing one would need other classical contexts to do so meaningfully. Are we really saying that worldtubes X and Y, etc., can simultaneously play the role of context for one another? Yes. It also helps to remember that the world of RBW is 4D-relational. The classical world does not emerge via some dynamical quantum process; both the classical and the quantum as we define them are co-equal/co-determining features of this world (Figure 4.12).

Carsten Held calls this kind of contextuality, "ontological contextuality." It "entails that system properties which we earlier thought to be intrinsic become relational in the sense that a system can only have these properties either if it has certain others, or if it is related to a certain measurement arrangement" [Held, 2013]. Thus, per RBW's model of objective reality, relata in the form of classical objects are co-fundamental; they are not "composed of" autonomous quantum systems.

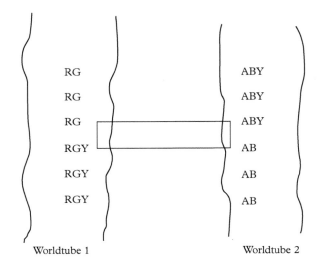

Figure 4.12 *Quantum exchange of energy–momentum. The property Y is associated with the source on the spacetimesource element (rectangle) shared by the worldtubes. As a result, property Y disappears from Worldtube 1 (Y Source) and reappears later at Worldtube 2 (Y detector) without mediation. That is, there is no third worldtube/line needed to explain the exchange of energy–momentum associated with property Y between Worldtube 1 and Worldtube 2. While these properties are depicted as residing in the worldtubes, they don't represent something truly intrinsic to the worldtubes, but are ultimately contextual/relational; that is, being the Source of Y only makes sense in the context of (in relation to) a "Y detector," and vice versa. The A, B, R, and G properties shown might be established with respect to classical objects not shown in this figure, for example.*

So again, like OSR, our view rejects the idea that objective reality is ultimately composed of things, that is, self-subsisting entities, individuals, or trans-temporal objects with intrinsic properties and "primitive thisness," haecceity, etc. In Einstein's terminology, particles do not have their own "being thus" [Einstein, 1948]. Again, this is not to say that relata don't exist in our model of objective reality, but it is a denial of their fundamentality in that model. Relations are primary while the relata are derivative, so relata are not fundamentally "built of relata." Relata inherit their individuality and identity from the structure of relations, so classical objects and their properties are secondary to relational structure.

Assuming relations are fundamental to relata is not unique to RBW. For example, Carlo Rovelli's relational interpretation of quantum mechanics [Rovelli, 1996] holds that a system's states or the values of its physical quantities as standardly conceived only exist relative to a division between a system and an observer or measuring instrument. As well, according to Rovelli's relational account, the appearance of determinate observations from pure quantum superpositions happens only relative to the interaction of the system and observer. Rovelli is rejecting absolutely determinate relata. Another closely related example is Mermin's Ithaca interpretation which tries to "understand quantum mechanics in terms of statistical correlations without there being any determinate correlata that the statistical correlations characterize" [Mermin, 1998]. According to Mermin, physics, for example, quantum mechanics, is about correlations and only correlations; "it's correlations all the way down." It is not about correlations between determinate physical records nor is it about correlations between determinate physical properties. Rather, physics is about correlations without correlata. On Mermin's view, correlations have physical reality and that which they correlate does not. Mermin claims that the physical reality of a system consists of the (internal) correlations among its subsystems and its (external) correlations with other systems, viewed together with itself as subsystems of a larger system. Both these views differ from our own in marked ways; for example, RBW is explicitly about the fundamentality of adynamical global constraints, it is explicitly psi-epistemic, it invokes ontological spatiotemporal contextuality, etc. But again, the point is that more and more theorists are in some fashion advocating for the fundamentality of contextuality and relations. All interpretational differences aside,[12] Rovelli states, and we agree:

> Einstein extended the notion of relativity to time: we can say that two events are simultaneous, only relative to a given motion . . . . Quantum mechanics extends this relativity in a radical way: *all* variable aspects of an object exist only in relation to other objects. It is only in interactions that nature draws the world.
>
> [Rovelli, 2017, p. 135]

> *Relationality.* The events of nature are always interactions. All events of a system occur in relation to another systems.
>
> [Rovelli, 2017, p. 137]

We should note here that in addition to foundational physicists Albert Einstein, John Wheeler, and Richard Feynman, Carlo Rovelli and his relational interpretation of QM was an essential influence on us, especially because he was among the first to hammer home that one could view relativity and QM as both pointing to the contextual nature of reality.

OSR alone already somewhat violates the dynamical (or mechanical) bias by rejecting things with intrinsic properties as the fundamental building blocks of reality—the world isn't fundamentally compositional. Ontological contextuality goes even further in this violation by claiming that classical objects such as measurement setups are co-fundamental with or irreducible to so-called quantum systems. That is, the RBW ontology, spacetimesource elements, is a strong violation of a compositional picture of realty, since the properties of these elements are inherited from their classical context as in the quantum contextualism advocated by Auffèves and Grangier [2014]. Furthermore, unlike ontological

---

[12] However, see [Rovelli, 2016] where he argues against realism about the wavefunction.

contextuality as typically conceived, the contextuality of RBW is spatiotemporal; for example, explaining the outcomes of QM experiments requires the entire spatiotemporal arrangement including the future boundary conditions. Thus, we call it spatiotemporal ontological contextuality. Indeed, the fundamental ontology of RBW are spacetimesource elements and these are 4D in nature and inherently contextual.

A good deal of the OSR literature focuses on philosophical concerns about scientific realism and intertheoretic relations, rather than being motivated by physics itself [Ladyman, 2007; Rickles, 2012]. There has also been much debate in the philosophical literature as to whether OSR provides any real help in resolving foundational issues of physics such as interpreting QM or in advancing physics itself. Consider Esfeld's claim, for example:

> OSR is not an interpretation of QM in addition to many worlds-type interpretations, collapse-type interpretations, or hidden variable-type interpretations. As the discussion of the arguments for OSR from QM . . . has shown, OSR is not in the position to provide on its own an ontology for QM, since it does not reply to the question of what implements the structures that it poses. In conclusion, after more than a decade of elaboration and debate on OSR about QM, it seems that the impact that OSR can have on providing an answer to the question of what the world is like, if QM is correct, is rather limited. From a scientific realist perspective, the crucial issue is the assessment of the pros and cons of the various detailed proposals for an ontology of QM, as it was before the appearance of OSR on the scene.
>
> [Esfeld, 2013, p. 30]

And this from Rickles and Bloom:

> While the basic idea defended here (a fundamental ontology of brute relations) can be found elsewhere in the philosophical literature on 'structural realism', we have yet to see the idea used as an argument for advancing physics, nor have we seen a truly convincing argument, involving a real construction based in modern physics, that successfully evades the objection that there can be no relations without first (in logical order) having things so related.
>
> [Rickles and Bloom, 2015, p. 104]

RBW is an answer to Esfeld's request in that we "provide on its own an ontology for QM." And, in response to Rickles and Bloom, as we show, RBW's analog to OSR has the potential to reground physics, dissolve current quagmires, and lead to new physics, just as they expect:

> Viewing the world as structurally constituted by primitive relations has the potential to lead to new kinds of research in physics, and knowledge of a more stable sort. Indeed, in the past those theories that have adopted a broadly similar approach (along the lines of what Einstein labeled 'principle theories') have led to just the kinds of advances that this essay competition seeks to capture: areas "where thinkers were

'stuck' and had to let go of some cherished assumptions to make progress." Principle theory approaches often look to general 'structural aspects' of physical behaviour over 'thing aspects' (what Einstein labeled 'constructive'), promoting invariances of world-structure to general principles.

[Rickles and Bloom, 2015, pp. 102–103]

Rickles and Bloom lament the fact that OSR has yet to be so motivated and further anticipate RBW almost perfectly when they state:

The position I have described involves the idea that physical systems (which I take to be characterized by the values for their observables) are exhausted by extrinsic or relational properties: they have no intrinsic, local properties at all! This is a curious consequence of background independence coupled with gauge invariance and leads to a rather odd picture in which objects and [spacetime] structure are deeply entangled. Inasmuch as there are objects at all, any properties they possess are structurally conferred: they have no reality outside some correlation. What this means is that the objects don't ground structure, they are nothing independently of the structure, which takes the form of a (gauge invariant) correlation between (non-gauge invariant) field values. With this view one can both evade the standard 'no relations without relata' objection and the problem of accounting for the appearance of time (in a timeless structure) in the same way.

[Rickles, 2012, p 8]

Indeed, as we show in the Foundational Physics for Chapter 4 per our AGC, gauge invariance is rooted in spatiotemporal relationalism.

4.10   A realist, psi-epistemic account

In terms of a QM interpretation, RBW is providing a realist, psi-epistemic account exactly as Leifer suggests: "If we are to maintain psi-epistemic explanations, then we instead need to look for retrocausal ontological models that posit a deeper reality underlying quantum theory that does not include the quantum state" [Leifer, 2014, p. 140]. However, the most fundamental underlying explanation is not so much retrocausal in the sense of information traveling from the future to the past, but adynamical in a global 4D perspective. Specifically, the AGC constrains the probability amplitude for spacetimesource elements, which are spatiotemporal 4D relations. In the Foundational Physics for Chapter 4, we will show how the path integral formalism of QM provides for the 4D distribution of spacetimesource elements in Mermin's experiment and specifically how the AGC underwrites the construct of the probability amplitude for the distribution of spacetimesource elements in the twin-slit experiment. Again, to be clear, a spacetimesource element is not *in* spacetime, it is *of* spacetime, even while a distribution of detector clicks is viewed in the spacetime context of the experimental equipment and process from initiation to termination.

Before continuing, we should point out that using a block universe interpretation of QM is supported by more than the Feynman path integral. Kaiser [1981], Bohr and Ulfbeck [1995], and Anandan [2003] all showed independently that the non-commutivity of the position and momentum operators in QM follows from the non-commutivity of the Lorentz boosts and spatial translations in SR, that is, the relativity of simultaneity. Kaiser writes:

> For had we begun with Newtonian spacetime, we would have the Galilean group instead of [the restricted Poincaré group]. Since Galilean boosts commute with spatial translations (time being absolute), the brackets between the corresponding generators vanish, hence no canonical commutation relations (CCR)! In the [$c \to \infty$ limit of the Poincaré algebra], *the CCR are a remnant of relativistic invariance where, due to the nonabsolute nature of simultaneity, spatial translations do not commute with pure Lorentz transformations.*

> [Kaiser, 1981, p. 706; Italics in original]

And, as we pointed out in chapter 2, the relativity of simultaneity motivates the block universe.

So, while our account takes the block universe seriously and uses future boundary conditions to explain experimental outcomes, it differs from some retrocausal accounts in many respects. First and foremost among the differences, we do not provide an account using dynamical explanation in the mechanical universe whereby the past determines the future yet information somehow travels from the future to the past and fundamental explanations are still in terms of dynamical equations of motion. For example, there are no waves coming from the future in our view. We believe that taken together, quantum theory and relativity are really telling us that dynamical explanation, the notion of fundamental entities being evolved in time by dynamical laws, is not the way to think about fundamental physics. Rather, we think it is worth pursuing the idea that an AGC might be fundamental to both quantum theory and relativity [Silberstein et al., 2013; Stuckey et al., 2016c].

Second, as we elaborated earlier, we also think it is worth pursuing the idea that what we treat as QM entities "emerge" from a fundamentally relational (contextual) basis, that is, 4D relations are fundamental, not entities with intrinsic properties and "primitive thisness." As we discussed, there is much in quantum theory that leads one in that direction, including entanglement and the indistinguishability of quantum particles. According to some theorists, if we move to quantum field theory (QFT) and quantum cosmology, things get worse for dynamism in this regard. For example, the "Unruh effect," named after Bill Unruh, is a well-known but counterintuitive prediction of QFT that with respect to the reference frame of an accelerating observer (in the relativistic sense of the word), "empty space" contains a gas of particles at a temperature proportional to the acceleration [Leinaas, 1991; Fulling, 2005; McCabe, 2007]. While not yet experimentally confirmed, it is claimed that an analogue under centripetal acceleration is observed in the spin polarization of electrons in circular accelerators. It is also claimed that the Unruh effect is necessary for consistency of the respective descriptions of observed phenomena, such as particle decay, in inertial and in accelerated reference frames.

Perhaps this should not be so counterintuitive given that in the Standard Model of particle physics generally, single-particle states for inertial reference frames in Minkowski spacetime are superpositions of eigenstates of the number operator for an accelerated class of reference frame. Therefore, presumably, single-particle states for an inertial reference frame are non-particle states for an accelerated reference frame. QFT characterizes a particle not simply as a property of an underlying quantum field unto itself, but as an inherently relational manifestation between a quantum field and a class of reference frame. Thus, the number of particles present is dependent upon the observer's state of motion and is therefore a relationship between the observer and the quantum field. In short, because the state of a free field is reference frame-dependent, the number of particles is also reference frame-dependent.

Third, because of our hunch about an AGC being fundamental, we based our account on the pathintegral formalism and seek a realist account with a single history. More specifically, our AGC is constructed in a modified version of lattice gauge theory (graphical version of QFT). Contrary to convention, lattice gauge theory is then assumed fundamental to QFT, whence QM. That is how our AGC ultimately underwrites QM. The twin-slit analysis in Foundational Physics for Chapter 4 will make this explicit.

Fourth, we wanted to provide not merely another interpretation of QM but a physical model that would also cover QFT (chapter 5) and provide the grounds for quantum gravity and unification (chapter 6).

Fifth, we sought a realist, psi-epistemic account of the quantum in a 4D setting without the need for realism about configuration space.

Sixth, in order to construct an account that was realist, psi-epistemic, without configuration space but only spacetime, consistent with special relativity (SR), with none of the problems associated with invoking paths, particle or field histories, waves, etc., we sought a characterization of the quantum in terms of unmediated interaction, that is, no "quantum worldlines."

While retrocausal accounts are proliferating [Corry, 2015], the path integral approach is proliferating among foundationalists [Sorkin, 2007a; Dowker, 2014; Wharton, 2014], belief in the fundamentality of relations as we discussed is proliferating [Ladyman, 2007], and psi-epistemic accounts are proliferating (though not realist ones) [Leifer, 2014], we know of no account that embodies all six of the aforementioned features and weaves them into a seamless package. The reasons for these six choices should be clear after reading this book, but suffice it to say they each have an important role to play.

While no account embodies all six of our desiderata we are certainly not alone in thinking in terms of adynamical global constraints, as Price and Wharton make clear:

In putting future and past on an equal footing, this kind of approach is different in spirit from (and quite possibly formally incompatible with) a more familiar style of physics: one in which the past continually generates the future, like a computer running through the steps in an algorithm. However, our usual preference for the computer-like model may simply reflect an anthropocentric bias. It is a good model for creatures like us, who acquire knowledge sequentially, past to future, and hence find it useful to update their predictions in the same way. But there is no guarantee that the principles on which the universe is constructed are of the sort that happens to be useful to creatures in our particular situation. Physics has certainly overcome such biases before—the Earth isn't the center of the universe, our sun is just one of many, there is no preferred frame of reference. Now, perhaps there's one further anthropocentric attitude that needs to go: the idea that the universe is as "in the dark" about the future as we are ourselves.

[Price and Wharton, 2013, p. 15]

The interpretations of QM that sail under the retrocausal banner are quite diverse, obviously. And, it only takes a quick perusal of the retrocausal literature to see that there is no universal agreement on what counts as retrocausal. For example, are any of the following necessary or sufficient?

1. The use of definite future boundary conditions.
2. The posit of a block universe.
3. Employing novel retro-time-evolved mechanisms such as waves from the future.
4. Explicitly rejecting Bell's statistical independence assumption.
5. Acknowledgement of time-symmetric dynamical laws.
6. Some truly robust or non-deflationary account of agent intervention or the future causing the past.

There are retrocausal accounts such as Kastner's Possibilist Transactional Interpretation [Kastner, 2013] that violate points 1 and 2, so these are not necessary. Nor are they sufficient, since neither the least action principle nor the block universe entails retrocausation. There are retrocausal accounts that violate point 3 such as RBW, so it is not necessary. Point 4 is not sufficient because Bell's "superdeterminism" exploits this loophole, but is not retrocausal. A superdeterministic world is one in which independence is violated via a past common cause—a common cause of one's choice of measurements and say the particle spin properties, in the case of Bell correlations. In short, superdeterminism is a conspiratorial theory with only past-to-future causation. Many acknowledge the time-symmetric nature of most dynamical laws and yet do not espouse retrocausation, so point 5 is not sufficient. And there are retrocausal accounts such as Wharton's "Lagrangian-only" approach [Wharton, 2014] that defend only relatively deflationary accounts of agent intervention and causation, so point 6 is not necessary. It seems that the only claim we can pin down as the necessary and sufficient condition for being a retrocausal interpretation of QM is that the account must, in some ontological and not merely formal sense, have the future determining the past or present as much as the past or present determines the future in some situations.

Evans tries to provide the basic package of necessary beliefs that combine to give retrocausality per the school of Price and Wharton:

> This then is the package of metaphysical ideas that combine to give a picture that is consistent with the possibility of retrocausality. We begin with two established metaphysical foundations in the block universe model of time and the interventionist account of causation. We then remove two potential obstacles originating in our ordinary temporal intuitions: we realise that we have no evidence to suggest our macroscopic asymmetric causal intuitions can be extrapolated to the microscopic realm and we realise that we do not necessarily have epistemic access to the past independent of our own future actions. With these obstacles gone, the emerging picture of a temporally and causally symmetric reality viewed from an epistemically limited vantage point concords well with the possibility of retrocausality. A significant aspect of this assembly of ideas is that none of the included elements are precluded by the known physical structure of our reality. Indeed, if anything, these elements are supported by the structure of at least one of our best physical theories: quantum mechanics.
>
> [Evans, 2015, p. 16]

But, again, conditions 2 (block universe) and 6 (robust interventionalism) are apparently not necessary for a retrocausal account in general.

So, looking at retrocausal accounts more generally it seems that there are two basic ways to go, one we call "time-evolved" or "retro-time-evolved" and the other we call global (4D). The former focuses on positing (relatively) new dynamical mechanisms to underwrite retrocausation and the latter takes a more global, adynamical approach. Cramer's retrocausal transactional interpretation was one of the first of this former sort in recent history [Cramer, 1986]. On this account the wave function is taken realistically and time-symmetrically. In the case of a simple EPR setup, we have an "offer-wave (function)" and a "retarded/confirmation-wave (function)" sent out from the point where the initial wave function (corresponding to the EPR state) is emitted (the Source) and the point where it is absorbed (the detectors). A "transaction" is completed once both "offer" and "retarded" waves meet and they bounce back and forth until all the boundary conditions are met. If one takes the talk of waves realistically then this would certainly be an example of an interpretation that adds a retrocausal mechanism to the block universe. But, as Cramer says himself, the backward-causal elements of his theory are "only a pedagogical convention," and that in fact "the process is atemporal" [Cramer, 1986, p. 661] (see also Silberstein et al., 2008). In Kastner's "possibilist" extension of the transactional interpretation [Kastner, 2013], she escapes the redundancy of adding a retrocausal mechanism to the block universe because the possibilist transactional interpretation abjures future boundary conditions. In the possibilist transactional interpretation, the offer waves and confirmation waves do not "live" in spacetime, but in possibility space. On her view the past is populated by empirical observations/actualized transactions, but the future is not actualized. It is filled with offer waves that have not yet arrived. Kastner calls this "space" of unactualized possibilities "pre-spacetime" and it has the properties of Hilbert space. Thus, Kastner dispenses with future boundary conditions and the block universe. In this book, we focus only on the subset of retrocausal accounts that make use of future boundary conditions. Since Kastner's possibilist transactional interpretation belongs to another branch of retrocausal accounts, we will not discuss it here. As will become clear, RBW is an attempt to do physics from a global (4D) point of view in the sense that we underwrite dynamical laws and causal patterns with an adynamical global constraint. That is, rather than trying to add some new mechanism within the block universe (such as waves from the future or possible futures) to account for how information from the future got to the emission event in the past, we step back and note that in a block universe the experimental process from initiation to termination, with everything in between, is all just "there."

These two approaches are largely mutually exclusive at least with respect to fundamental physical models of retrocausation. However, there is some room for compromise. Take, for example, the Price and Wharton school. In Price's Helsinki toy model paper he "shows how something that 'looks like' retrocausality can emerge from global constraints on a very simple system of 'interactions', when the system in question is given a natural interpretation in the light of familiar assumptions about experimental intervention and observation" [Price, 2008, p. 2] And Wharton's "Lagrangian-only" approach sets the Lagrangian density equal to zero as an adynamical global constraint [Wharton, 2014]. He differs from RBW in that his approach is mediated (by classical field configurations), but his goal is that these field configurations will not satisfy a differential equation, that is, they will only satisfy a least action principle. Price, on the other hand, talks about dynamical and causal explanations as "perspectival" [Price, 2005; Price and Weslake, 2008] from within the block universe and he champions a deflationary interventionist or manipulationist account of causation just as Evans notes. So, while the Price and Wharton school of retrocausation does not add new retro-dynamics to the universe such as the Two-State Vector Formalism [Aharonov et al., 1964; Aharonov and Vaidman, 1990], it does heavily emphasize the interventionist agent-focused account of causation, however deflationary it may be. The point is that while Price and Wharton are squarely in the global 4D camp, they think it is important to recognize these causal regularities and counterfactuals.

It is reasonable to ask, given their allegiance to the 4D camp, why Price and Wharton choose to label themselves retrocausal as opposed to adynamical/global constraint? Other people have raised this concern; for example, Corry says to invoke the notion of a "single, indivisible non-local event" is to "deny a causal explanation" [Corry, 2015]. Here is how Kastner provocatively states the problem:

> Thus, Cramer is faced with a dilemma that, in the opinion of this reviewer, he has not adequately dealt with: if all the processes are wholly in spacetime, then he has a block world. In that picture, the dynamical interplay of offers and confirmations is just a story grafted on that does not describe anything real, since all the spacetime events are already there.
>
> [Kastner, 2016]

We assume their thinking is thus: if one has an interventionist account of causation (robust or deflationary), then a "global constraint" model becomes retrocausal because one's choice of a future event is correlated with an earlier event. So, there is nothing more to retrocausation than global constraint plus our ability to intervene. Withholding the label of causation might suggest that there is something more required for causation which, given the package of beliefs outlined by Evans, both Price and Wharton deny. An agent can choose a different measurement apparatus, which (via the global constraint) determines the "initial" likelihood of actual outcomes. Thus, Bell's statistical independence assumption is explicitly violated, because the final measurement geometry is an external "choice of the agent," and it constrains the past.

If we have properly captured the thinking of Price and Wharton, then RBW counts as retrocausal. Regarding the Price and Wharton program, an anonymous referee for Stuckey et al., 2015 wrote: "I do not see how anything truly 'retrocausal', in a dynamical sense, can occur given global time-symmetric constraints on spacetime. The authors seem to me to be too charitable here, a future boundary condition implies an adynamical block world, in which talk of dynamics or intervention is superfluous at best, and inconsistent at worst." Given how deflationary their notion of causation and intervention are, we can appreciate the superfluous charge. We certainly agree with their thinking in these matters and are happy to fly the retrocausal flag. Perhaps the biggest difference between RBW and Price and Wharton is just a matter of emphasis. From the very beginning, we have chosen to emphasize not the agent's perspective, not the causal or dynamical perspective, but instead the global 4D perspective. Thus, we have chosen to focus on constructing fundamental physics based on an adynamical global constraint. Of course, our decision to do so has certainly made it harder to sell RBW to those squarely ensconced in dynamism. We must also note that Price and Wharton no doubt do not share all the preceding six features we listed as essential to RBW.

But most importantly, we share the same basic goal of coming up with a physical model of adynamical explanation from the global 4D perspective that underwrites dynamical laws and causal regularities. And we should also note our extreme debt of gratitude to the notable Huw Price and his book *Time's Arrow and Archimedes' Point* [Price, 1996]. After all, RBW's "God's-eye view" is just Price's "Archimedean view" on steroids. As is obvious throughout, our gratitude also extends to Ken Wharton and his work on the Lagrangian schema.

Foundational Physics for Chapter 4

In the main thread, we introduced delayed choice, Bell's inequality, no counterfactual definiteness, and entanglement using experiments from Zeilinger, Hardy, and Mermin. We showed that the NSU leads to all the mysteries of QM and our Relational Blockworld (RBW) approach to the LSU easily resolves them. We concluded with a synopsis of how RBW addresses the interpretational issues of QM. In this foundational physics thread, we will:

- Introduce the Schrödinger equation.
- Provide amplitudes from the Feynman path integral approach to the Mermin experiment.
- Mathematically explain RBW's adynamical global constraint (AGC).
- Graphically articulate the concepts of space, time, and diachronic objecthood.
- Show how the previous two items follow from the boundary of a boundary principle.
- Explain how gauge invariance and divergence-free sources are natural consequences of the AGC.
- Show how the AGC underwrites Rovelli's claim that gauge "reveals the relational structure of our world."
- Provide a detailed analysis of the twin-slit experiment per RBW.

4.11   The Schrödinger equation

As we said in the main thread, the wave function that is used to produce a probability amplitude which is used to produce a probability (by the Born rule) is found by solving the Schrödinger equation. In its simplest form, we have

$$\hat{H}\psi = i\hbar\frac{\partial}{\partial t}\psi \qquad (4.1)$$

where $\hbar$ is Planck's constant h divided by 2π and $\hat{H}$ is the Hamiltonian differential operator constructed from the sum of kinetic ($\frac{p^2}{2m}$) and potential energies ($V(\vec{r})$). In 1D space, $p \to i\hbar\frac{\partial}{\partial x}$ and $x \to x$ so the Schrödinger equation is

$$\left[\frac{-\hbar^2}{2m}\frac{\partial^2}{\partial x^2} + V(x)\right]\psi = i\hbar\frac{\partial}{\partial t}\psi \qquad (4.2)$$

$\psi(x,t)$ is then found for various $V(\vec{r})$ and particle masses m. The simplest example is the time-independent particle in a box. Let the box extend from $x = 0$ to

$x = L$ and suppose the wave function cannot leak out of the box, that is, $V(x) = 0$ for $0 \le x \le L$ (inside the box) and $V(x) = \infty$ everywhere else (outside the box). Thus, we just need to solve Schrödinger's equation inside the box and let $\psi = 0$ everywhere outside the box. Our differential equation is

$$\frac{\hbar^2}{2m} \frac{d^2\psi}{dx^2} + E\psi = 0 \tag{4.3}$$

where E is the energy of the particle with mass m inside the box. The general solution to this equation is

$$\psi = A \sin\left(\sqrt{\frac{2mE}{\hbar^2}}x\right) + B \cos\left(\sqrt{\frac{2mE}{\hbar^2}}x\right) \tag{4.4}$$

where A and B are constants to be determined from the boundary conditions $\psi(0) = 0$ and $\psi(L) = 0$. $\psi(0) = 0$ implies $B = 0$ and $\psi(L) = 0$ implies $\sin\left(\sqrt{\frac{2mE}{\hbar^2}}L\right) = 0$ ($A = 0$ also works, but that simply says $\psi(x) = 0$ meaning there is no particle in the box). This means $\left(\sqrt{\frac{2mE}{\hbar^2}}L\right) = n\pi$ where $n = 1, 2, 3, \dots$, that is, E is quantized:

$$E_n = \frac{n^2\pi^2\hbar^2}{2mL^2} \tag{4.5}$$

The wave function is $\psi(x) = A \sin\left(\frac{n\pi x}{L}\right)$ where A is determined by requiring

$$\int_0^L |\psi|^2 \, dx = 1 \tag{4.6}$$

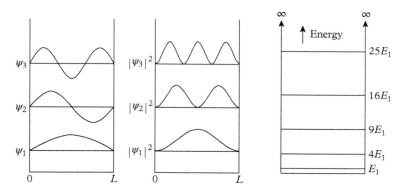

Figure 4.13 $\psi(x)$ and $\psi^2(x)$ for $n = 1$, $n = 2$, and $n = 3$ of the particle in a box. Also shown is the energy of each level relative to the ground state, E_1, for $n = 1, 2, 3, 4, 5$.

That is, the sum of the probabilities is 1 ($\psi(x)$ is said to be "normalized"). Various $\psi(x)$ and $\psi^2(x)$ are plotted in Figure 4.13. Notice this is very different than the result from classical mechanics whereby the particle would simply bounce back and forth between the two walls at constant speed. In that case the probability per unit length of finding the particle would be uniform throughout the box, that is, $\psi^2(x) = $ a constant. Also, in classical Newtonian mechanics, it is possible to have the particle simply sitting at rest in the box, so $E = \frac{1}{2}mv^2 = 0$. But in QM, however, the lowest energy (called the "ground state") is $E_1 = \frac{\pi^2 \hbar^2}{2mL^2}$.

4.12   The Mermin experiment explained via RBW

Again, the corresponding Lagrangian Schema computational approach for determining the distribution of spacetimesource elements in their classical context of the block universe is the pathintegral formalism. Using the method of Sinha and Sorkin [1991] or Wharton and colleagues [2011] for constructing probability amplitudes via the path integral, the spatiotemporally global nature of the process becomes clear, so we now analyze Mermin's experiment [Mermin, 1981] using this formalism.

Recall that each experimentalist in Mermin's experiment has three possible settings (1, 2, or 3) for their detector and there are two possible outcomes (R or G). Let the outcomes R and G correspond to spin up (u) and down (d) along a given Stern–Gerlach (SG) magnet orientation (Figure 4.14), and let the possible SG magnet orientations be given by Figure 4.15.

The quantum state that describes the entangled pair of particles is the correlated state $\psi = \frac{1}{\sqrt{2}}(|\,uu\rangle + |\,dd\rangle)$ (which can be constructed in the undergraduate physics lab [Dehlinger and Mitchell, 2002]), and the probability amplitudes are

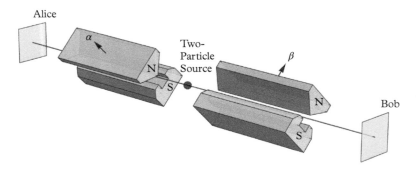

Figure 4.14 *Alice and Bob making spin measurements with their Stern–Gerlach (SG) magnets and detectors.*

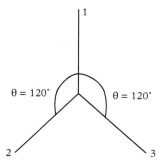

Figure 4.15 *Possible orientations of the SG magnets in Mermin's experiment.*

$$A_{uu} = \frac{1}{2\sqrt{2}}\left(e^{i\alpha} + e^{i\beta}\right) \tag{4.7}$$

$$A_{dd} = \frac{1}{2\sqrt{2}}\left(-e^{i\alpha} - e^{i\beta}\right) \tag{4.8}$$

$$A_{ud} = \frac{1}{2\sqrt{2}}\left(e^{i\alpha} - e^{i\beta}\right) \tag{4.9}$$

$$A_{du} = \frac{1}{2\sqrt{2}}\left(-e^{i\alpha} + e^{i\beta}\right) \tag{4.10}$$

for SG orientation angles α and β. One can see that each probability amplitude is spatiotemporally holistic in that it is the probability amplitude for a Source emission event with its corresponding orientations of the two SG magnets (α and β) and four outcomes (uu, dd, ud, or du). Thus, one may view this as the first step in the computation of the distribution function for the various spacetimesource elements of Mermin's experiment in their classical context of the block universe (Figure 4.16). Again, probability here refers to a distribution of experimental configurations and outcomes in the block universe (Figure 4.17).

In order to compute the overall $\frac{1}{2}$ uu and dd correlated outcomes that violate Bell's inequality (greater than $\frac{5}{9}$ for Mermin's experiment as shown in chapter 1), we square the probability amplitudes (Born rule) to obtain the following probabilities:

$$P_{uu} = |A_{uu}|^2 = |A_{dd}|^2 = P_{dd} = \frac{1}{2}\cos^2\left(\frac{\alpha - \beta}{2}\right) \tag{4.11}$$

and

$$P_{ud} = |A_{ud}|^2 = |A_{du}|^2 = P_{du} = \frac{1}{2}\sin^2\left(\frac{\alpha - \beta}{2}\right) \tag{4.12}$$

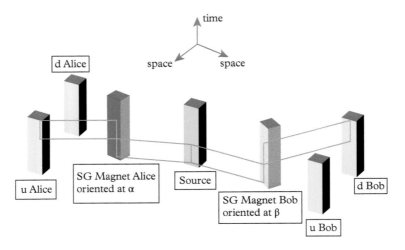

Figure 4.16 *The four empty rectangles connecting the Source emission event to the Stern–Gerlach (SG) magnets and Bob's down (d) outcome and Alice's up (u) outcome depict a spacetimesource element for this particular trial in Mermin's experiment.*

The 4D pattern that needs to be explained is the set of experimental trials (initiation to termination to include outcomes) in which the 0° relative SG orientations (11, 22, 33) always produce like outcomes (50% uu and 50% dd), that is, the probability of like outcomes is 1 in $\frac{1}{3}$ of the trials, and the 120° relative SG orientations (12, 13, 23, 21, 31, 32) only produce like outcomes 25% of the time (12.5% uu and 12.5% dd), that is, the probability of like outcomes is $\frac{1}{4}$ in the other $\frac{2}{3}$ of the trials. That gives us the overall $\frac{1}{2}$ agreement, that is, $(1)\left(\frac{1}{3}\right) + \left(\frac{1}{4}\right)\left(\frac{2}{3}\right) = \frac{1}{2}$, in violation of Bell's inequality "greater than $\frac{5}{9}$" found using Mermin instruction sets corresponding to counterfactual definiteness. Thus, the instruction sets fail to produce the observed correlations (4D pattern) for this experiment, thereby producing a "quantum myster[y] for anybody." The spacetimesource element (Figure 4.16) for any particular trial (which includes detector settings and outcomes) does not contain reference to unrealized measurements and, therefore, has no need for instruction sets or counterfactual definiteness. In such an adynamical explanation, instruction sets are superfluous. The ultimate explanation of the observed distribution of the spacetimesource elements, that is, the 4D pattern, uses a combination of spatiotemporal ontological contextuality and adynamical global constraints (as did Weinstein [Weinstein, 2015], see following). In this case, the fundamental adynamical explanation does not harbor a less fundamental dynamical counterpart and that, again, is the source of the conundrum. We note in passing that since RBW's adynamical explanation does not employ worldlines, superluminal or otherwise, there is no possibility of conflict with SR, that is, no nonlocality in the SR sense.

Time

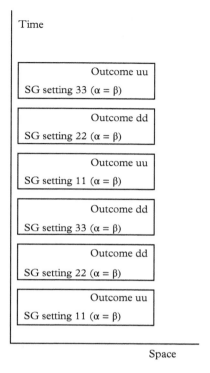

SG setting 33 ($\alpha = \beta$)	Outcome uu
SG setting 22 ($\alpha = \beta$)	Outcome dd
SG setting 11 ($\alpha = \beta$)	Outcome uu
SG setting 33 ($\alpha = \beta$)	Outcome dd
SG setting 22 ($\alpha = \beta$)	Outcome dd
SG setting 11 ($\alpha = \beta$)	Outcome uu

Space

Figure 4.17 *A spacetime depiction of six trials of the Mermin experiment with its two possible outcomes uu and dd for the three possible $\alpha = \beta$ settings. The Feynman path integral has assigned the probability $\frac{1}{2}$ to outcomes uu and dd per Eq. 4.11 when $\alpha = \beta$. Of course, that means the probability for outcomes ud and du must be zero for $\alpha = \beta$ settings, and that is given by Eq. 4.12. Each of the rectangles in this figure corresponds to a particular spacetimesource element in the Mermin experiment as in Figure 4.16 with the ud result shown there replaced with uu or dd and $\alpha = \beta$.*

4.13   The adynamical global constraint of RBW

This section and the next will be formally challenging, but one doesn't have to delve deeply into the mathematical details to appreciate the manner in which our AGC is fundamental to QM and how it contains a wide range of implications beyond the quantum. Simply skimming the mathematical details and reading the surrounding text should suffice to convey these points. We include the mathematical details to provide depth for the reader with a background in physics. With that caveat, let us begin.

Given this new adynamical, graphical, contextual ontology, we propose a commensurate break with narrative explanation, that is, a break with the continuous evolution of the state of objects with intrinsic properties. While it is true that least action principles have been around for a long time, some assume these methods are formal tricks and not fundamental to dynamical equations. As Ballentine wrote:

> Lastly, we raise the question of the physical status of the infinity of Feynman paths (as the possible histories are often called). Does the system really traverse all paths simultaneously? Or does it sample all paths and choose one? Or are these Feynman paths merely a computational device, lacking any physical reality in themselves? In the case of imaginary time path integrals it is clear that they are merely a computational device. This is most likely also true for real time path integrals, although other opinions no doubt exist.
>
> [Ballentine, 1998, p. 123]

To review, the action S is the Lagrangian L integrated over the time interval in question,

$$S = \int L \, dt \tag{4.13}$$

where L is the difference between the system's kinetic and potential energies, so S is a spatiotemporally global quantity. The dynamical laws of the Newtonian Schema follow for each infinitesimal time dt in the Lagrangian Schema when one demands that $\delta S = 0$, that S is extremal.

While our adynamical approach employs mathematical formalism akin to dynamical theories, for example, lattice gauge theory, we redefine what it means to "explain" something in physics. Rather than finding a rule for time-evolved entities per Carroll, the AGC leads to the self-consistency of a graphical spacetime metric and its relationally defined sources. While we do talk about "constructing" or "building" spatiotemporal objects in our view, we are not implying any sort of "evolving blockworld" as in causet dynamics [Sorkin, 2007b]. Our use of this terminology is merely in the context of a computational algorithm. So, one might ask, for example, "why does link X have metric G and stress–energy tensor T?" Recall, for example, that the metric and stress–energy tensor on the spacetime manifold M are consistent with each other, that is, the pair together satisfy Einstein's equations.

Weinstein among others anticipated adynamical global constraint explanation when he wrote:

> What I want to do here is raise the possibility that there is a more fundamental theory possessing nonlocal adynamical constraints that underlies our current theories. Such a theory might account for the mysterious nonlocal effects currently described, but not explained, by quantum mechanics, and might additionally reduce the extent to which cosmological models depend on finely tuned initial data to explain the large scale correlations we observe. The assumption that spatially separated physical systems are entirely uncorrelated is a parochial assumption borne of our experience with the everyday objects described by classical mechanics. Why not suppose that at

certain scales or certain epochs, this independence emerges from what is otherwise a highly structured, nonlocally correlated microphysics?

[Weinstein, 2015, p. 2]

As he explains, every extant fundamental theory of physics assumes the non-existence of such nonlocal constraints:

> Despite radical differences in their conceptions of space, time, and the nature of matter, all of the physical theories we presently use, non-relativistic and relativistic, classical and quantum, share one assumption: the features of the world at distinct points in space are understood to be independent. Particles may exist anywhere, independent of the location or velocity of other particles. Classical fields may take on any value at a given point, constrained only by local constraints like Gauss's law. Quantum field theories incorporate the same independence in their demand that field operators at distinct points in space commute with one another. The independence of physical properties at distinct points is a theoretical assumption, albeit one that is grounded in our everyday experience. We appear to be able to manipulate the contents of a given region of space unrestricted by the contents of other regions. We can arrange the desk in our office without concern for the location of the couch at home in our living room.

[Weinstein, 2015, p. 2]

As stated earlier, graph theory is the optimal formalism for RBW, so we have formulated RBW on a graph by modifying lattice gauge theory (discrete counterpart for QFT) and Regge calculus (discrete counterpart for GR). In chapter 6, we will show how modified lattice gauge theory explains our approach to unification and quantum gravity. We will also explain how a direct implication of our quantum gravity (modified Regge calculus) explains the dark energy phenomenon without a cosmological constant. Here we will use lattice gauge theory to introduce our AGC. We begin with the coupled masses of Figure 4.18 and their Lagrangian

$$L = \frac{1}{2}m\dot{q}_1^2 + \frac{1}{2}m\dot{q}_2^2 - \frac{1}{2}k\left(q_1 - q_2\right)^2 \qquad (4.14)$$

In the Lagrangian formulation of QFT, one computes empirical predictions using the "transition amplitude." Instead of integrating over various spacetime paths as in Feynman's path integral, one integrates over all possible field configurations in

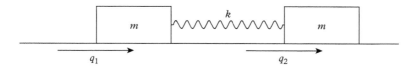

Figure 4.18 *Coupled masses.*

spacetime (or on the graph of lattice gauge theory). The transition amplitude in this case is taken from Zee, 2003, p. 173:

$$Z = \int Dq(t)\exp\left[\frac{i}{\hbar}\int dt\left(\frac{1}{2}m\dot{q}_1^2 + \frac{1}{2}m\dot{q}_2^2 - \frac{1}{2}k\,(q_1 - q_2)^2 + J_1 q_1 + J_2 q_2\right)\right] \quad (4.15)$$

for source J. The discrete version of this transition amplitude on the graph of Figure 4.19 is

$$Z = \int \cdots \int dQ_1 \ldots dQ_6 \exp\left[\frac{i}{\hbar}\left(\frac{1}{2}\vec{Q}\cdot\mathbf{K}\cdot\vec{Q} + \vec{J}\cdot\vec{Q}\right)\right] \quad (4.16)$$

where $(Q_1, Q_2, Q_3) = (q_1(t_1), q_1(t_2), q_1(t_3))$, $(Q_4, Q_5, Q_6) = (q_2(t_1), q_2(t_2), q_2(t_3))$, $\vec{Q} = (Q_1, Q_2, Q_3, Q_4, Q_5, Q_6)$, $\vec{J} = (J_1(t_1), J_1(t_2), J_1(t_3), J_2(t_1), J_2(t_2), J_2(t_3))$, and

$$\mathbf{K} = \begin{pmatrix}
\left(\frac{m}{\Delta t} - k\Delta t\right) & -\frac{m}{\Delta t} & 0 & k\Delta t & 0 & 0 \\
-\frac{m}{\Delta t} & \left(\frac{2m}{\Delta t} - k\Delta t\right) & -\frac{m}{\Delta t} & 0 & k\Delta t & 0 \\
0 & -\frac{m}{\Delta t} & \left(\frac{m}{\Delta t} - k\Delta t\right) & 0 & 0 & k\Delta t \\
k\Delta t & 0 & 0 & \left(\frac{m}{\Delta t} - k\Delta t\right) & -\frac{m}{\Delta t} & 0 \\
0 & k\Delta t & 0 & -\frac{m}{\Delta t} & \left(\frac{2m}{\Delta t} - k\Delta t\right) & -\frac{m}{\Delta t} \\
0 & 0 & k\Delta t & 0 & -\frac{m}{\Delta t} & \left(\frac{m}{\Delta t} - k\Delta t\right)
\end{pmatrix} \quad (4.17)$$

In RBW, we call $\mathbf{K}$ the "relations matrix" because the rows of $\mathbf{K}$ sum to zero. For example, if we were describing the line of people [Alice, Bob, Charlie, David], we would say, "Alice is in front of Bob" and "Bob is behind Alice and in front of Charlie." Assigning a numerical value of $+1$ to "in front of" and -1 to "behind," we see that adding the statements describing the locations of all four people gives a result of zero. This "summing to zero" happens because we have a self-referential, relational description of all four people. Mathematically we write $\mathbf{K}\cdot[111111]^T = 0$ and say that $[111111]^T$ is a non-null (not zero) eigenvector of $\mathbf{K}$ with eigenvalue zero. Thus, $\mathbf{K}$ is said to possess a non-trivial null space in contrast to the fact that all matrices multiplied by the zero (null) vector $[0,0,\ldots,0]$ will give a result of zero (this is the trivial null space). The complement to that non-trivial null space (the vectors $[\ldots]$ that satisfy $\mathbf{K}\cdot[\ldots]^T \neq 0$) is called the "row space" of $\mathbf{K}$. That $\mathbf{K}$ possesses a non-trivial null space is the graphical counterpart to gauge invariance and restricting the transition amplitude Z integral to the row space of $\mathbf{K}$ is the graphical counterpart of Fadeev–Popov gauge fixing [Zee, 2003, p. 168–70]. Since $\vec{J}$ also appears in the integral for Z, we need it to reside entirely in the row space of $\mathbf{K}$ which means $\vec{J}\cdot[111111]^T = 0$, so that the components of $\vec{J}$ also sum to zero, that is, $\vec{J}$ is said to be "divergence-free." This is a necessary, but not sufficient, criterion for a conserved exchange of energy–momentum for a given spacetimesource element. Therefore, we also require that each spacetime-source element represents a conserved exchange of energy–momentum. This is our proposed AGC for the construct of $\mathbf{K}$ and $\vec{J}$ associated with a fundamental spacetimesource element.

This AGC is motivated by fundamental notions of space, time, and the time-evolved *things* of dynamical narration, that is, "diachronic entities." Indeed, there is no principle which dictates the construct of diachronic entities fundamental to the formalism of dynamics in general; these objects are "put in by hand" throughout physics. When Albrecht and Iglesias [2012] allowed time to be an "internal variable" after quantization, as in the Wheeler–DeWitt equation [Jaroszkiewicz, 2016, pp. 93–4], they found "there is no one set of laws, but a whole library of different cosmic law books" [Siegfried, 2008, p. 28]. They called this the "clock ambiguity." In order to circumvent this "arbitrariness in the predictions of the theory" they proposed that "the principle behind the regularities that govern the interaction of entities is . . . the idea that individual entities exist at all" [Siegfried, 2008, p. 28]. Albrecht and Iglesias characterize this as "the central role of quasi-separability" [Albrecht and Iglesias, 2012, p 67].

Since RBW is a form of spatiotemporal ontological contextuality, we must decompose the worldlines of "individual entities" relationally, thereby explaining "the central role of quasiseparability." To see how this is done, we point out that the AGC of RBW is based on the "boundary of a boundary principle," that is, $\partial\partial = 0$, which underwrites GR and electromagnetism [Misner et al., 1973, p. 364; Wise, 2006] (Figure 4.20). Indeed, $\mathbf{K}$ can be constructed from boundary operators on the graph, that is, $\mathbf{K} = \partial_1\partial_1^T$ where $\partial_1\partial_2 = 0$ is the graphical counterpart to $\partial\partial = 0$ [Stuckey et al., 2016c]. A close inspection at what $\partial_1\partial_2 = 0$ is saying about the graphical entities shows that it is merely repeating what we have already stated: "Alice is in front of Bob" means "Bob is behind Alice." In this context, it is saying that if an oriented link goes from node 1 to node 2 as part of an oriented plaguette, then "out of node 1" cancels "into node 2" in a global sum. So, we propose that this topological tautology underwrites the trans-temporal identity employed tacitly in all dynamical theories. Our discrete (graphical) starting point provides a topological basis for sources $\vec{J}$, space

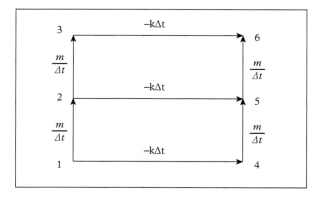

Figure 4.19 *Six-node graph for coupled masses.*

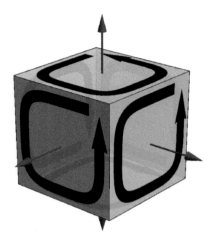

Figure 4.20 *The boundary of a boundary is zero. The oriented plaquettes bound the cube and the directed edges bound the plaquettes. As one can see from the figure, every edge has oppositely oriented directions that cancel out. Thus, the boundaries of the plaquettes (the edges), which bound the cube, sum to zero.*

and time. Clearly, the process by which the worldlines for two sources in space-time is modeled using $\vec{J}$ on the graph of Figure 4.19 is an organization of the set $\vec{J} = (J_1(t_1), J_1(t_2), J_1(t_3), J_2(t_1), J_2(t_2), J_2(t_3))$ on two levels: there is the split of the set into two subsets $(J_1(t_1), J_1(t_2), J_1(t_3))$ and $(J_2(t_1), J_2(t_2), J_2(t_3))$, one for each source, and there is the ordering t_i over each subset. The split represents space, the ordering represents time, and the result is relational, diachronic objecthood. In this sense, space, time, and sources are relationally and inextricably co-constructed in our formalism. Consequently, we believe the articulation of the otherwise tacit construct of dynamical entities, that is, "the idea that individual entities exist at all," has a mathematical counterpart fundamental to the action, viz., the boundary of a boundary principle, $\partial\partial = 0$, that encodes "the central role of quasiseparability" at the fundamental level. This is in accord with Toffoli's belief that there exists a mathematical tautology fundamental to the action:

> Rather, the motivation is that principles of great generality must be by their very nature trivial, that is, expressions of broad tautological identities. If the principle of least action, which is so general, still looks somewhat mysterious, that means we still do not understand what it is really an expression of–what it is trying to tell us.
>
> [Toffoli, 2003, p 380]

So, in summary, that **K** possesses a non-trivial null space is the graphical equivalent of gauge invariance and restricting $\vec{J}$ to the row space of **K** provides a natural gauge fixing, that is, restricting the integral of the transition amplitude to the row space of **K**. That **K** possesses a non-trivial null space also means the determinant of **K** is zero, so the set of vectors constituting the rows of **K** is not linearly independent. That some subset of these row vectors is determined by its complement follows from having the graphical set relationally constructed. Thus, divergence-free $\vec{J}$ follows from relationally defined **K** as a direct result of our AGC, which follows from the geometric tautology $\partial\partial = 0$ encoding the construct of diachronic entities necessary for classical physics. Consequently, we agree with Rovelli that "gauge is ubiquitous. It is not unphysical redundancy of our mathematics. It reveals the relational structure of our world" [Rovelli, 2013, p. 7]. As it turns out, the manner in which the AGC is applied to the construct of the relations matrix **K** for the Schrödinger action [Stuckey et al., 2015] is the same manner in which it can be applied to the construct of **K** for the quadratic form in the Klein–Gordon, Dirac, Maxwell, and Einstein–Hilbert actions, with extension to the Standard Model of particle physics [Stuckey et al., 2016c]. We will say more about that and its implications for unification in chapter 6. Now, we will bring the AGC to bear on the twin-slit experiment.

4.14 RBW analysis of the twin-slit experiment

Finally, in order to illustrate the explanatory function of the AGC, we now apply it to the quintessential foundations example, the twin-slit experiment. As will quickly become apparent, our modified lattice gauge theory approach is computational overkill in this context, but it provides an excellent illustration of how the AGC for spacetimesource elements ultimately underwrites QM. The computation is in three parts and the goal is to produce a non-relativistic, source-to-source QFT probability amplitude ψ for the spacetimesource element in the twin-slit experiment per modified lattice gauge theory. First, we use the transition amplitude for the Klein–Gordon action in the non-relativistic limit to produce a propagator $D(x-x')$ between point sources from the generating function $W(J)$. Next, we relate $D(x-x')$ to the probability amplitude ψ of the Schrödinger equation, even though the Schrödinger equation is homogeneous (has no source terms). Lastly, we discretize the transition amplitude of the non-relativistic Klein–Gordon action with source terms and use the AGC to find our modified lattice gauge theory counterpart to $W(J)$, and thus ψ, for the spacetimesource element. A modification to the discretization process is required by the AGC since there is an undifferenced (non-relational) term ψ^* in the non-relativistic Klein–Gordon action.[13] The AGC also tells us which eigenmode of our difference matrix is relevant. Essentially, the second and third parts justify and explain our use of the propagator $D(x-x')$ between point sources in non-relativistic QFT in computing the QM probability amplitude ψ for the spacetimesource element of the twin-slit experiment.

---

[13] This non-relational term ψ^* doesn't appear in the fully relativistic Klein–Gordon action. It only appears in the conversion to its non-relativistic form.

The spacetimesource element for an exchange of mass m in the context of a pair of slits is shown in Figure 4.7. The goal of this computation is to obtain the amplitude for that spacetimesource element, component by component, and plot the resulting intensity as a function of angular displacement on the detector for some mass, slit spacing, Source-to-slit distance, and slit-to-detector distance. As will become evident in the analysis following, the construct of the Schrödinger action is related to the previous example of coupled mechanical oscillators by virtue of the shared quadratic form of their actions, that is, we are not modeling quantum exchanges literally as coupled mechanical oscillators. Thus, the manner in which the AGC is applied to the construct of **K** for the Schrödinger action here is the same manner in which it can be applied to the construct of **K** for the quadratic form in the Klein–Gordon, Dirac, Maxwell, and Einstein–Hilbert actions, with extension to the Standard Model of particle physics. Let us begin.

The non-relativistic limit of the Klein–Gordon equation gives the free-particle Schrödinger equation by factoring out the rest mass contribution to the energy E, assuming the Newtonian form for kinetic energy, and discarding the second-order time derivative [Zee, 2003, p. 172]. To illustrate the first two steps, plug $\phi = Ae^{i(px-Et)/\hbar}$ into the Klein–Gordon equation and obtain $(-E^2 + p^2c^2 + m^2c^4) = 0$, which tells us that E is the total relativistic energy. Now plug $\psi = Ae^{i(px-Et)/\hbar}$ into the free-particle Schrödinger equation and obtain $\frac{p^2}{2m} = E$, which tells us that E is only the Newtonian kinetic energy. Thus, we must factor out the rest energy of the particle, that is, $\psi = e^{imc^2 t/\hbar}\phi$, assume the low-velocity limit of the relativistic kinetic energy, and discard the relevant term from our Lagrangian density (leading to the second-order time derivative) in going from ϕ of the Klein–Gordon equation to ψ of the free-particle Schrödinger equation. We will simply outline the details here. Again, these can be skimmed if one has no interest in the mathematical formalism.

Overall, we start with a transition amplitude from quantum field theory to get our generating function whence the propagator that gives us the QM probability amplitude ψ. The transition amplitude for the Klein–Gordon equation is

$$Z(J) = \int D\phi \exp\left[\frac{i}{\hbar}\int d^4x\left(\frac{1}{2}(\partial\phi)^2 - \frac{1}{2}\overline{m}^2\phi^2 + J\phi\right)\right] \quad (4.18)$$

which in $(1+1)$D is

$$Z(J) = \int D\phi \exp\left[\frac{i}{\hbar}\int dxdt\left(\frac{1}{2}\left(\frac{\partial\phi}{\partial t}\right)^2 - \frac{c^2}{2}\left(\frac{\partial\phi}{\partial x}\right)^2 - \frac{1}{2}\overline{m}^2\phi^2 + J\phi\right)\right] \quad (4.19)$$

where $\overline{m} = \frac{mc^2}{\hbar}$. Making the changes described above with $\psi = e^{i\overline{m}t}\sqrt{\overline{m}}\phi$ gives the non-relativistic Klein–Gordon transition amplitude corresponding to the free-particle Schrödinger equation [see Zee, 2003, p. 173]:

$$Z(J) = \int D\psi \exp\left[\frac{i}{\hbar}\int dxdt\left(i\psi^*\left(\frac{\partial\psi}{\partial t}\right) - \frac{c^2}{2\overline{m}}\left(\frac{\partial\psi}{\partial x}\right)^2 + J\psi\right)\right] \quad (4.20)$$

Now integrate the second term by parts and obtain

$$Z(J) = \int D\psi \exp\left[\frac{i}{\hbar} \int dxdt \left(i\psi^*\left(\frac{\partial \psi}{\partial t}\right) + \frac{\hbar}{2m}\psi^*\frac{\partial^2 \psi}{\partial x^2} + J\psi\right)\right] \qquad (4.21)$$

This gives

$$Z(J) = \int D\psi \exp\left[\frac{i}{\hbar} \int dxdt \left(\frac{1}{2}\psi^*K\psi + J\psi\right)\right] \qquad (4.22)$$

where

$$K = \left(2i\frac{\partial}{\partial t} + \frac{\hbar}{m}\frac{\partial^2}{\partial x^2}\right) \qquad (4.23)$$

Discretizing this on a four-node graph, we find that **K** for any of the rectangular sections of the spacetimesource element depicted in Figure 4.7 is given by

$$\mathbf{K} = \begin{pmatrix} \left(\frac{2i}{\Delta t} - \frac{\hbar}{m\Delta x^2}\right) & -\frac{2i}{\Delta t} & \frac{\hbar}{m\Delta x^2} & 0 \\ -\frac{2i}{\Delta t} & \left(\frac{2i}{\Delta t} - \frac{\hbar}{m\Delta x^2}\right) & 0 & \frac{\hbar}{m\Delta x^2} \\ \frac{\hbar}{m\Delta x^2} & 0 & \left(\frac{2i}{\Delta t} - \frac{\hbar}{m\Delta x^2}\right) & -\frac{2i}{\Delta t} \\ 0 & \frac{\hbar}{m\Delta x^2} & -\frac{2i}{\Delta t} & \left(\frac{2i}{\Delta t} - \frac{\hbar}{m\Delta x^2}\right) \end{pmatrix} \qquad (4.24)$$

The eigenvalues are $\left(0, \frac{4i}{\Delta t}, -\frac{2\hbar}{m\Delta x^2}, \frac{4i}{\Delta t} - \frac{2\hbar}{m\Delta x^2}\right)$ with eigenvectors $(1,1,1,1)$, $(-1,1,-1,1)$, $(-1,-1,1,1)$, and $(1,-1,-1,1)$, respectively. This is a Hadamard structure that we see repeated in **K** for the Klein–Gordon and Dirac equations [Stuckey et al., 2016c]. These eigenvectors correspond to the four modes shown in Figures 4.21, 4.22, 4.23, and 4.24 and, as we explain in the captions, only the mode of Figure 4.24 satisfies the AGC.

In Stuckey et al. [2016c], we extend this result to find **K** for the Klein–Gordon, Dirac, Maxwell, and Einstein–Hilbert actions, with extension to the U(1) and SU(N) interactions for the Standard Model of particle physics. Strictly speaking,

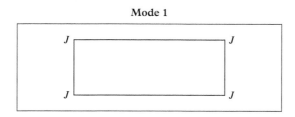

Mode 1

Figure 4.21 *There is no spatial or temporal variation in $\vec{J}$, so $\vec{J}$ is not divergence-free and therefore does not reside in the row space of **K**. This source does not satisfy the AGC.*

Mode 2

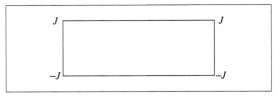

Figure 4.22 *There is only temporal variation in $\vec{J}$. While $\vec{J}$ resides in the row space of K and is therefore divergence-free in the mathematical sense, it is not conserved within the element. Therefore, this source does not satisfy the AGC.*

Mode 3

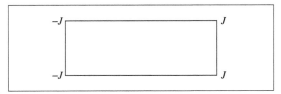

Figure 4.23 *There is only spatial variation in $\vec{J}$. While $\vec{J}$ resides in the row space of K and is conserved within the element, it does not represent an interaction. Therefore, this source does not satisfy the AGC.*

Mode 4

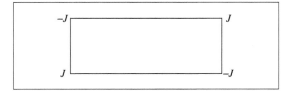

Figure 4.24 *There is both spatial and temporal variation in $\vec{J}$, which resides in the row space of K, is conserved in the element, and represents an interaction. This source satisfies the AGC.*

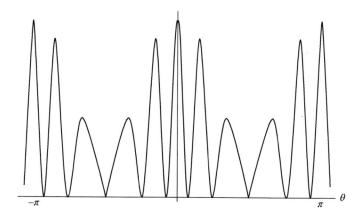

Figure 4.25 *Intensity versus angular displacement in radians for electrons with* λ = *72.8 μm, slit separation of 1.00 mm, screen-to-detector distance of 50.0 cm, and Source-to-slits distance of 50.0 cm.*

when finding the gradient of a vector field on the graph as with the Dirac operator, we need to specify a means of parallel transport. So, in our view and that of lattice gauge theory, local gauge invariance is seen as a modification to the matter field gradient on the graph required by parallel transport (SU(N) for N vectors being differenced between nodes). The AGC applies to the fundamental spacetimesource element in all these cases.

The amplitude for each component of the spacetimesource element in Figure 4.7 is

$$A(x, t, m) = -\frac{1}{4}\sqrt{\frac{m}{\pi \hbar t}}\left[iC\left(\sqrt{\frac{mx^2}{\pi \hbar t}}\right) + S\left(\sqrt{\frac{mx^2}{\pi \hbar t}}\right)\right]\exp\left(\frac{imx^2}{2\hbar t}\right) \quad (4.25)$$

where $C(z) = \int_0^z \cos\left(\frac{\pi}{2}u^2\right)du$ and $S(z) = \int_0^z \sin\left(\frac{\pi}{2}u^2\right)du$ are Fresnel integrals. Now to construct the amplitude A_{total} for the entire spacetimesource element in Figure 4.7 for an outcome in the twin-slit experiment we have

$$A_{total} = A(x_1, t_1, m)A(x_3, t_3, m) + A(x_2, t_2, m)A(x_4, t_4, m) \quad (4.26)$$

where x_1 and t_1 are the distance and time ("classical context") from Source to Slit 2, x_2 and t_2 are the distance and time from Source to Slit 1, x_3 and t_3 are the distance and time from Slit 2 to the Detection Event, and x_4 and t_4 are the distance and time from Slit 1 to the Detection Event. For an electron traveling at 10.0 m/s through the device (dynamical language) we obtain the plot in Figure 4.25.

5

Relational Blockworld and Quantum Field Theory

*Quantum logic is not just bizarre, it is also beautiful in its mathematical per-
fection. If you come to think of it, we could not have wished for a better tool for
doing calculations. Combining it with other physical principles such as special relativity opens up a grandiose mathematical scheme called quantum field
theory, allowing us to describe the most elementary constituents of matter in
a synthesis of simplicity and complexity. Numerous experiments have shown
that this really does describe the world we live in. So, quantum mechanics is
here to stay, but we still have to learn how it emerged from simple logical principles. The quantum world should not be looked upon as alien, but one where
straight and natural logic rules.*

Nobel Laureate Gerard 't Hooft ['t Hooft, 2016, p. x]

5.1   Introduction

Having dealt with the conundrums of quantum mechanics (QM), we now bring
our version of God's-eye physics to bear on particle physics as per quantum field
theory (QFT). As we explained in chapter 4, data for *fundamental* physics are
individual detector clicks, not the worldlines for particles (formed from a series of
detector clicks) in a high-energy physics detector. As we will see in this chapter,
particle physics as per QFT is in the business of classifying spatiotemoral relationships between sets of detector clicks, that is, the worldlines of particles. Thus,
even though the Lagrangian Schema is widely used in particle physics, it is understood and approached dynamically. We will argue here that, as with the puzzles,
problems, paradoxes, and conundrums of GR and QM, this Newtonian Schema
Universe (NSU) thinking is responsible for the stalemate particle physics finds itself in today. We will then introduce a new Langrangian Schema Universe (LSU)
path forward via Relational Blockworld (RBW) that we will explain in chapter 6.

The issues of QFT are formally non-trivial. Entire books have been dedicated to them, so our accounts here are necessarily abbreviated and focused; for
example, we are only addressing the application of QFT to particle physics. In
Philosophy of Physics for Chapter 5, we will deal with many of the interpretational issues of QFT such as gauge invariance, gauge fixing, the Aharonv–Bohm
effect, and others. In Foundational Physics for Chapter 5, we show that classical

Beyond the Dynamical Universe. Michael Silberstein, W.M. Stuckey and Timothy McDevitt,
Oxford University Press (2018). © Michael Silberstein, W.M. Stuckey and Timothy McDevitt.
DOI 10.1093/oso/9780198807087.001.0001

field theory is satisfied by the extremum field configurations of QFT and introduce gauge fields. This is analogous to spacetime paths of classical mechanics being the extremum paths in the path integral formulation of QM.

While the main threads up to this point were mostly devoid of formalism, the main thread for chapter 5 revisits the introductory physics of special relativity and simple harmonic motion, since "even after some 75 years, the whole subject of quantum field theory remains rooted in this harmonic paradigm" [Zee, 2003, p. 5]. We also display and discuss conceptually the Central Identity of Quantum Field Theory. Before introducing QFT as a concept, we present an abbreviated history of unification leading to its development. Finally, we will provide our account of what particles or "excitations of a field" are in RBW, and how that might at least provide a coherent ontology for QFT, and an account of what QFT is actually doing.

5.2   A brief history of unification

Although we could begin earlier, we will abbreviate the story by beginning with the unification of electricity and magnetism in the 1800s. It was known that a magnetic field was created around a conducting wire when an electric field created a current in that wire. And, changing the magnetic field flow through a loop of wire (a changing magnetic field flux) created an electric field in the wire. These facts were captured by Ampere's Law and Faraday's Law, respectively. Maxwell then asked "what happens to Ampere's Law if I put a capacitor in the circuit?" The problem is that the capacitor introduces a small gap in the circuit where no charge could flow, that is, a small space where there is no current. Yet, a magnetic field was still produced around the wire as the capacitor charged and discharged. He reasoned that the electric field flux in the empty space of the gap, that is, in the capacitor, must produce a current of sorts. He called this the "displacement current" and added it to Ampere's Law (so it is now named the Ampere–Maxwell Law).

This new form of Ampere's Law in which a changing electric field flux allows for a magnetic field-producing displacement current in otherwise empty space, plus Faraday's Law, combined to predict waves of co-creating electric and magnetic fields moving through space at the speed of light. Maxwell's equations[1] had unified the electric and magnetic fields thereby predicting electromagnetic radiation. From low to high frequency, electromagnetic radiation is called radio waves, microwaves, infrared radiation, light, ultraviolet radiation, X-rays, and γ-rays. Although electromagnetism doesn't have to be thought of mechanically, Maxwell created it using the familiar mechanical paradigm[2] and it is widely understood

---

[1] Besides Faraday's Law and the Ampere–Maxwell Law, Maxwell's four equations include Gauss's Law for the electric and magnetic fields.

[2] Again, Maxwell believed "all physics research should advance on strict mechanical principles" [Goodstein, 1985, Episode 39].

mechanically today. That is, the exchange of energy–momentum via electromagnetic radiation is understood to be mediated by an oscillating electromagnetic field between sources.[3]

While Maxwell's equations unified electric and magnetic phenomena, they created a rift with the Galilean-invariant Newtonian mechanics because Maxwell's equations are Lorentz-invariant.[4] In other words, as we explained in chapter 2, according to Lorentz transformations the spatial distance and time between events can both vary between moving frames for the same events. But, according to Galilean transformations, everyone measures the same spatial distance and time between the same events. These are mutually exclusive facts, so at least one of Maxwell's equations or Newtonian mechanics had to be "incomplete" and special relativity (SR) showed that it was Newtonian mechanics that was only an approximation to relativistic mechanics. For example, the SR kinetic energy for an object of mass m moving at speed v is $(\gamma - 1)mc^2$, that is, the total energy γmc^2 minus the rest energy mc^2. Expanding $\gamma = \sqrt{\frac{1}{1-\frac{v^2}{c^2}}}$ in $(\gamma - 1)mc^2$ to first order in $\frac{v^2}{c^2}$, that is, $\gamma \approx 1 + \frac{1}{2}\frac{v^2}{c^2}$, gives the familiar Newtonian kinetic energy $\frac{1}{2}mv^2$. So, Newtonian mechanics is understood to be a small-speed approximation to SR mechanics. Thus, Einstein had unified mechanics and electromagnetism.

About the same time Einstein was producing SR, Planck was producing the first quantum mechanics (QM) to explain blackbody radiation, that is, that all objects emit electromagnetic radiation characteristic of their temperature. In order to derive his blackbody radiation equation, Planck modeled matter as oscillators that could only vibrate at discrete (quantum) frequencies with energy $E = hf$ where h is Planck's constant and f is the frequency of the oscillator. Einstein then reasoned that the blackbody radiation must therefore be emitted and absorbed in discrete bundles called "photons" with energy $E = hf$. This explained the photoelectric effect[5] and won Einstein the Nobel Prize in Physics. Apparently, QM trumps electromagnetism (a theory of classical physics), but notice that QM explanation is still very mechanical.

---

[3] Recall, "sources" are associated with emission and absorption events for all forms of quantum energy–momentum exchange.

[4] For example, the Lorentz force $q\vec{v} \times \vec{B}$ can be obtained from the Coulomb force $q\vec{E}$ by a Lorentz transformation. This makes sense, since in the frame where the charge Q producing $\vec{E}$ and the charge q are both at rest, we should see only the Coulomb force. But, in a frame moving relative to that rest frame, the moving charge Q constitutes a current which should therefore produce a magnetic field. Then, the charge q is moving in Q's magnetic field, so we should see a Lorentz force.

[5] Electrons emitted from the surface of a metal when illuminated by electromagnetic radiation had a maximum kinetic energy depending on the frequency of the impinging electromagnetic radiation independent of the intensity of the radiation. There was no reason in classical thinking to expect that since increasing the intensity of the radiation at fixed frequency increased the energy per unit time responsible for ejecting electrons. So, why wouldn't we see more energetic electrons? However, if an increased intensity at a given frequency meant more photons while the energy of each photon was not increased, then we would expect more ejected electrons at higher intensity, but their max kinetic energy would not change since each electron was ejected by one photon.

The result of the photoelectric effect is that electromagnetic radiation is wave-like according to classical physics (generates interference patterns) while it is particle-like in quantum physics (photoelectric effect). Is there some unifying theme? An early, affirmative answer to this question was given by de Broglie. Since Planck's constant h has units of angular momentum, Bohr had explained the spectrum of the hydrogen atom by assuming the electron could only orbit the nucleus with discrete (quantized) angular momentum, $L = \frac{nh}{2\pi} \equiv n\hbar$ where $n = 1, 2, 3, \ldots$. de Broglie reasoned that since electromagnetic radiation acts like waves in classical physics and like particles in quantum physics, perhaps matter does the converse. That is, matter is typically understood as particle-like (e.g., electrons) in classical physics, so perhaps it has wave-like characteristics in quantum physics. By combining SR's equation $E = mc^2$ for particles with QM's equation $E = hf = \frac{hc}{\lambda}$ for waves and setting $c = v$, de Broglie obtained a wavelength for matter bearing his name $\lambda = \frac{h}{mv}$. Next, he replaced the electron as a particle in orbit around the proton of a hydrogen atom with the electron's de Broglie wavelength and assumed the orbit must be an integer multiple of the wavelength, so as to interfere with itself constructively. We have $L = rmv$ and $mv = \frac{h}{\lambda}$, so assuming the orbital circumference $2\pi r = n\lambda$ gives $L = n\hbar$ which explains Bohr's otherwise ad hoc assumption via de Broglie's nascent unification of SR and QM. Notice that while the electron as a particle was replaced by a standing wave, we still have a very mechanical view of the hydrogen atom.

Moving to atoms heavier than hydrogen with multiple protons in the nucleus, we find a problem for the dynamical worldview, that is, why do protons stick together in the nucleus? Why doesn't electrostatic repulsion (Coulomb force) drive them apart? Clearly there must be another force acting in the nucleus that is stronger than the Coulomb force that holds protons (and neutrons) together. The Coulomb force is from classical physics where the electromagnetic field is continuous, so maybe quantum physics trumps classical physics again via a force based on a discrete entity, that is, a particle. This force must be short range or it would pull all atoms in matter together into one giant nucleus. Yukawa postulated that since the electromagnetic force is mediated by massless photons over infinite range (quantum view) perhaps a particle mediating a short-range force would have non-zero mass. Now the spontaneous emission of a force particle by a nucleon (proton or neutron) would temporally violate the conservation of energy for the nucleus, but Heisenberg's Uncertainty Principle, $\Delta E \Delta t \geq \frac{\hbar}{2}$, says this would be allowed as long as the rest energy E_p of the force particle is less than the uncertainty in the energy of the nucleus ΔE. This sets an upper limit on E_p of ΔE. Plugging this upper limit into Heisenberg's Uncertainty Principle we obtain an equality $E_p = \frac{\hbar}{2\Delta t}$ based on the size of the uncertainty Δt in the time for the particle exchange to occur. Assuming the particle is exchanged over the typical diameter of a nucleus $d = 10^{-15}$ m (a femtometer) at the fastest possible speed c gives a lower bound for Δt giving an upper bound for E_p. Using SR's equation $E = mc^2$ Yukawa solved for the mass of his force particle $E_p = mc^2 = \frac{hc}{2d} = 100$ MeV (the masses of particles are typically given as their rest energies, mc^2). Again, an early unification of SR and QM "saved the appearances" for the mechanical paradigm and in doing so began the enterprise of particle physics.

While Yukawa's particle was subsequently discovered (and named the "pion"), it is not the particle mediating the "strong nuclear force" that binds nucleons per the Standard Model of particle physics. That particle is the "gluon" of a theory called "quantum chromodynamics," which is couched in quantum field theory (QFT), the ultimate unification of SR and QM, as we will describe conceptually in the following. Gell-Mann's quantum chromodynamics was introduced to simplify the classification of hundreds of particles discovered in particle physics experiments through the 1950s. He assumed that nucleons and myriad other particles called "hadrons" were composed of "quarks." Today, particles of matter (or non-force particles) are called "fermions" and come in two types, "quarks" and "leptons." The distinction is that quarks interact via the strong force while leptons do not. Currently, both quarks and leptons (an electron is a lepton, for example) are believed to be point particles; that is they are not further decomposable, they are fundamental particles. Particles of force are called "bosons." Thus, according to particle physics, all phenomena can be explained by the exchange of bosons by fermions and we see that mechanistic thinking still pervades modern physics.

The set of fundamental particles in the Standard Model (Figure 5.1) unifies a large number of phenomena, but particle physicists consider it to be too complicated to be a fundamental theory. There are six quarks, six leptons, four force particles, a mass-endowing particle (Higgs boson), and several conservation laws[6]

Standard Model of Particle Physics	
Fermions	
Quarks	Leptons
Up	Electron
Down	Electron Neutrino
Charm	Muon
Strange	Muon Neutrino
Top	Tau
Bottom	Tau Neutrino
Gauge Bosons	
Photon	Electromagnetic Force
Gluon	Strong Nuclear Force
Weak Bosons	Weak Nuclear Force
Graviton	Gravitational Force
Scalar Boson	
Higgs	Endows Mass

Figure 5.1 *The Standard Model of particle physics. Strictly speaking, gravity is not included in the Standard Model, since it has not been quantized as yet. This table depicts the expectation that it will be quantized eventually.*

[6] Besides conservation of energy, momentum, and charge, we have conservation of baryon number, electron lepton number, muon lepton number, tau lepton number, and strangeness, for example. Baryons are hadrons made of three quarks while mesons are hadrons made of two quarks.

needed to account for particle physics phenomena. According to QFT, the Standard Model contains about 20 adjustable parameters that cannot be accounted for otherwise. Further, Glashow, Salam, and Weinberg won the 1979 Nobel Prize in Physics for unifying the quantum electromagnetic force (quantum electrodynamics) and weak nuclear force[7] in what is called the "electroweak force." The gauge group (see Foundational Physics for Chapter 5) responsible for this force is U(1) x SU(2) and the gauge group for quantum chromodynamics is SU(3), so the Standard Model is often written as U(1) x SU(2) x SU(3). However, no one has been able to unify the electroweak force and strong nuclear force further (called Grand Unification), despite efforts spanning 40 years. Attempts to do so have produced predictions such as the decay of the proton and the existence of a magnetic monopole, neither of which has been observed.

Finally, there is the matter of the weakest force, gravity. Some attempts to unify the forces have skipped Grand Unification and proceeded straight to Super Unification, that is, unification of all four forces, such as in string theory. While string theorists have claimed great progress in the 40 or more years of their efforts, the theory fails on a number of fronts, as we described in chapter 1. Even if one doesn't believe gravity should be unified with the other fundamental forces, everyone admits it needs to be quantized,[8] as we will explain in chapter 6. So, while QFT successfully unifies SR and QM, gravity (as described by GR) has eluded all efforts at quantization (although loop quantum gravity theorists may disagree).

This is the impasse of unification (and quantum gravity) that has existed for almost 40 years despite the efforts of many brilliant physicists over that time. As we saw in this brief history, the mechanical paradigm of the dynamical universe has dominated development of modern physics. Even though the theory (QFT) that underwrites the Standard Model is often couched in the Lagrangian Schema, it is understood dynamically as we will see. After introducing QFT, we will argue once again that it is perhaps time to move "beyond the dynamical universe" in order to advance physics.

5.3   Abridged introduction to quantum field theory

Since QFT is not typically taught in undergraduate programs, we present a short introduction here. We begin with a review of the harmonic oscillator, since QFT can be understood as an array of coupled harmonic oscillators.

When a mass m is displaced slightly from stable equilibrium, it is subject to a linear restoring force and oscillates about equilibrium in a sinusoidal fashion called simple harmonic motion (Figure 5.2). The frequency/period of oscillation is related to m and the constant of proportionality k between the restoring force F

---

[7] This force is responsible for nuclear decay.
[8] This is called "quantum gravity."

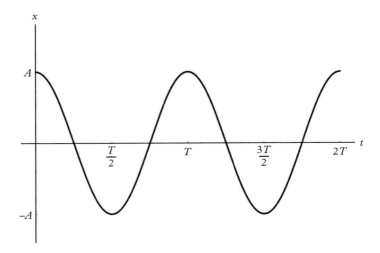

Figure 5.2 *Simple harmonic motion.*

and the distance $x(t)$ from equilibrium that m is displaced at any given time t. That is, Newton's second law of motion $F = m\frac{d^2x}{dt^2}$ with $F = -kx$ gives $x(t) = A\cos(\sqrt{\frac{k}{m}}t)$ where $A = x(0)$ is the oscillation amplitude and $2\pi\sqrt{\frac{m}{k}}$ is the period of oscillation (Figure 5.2). The potential energy is $\frac{1}{2}kx^2$.

Now consider a pair of masses m coupled by a spring of spring constant k with their displacements from equilibrium labeled q_1 and q_2 (Figure 4.18). Putting this configuration on a spacetime graph (Figure 4.19) you can write the discrete Lagrangian (kinetic energy minus potential energy, recall Foundational Physics for Chapter 1) as $\frac{1}{2}\vec{Q} \cdot \mathbf{K} \cdot \vec{Q}$ where the components of $\vec{Q} = (q_1(t_1), q_1(t_2), q_1(t_3), q_2(t_1), q_2(t_2), q_2(t_3))$ reside on the nodes (1,2,3,4,5,6), respectively, and $\mathbf{K}$ is a matrix containing terms shown on the links of Figure 4.19. In other words, $\mathbf{K}$ is a difference matrix (counterpart to a differential operator in continuous spacetime) responsible for the discrete counterparts of the velocities $\frac{dq_i}{dt}$ needed for the kinetic energy, as well as discrete counterparts of displacements needed for the potential energy. Notice that now both the kinetic energy and spring potential energy are contained in a single term that looks like the potential energy alone for a simple harmonic oscillator. Since $\mathbf{K}$ is taking differences between the "field" $\vec{Q}$ on the nodes, it is relational and therefore it contains redundancies.

For example, as we pointed out in Foundational Physics for Chapter 4, if we were describing the line of people (Alice, Bob, Charlie, David), we would say, "Alice is in front of Bob" and "Bob is behind Alice and in front of Charlie." Assigning a numerical value of +1 to "in front of" and −1 to "behind," we see that adding the statements describing the locations of all four people gives a result of zero. This "summing to zero" happens because we have a self-referential, relational description of all four people. Since **K** is a "relations matrix" in this same sense, its rows sum to zero. We write $\mathbf{K} \cdot [111111]^T = 0$ and say that $[111111]^T$ is a non-null (not zero) eigenvector of **K** with eigenvalue zero. Thus, **K** is said to possess a non-trivial null space in contrast to the fact that all matrices multiplied by the zero (null) vector $[0,0,\dots,0]$ will give a result of zero (this is the trivial null space). The complement to that non-trivial null space (the vectors $[\dots]$ that satisfy $\mathbf{K} \cdot [\dots]^T \neq 0$) is called the "row space" of **K**.

Creating a grid of coupled masses throughout space and writing the action S as a field Lagrangian density integrated in spacetime, instead of a particle Lagrangian integrated in time alone, gives us a Lorentz-invariant S. This is how QFT incorporates SR. We then exponentiate this relativistic S and integrate over all field configurations to get the transition amplitude Z for QFT. This is analogous to exponentiating the non-Lorentz invariant particle S and integrating over all paths to produce the propagator for QM (Feynman path integral). This is how QFT incorporates QM. Indeed, $Z(J)$ is used to produce the probabilities for QFT, so $Z(J)$ is the QFT counterpart to the Feynman path integral of QM. In QM, we fixed the endpoints of the path while in QFT we fix the locations of Sources and sinks of energy–momentum contained in the vibrating field. These Sources and sinks are called "sources" and denoted by J and the quanta of energy emitted and absorbed by J are particles. Thus, Lancaster and Blundell [2014, p. 19] pointed out that QM could be thought of as treating particles like waves (so-called first quantization) while QFT could be thought of as treating waves like particles (so-called second quantization).

The Central Identity of Quantum Field Theory is [Zee, 2003, p. 167]

$$Z(J) = \int D\phi \, e^{-\frac{1}{2}\phi \cdot \mathbf{K} \cdot \phi - V(\phi) + J \cdot \phi} = e^{-V\left(\frac{\delta}{\delta J}\right)} e^{\frac{1}{2}J \cdot \mathbf{K}^{-1} \cdot J} \qquad (5.1)$$

where ϕ is a vector of all fields and the relations matrix **K** accounts for the quadratic terms in the Lagrangian (e.g., as was done in Foundational Physics for Chapter 4). The rest of the terms are written $V(\phi)$ and finally ϕ is coupled to the source vector in the last term. The exponentiated derivative $V(\frac{\delta}{\delta J})$ is used to differentiate the solution of the free field solution $(e^{\frac{1}{2}J \cdot \mathbf{K}^{-1} \cdot J})$ which gives the probability amplitude in series form where the terms in the series represent different ways to get the energy from Source to sink. Again, just as the Feynman path integral for QM produces a probability amplitude by summing over all particle paths from Source to sink, QFT gives a probability amplitude by summing over all field configurations resulting in an energy exchange from Source to sink. The main difference is that QM has a fixed number of particles while the different field configurations of QFT can correspond to different numbers of particles. Indeed, the Hamiltonian approach to particle physics per algebraic QFT is that of particle creation and annihilation via creation and annihilation operators acting on the ground state of the fields.

So, just as one can decompose the QM wave function in the eigenbases of different operators to find its spectra of observables, one can decompose the fields of

QFT in the different particles of what is called n-particle Fock space. The ontology of QFT assumes space is filled with all types of (interacting) quantum fields and what we observe are the quantum excitations of those quantum fields. As Zee states, "[t]hat's just about all there is to quantum field theory" [Zee, 2003, p. 85].

5.4   An adynamical approach to particle physics

We have seen that the history of modern physics is a story in the dynamical universe. This story has culminated in our most sophisticated theory of quantum physics (QFT) being understood very dynamically/mechanically, even in its Lagrangian Schema formulation. Perhaps this is keeping us from advancing modern physics? How could an adynamical approach help?

As we explained in chapter 4, according to RBW the most fundamental data are individual detector clicks. The distribution of these clicks in their classical context is given by an adynamical global constraint. Clearly particle physics per QFT is about worldlines in spacetime (a series of clicks) rather than the individual clicks per se, so RBW is certainly a further ontological decomposition in that sense. Here is how we view particle physics according to RBW.

In RBW, the entire process of a particle physics experiment is viewed as a relationship between the accelerator beam (Source) and detector (sink). With this picture in mind, we can say simply what we are proposing is as follows: the discrete graphical approach to QFT (lattice gauge theory) is fundamental to QFT, not the converse per conventional thinking. Once one accepts this premise, it is merely a matter of degree to have sources connected by large graphical links (direct action), which is the basis for our explanation of the twin-slit experiment (chapter 4 foundational physics thread) and our modified Regge calculus approach to dark energy (chapter 6). In this approach, the role of the field in quantum physics is to designate the relative spatiotemporal locations of discrete experimental outcomes in a classical context that evidence the transfer of energy–momentum. There is no graphical counterpart to "quantum systems" traveling through space as a function of time from Source to sink to "cause" detector clicks. There are only various distributions of spacetimesource elements representing experimental Sources and sinks in the classical context of beam splitters, mirrors, particle sources, detectors, etc., in the given experiment from initiation to termination. This implies the empirical goal at the fundamental level is to tell a unified story about detector events at the level of individual clicks—how they are distributed in space (e.g., interference patterns, interferometer outcomes, spin measurements), how they are distributed in time (e.g., click rates, coincidence counts), how they are distributed in space and time (e.g., particle trajectories), and how they generate more complex phenomena (e.g., photoelectric effect, superconductivity). Thus in RBW, particle physics according to QFT is in the business of characterizing large sets of detector data, that is, all the individual clicks.

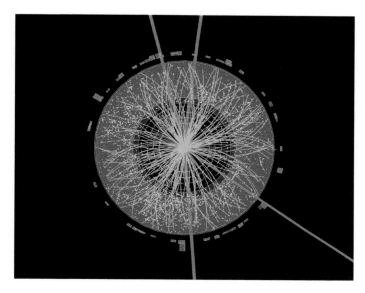

Figure 5.3 *The event was recorded by ATLAS on 18-May-2012, 20:28:11 CEST in run number 203602 as event number 82614360 http://cds.cern.ch/record/1459495. Reproduced by permission of CERN.*

A particle physics "event" contains "approximately 100,000 individual measurements of either energy or spatial information" [Frisch, 1993] (Figure 5.3). Individual detector clicks (called "hits in the tracking chamber") are first localized spatially (called "preprocessing"), then associated with a particular track (called "pattern recognition"). The tracks must then be parameterized to obtain dynamical characteristics (called "geometrical fitting") [Fernow, 1986, §1.7.1, 1.7.2, and 1.7.3]. This is the process by which worldlines are inferred from particle physics detectors and they constitute what we mean by time-evolved "classical objects." Each worldline can be deduced click by click in temporal succession using $\psi(x, t)$, as shown by Mott for alpha particles in a cloud chamber [Mott, 1929]. Therefore, a probability amplitude could be computed for each worldline using spacetimesource elements click by click with each click providing empirical evidence of an otherwise unobservable, underlying spacetimesource element. However, as shown by Mott, after the first click the remaining clicks follow a classical trajectory with high probability, so the only quantum computation needed (practically speaking) is for the probability amplitude of the spacetimesource element of the set of first clicks, that is, the first click for each worldline in the collection. And, the properties (mass, charge, momentum, energy, baryon number, electron lepton number, etc.) for that spacetimesource element would simply be the properties of the subsequent worldlines (particles) defined relationally in the context of the accelerator Source and particle detector per spatiotemporal ontological contextuality (as described in Philosophy of Physics for Chapter 4).

Accordingly, a particle physics detector event is one giant interference pattern (interference as in RBW), and the way to understand a particular pattern involving thousands of clicks can only realistically be accomplished by parsing an event into smaller subsets, and the choice of subsets is empirically obvious—spacetime trajectories. These trajectories are then characterized by mass, spin, and charge. According to RBW's adynamical explanation, the colliding beams in the accelerator and the detector surrounding the "collision point" form the graphical input that, in conjunction with RBW's adynamical global constraint, dictate the spacetime distribution of spacetimesource element configurations responsible for particle trajectories. This is exactly what QFT does for particle physics according to RBW. Since the individual trajectories are themselves continuous, QFT uses propagators in continuous spacetime which entails an indenumerably infinite number of locations for both clicks and interaction vertices.

This severely undermines the dynamical picture of perturbations moving through a continuum medium (naive field) between sources; it undermines the naive notion of a particle as traditionally understood. In fact, the typical notion of a particle is associated with the global particle state of n-particle Fock space and according to Colosi and Rovelli, "the notion of global particle state is ambiguous, ill-defined, or completely impossible to define" [Colosi and Rovelli, 2009, p. 14]. What we mean by "particle" is a collection of detector hits forming a spacetime trajectory resulting from a collection of adynamically constrained spacetimesource elements in the presence of colliding beams and a detector. And this doesn't entail the existence of an object with intrinsic properties, such as mass and charge, moving through the detector to cause the hits.

Our view of particles agrees with Colosi and Rovelli on two important counts. First, that particles are best modeled by local particle states rather than n-particle Fock states computed over infinite regions, squaring with the fact that particle detectors are finite in size and experiments are finite in time. The advantage to this approach is that one can unambiguously define the notion of particles in curved spacetime as excitations in a local (flat) region, which makes it amenable to Regge calculus (more on that in chapter 6). Second, this theory of particles is much more compatible with the quantum notion of complementary observables in that every detector has its own Hamiltonian (different sized graph with different properties), and therefore its own particle basis (unlike the unique basis of Fock space). According to Colosi and Rovelli, "[i]n other words, we are in a genuine quantum mechanical situation in which distinct particle numbers are complementary observables. Different bases that diagonalize different HR [Hamiltonian] operators have equal footing. Whether a particle exists or not depends on what I decide to measure" [Colosi and Rovelli, 2009].

Thus, in our view, particles simply describe how detectors and Sources are relationally/contextually co-defined via RBW's adynamical global constraint. Of course, this has consequences for unification as conventionally understood and that will be discussed in chapter 6 along with RBW's astrophysical implications for dark energy and dark matter. The bottom line is that, as with the puzzles, problems, paradoxes, and conundrums of GR and QM, the impasse of particle physics according to QFT may well be due to adherence to the dynamical universe and an entirely new view of future physics is available via adynamical thinking. To wit, in the philosophy of physics thread for this chapter many of the key interpretative problems and conundrums unique to QFT are resolved by taking the approach outlined here in the main thread for chapter 5.

Philosophy of Physics for Chapter 5

5.5   Introduction

Some have stressed that the conceptual problems besetting QM remain the central concerns [Cushing, 1988], while others stress that QFT exacerbates some of the interpretive problems of QM and possesses foundational problems all its own [Healey, 2007; Kuhlmann, 2009]. Some (especially physicists) have stressed that QFT is the greatest and most explanatory intellectual achievement of modern science [Gross, 1999; Oerter, 2006], while others believe QFT is "much more a set of formal strategies and mathematical tools than a closed theory" [Kuhlmann, 2009]. Of course, on both counts both sides are right. In addition to the problems of QM, an interpretation must address concerns unique to QFT, for example, notorious problems with particle and field ontologies and renormalization, how to interpret gauge invariance and the Aharonov–Bohm effect, the problem of inequivalent representations, and explaining the effectiveness of the interaction picture and perturbation theory in light of Haag's theorem.

As for progress in this area, Healey notes, "no consensus has yet emerged, even on how to interpret the theory of a free, quantized, real scalar field" [Healey, 2007, p. 203]. And, "There is no agreement as to what object or objects a quantum field theory purports to describe, let alone what their basic properties would be" [Healey, 2007, p. 221]. As Kuhlmann notes:

> In conclusion one has to recall that one reason why the ontological interpretation of QFT is so difficult is the fact that it is exceptionally unclear which parts of the formalism should be taken to represent anything physical in the first place. And it looks as if that problem will persist for quite some time.
>
> [Kuhlmann, 2009]

Those who emphasize the incompleteness of QFT over its successes often focus on the many ad hoc and, for some, troubling "fixes" involved in the practice of QFT. For example, since QFT is independent of overall factors in the transition amplitude, such factors are simply "thrown away" even when these factors are infinity as is the case when the volume of the gauge symmetry group in Fadeev–Popov gauge fixing is infinite [Zee, 2003, p. 170], as Dirac complained:

> Hence most physicists are very satisfied with the situation. They say: "Quantum electrodynamics is a good theory, and we do not have to worry about it any more." I must say that I am very dissatisfied with the situation, because this so-called "good theory" does involve neglecting infinities which appear in its equations, neglecting them in an arbitrary way. This is just not sensible mathematics. Sensible mathematics involves neglecting a quantity when it turns out to be small—not neglecting it just because it is infinitely great and you do not want it!
>
> [Dirac, 1978, p. 36]

And, in the process of renormalization one must "tweak" parameters in the Lagrangian so they remain finite under regularization ['t Hooft, 2007, p. 712].

QFT has triumphed empirically, but virtually all physicists agree that it is not a fundamental theory because it does have a limited domain of applicability, viz., it does not deal with particle interactions at ranges where gravity becomes important. It might be that the Standard Model of particle physics plus the gravitational field is fundamental [Weinberg, 2009], but most physicists assume there exists an underlying, unified theory which would naturally justify the ad hoc fixes employed in QFT and tell us how to handle the particle interactions where gravity is deemed relevant [Cao, 1999, §7].

Clearly, QFT is in need of more philosophical attention. Consequently in this philosophy of physics thread, we use RBW to provide a brief account of gauge invariance, gauge fixing, and the Aharonov–Bohm effect per the adynamical global constraint (AGC) that illustrates our spatiotemporal ontological contextuality alternative to problematic particle and field ontologies. Additionally, we use RBW to explain QFT's need for regularization and renormalization, and largely discharge the problems of Poincaré invariance in a graphical approach, inequivalent representations, and Haag's theorem.

5.6   Gauge invariance, gauge fixing, and the Aharonov–Bohm effect

As one adds different fields interacting in Lorentz-invariant fashion, the integral for $Z(J)$ becomes highly non-trivial and must be solved via various approximations. It is this computation of $Z(J)$ that leads to the issues of QFT.

For example, the solution of $Z(J)$ shown in the Central Identity of Quantum Field Theory [Zee, 2003, p. 167]

$$Z(J) = \int D\phi\, e^{-\frac{1}{2}\phi\cdot K\cdot\phi - V(\phi) + J\cdot\phi} = e^{-V\left(\frac{\delta}{\delta J}\right)} e^{\frac{1}{2}J\cdot K^{-1}\cdot J} \tag{5.2}$$

requires inverting K, but K doesn't have an inverse because its determinant is zero since it possesses a non-trivial zero eigenvalue. This is overcome by restricting the $Z(J)$ integral to the row space of K, that is, that region of K with non-zero eigenvalues. That K possesses a non-trivial null space is the graphical counterpart to gauge invariance and restricting the $Z(J)$ integral to the row space of K is the graphical counterpart of Fadeev–Popov gauge fixing [Zee, 2003, p. 168–70]. Since $\vec{J}$ also appears in the integral for $Z(J)$, we need it to reside entirely in the row space of K which means $\vec{J} \cdot [111111]^T = 0$, so that the components of $\vec{J}$ also sum to zero: $\vec{J}$ is said to be "divergence free." Recall, this same constraint applies to the stress-energy tensor in GR where it represents the local conservation of energy–momentum on the spacetime manifold M. Likewise here, divergence-free $\vec{J}$ is a necessary requirement for the conservation of energy–momentum (though not sufficient, as we showed with the twin-slit experiment in Foundational Physics for Chapter 4). Notice as per lattice gauge theory that gauge invariance is a natural consequence of relationalism. Consequently, we agree with Rovelli that "gauge is ubiquitous. It is not unphysical redundancy of our mathematics. It reveals the relational structure of our world" [Rovelli, 2013, p. 7]. Thus, we see that gauge invariance and gauge fixing are necessarily germane to RBW's spatiotemporal ontological contextuality.

Spatiotemporal contextuality also helps explain the Aharonov–Bohm effect (Figure 5.4). Here a spatially localized magnetic field $\vec{B}$ (contained in a long solenoid, for example) is added to the twin-slit experiment with electrons (Figure 4.7) as shown in Figure 5.4. The resulting interference pattern is shifted one way or the other depending on whether $\vec{B}$ is into or out of the page. Viewed classically this is somewhat mysterious, since the electron paths (Slit 1 or Slit 2 to Detection Event) don't traverse the region containing $\vec{B}$ (interior of the solenoid). Thus, as with the twin-slit experiment proper, an attempt to explain the Aharonov–Bohm fringe shift is difficult if one assumes each electron takes one path or the other. The interaction, however, involves the vector field $\vec{A}$, where $\vec{B} = \nabla \times \vec{A}$, since the momentum for the electron in the Lagrangian becomes $m\vec{v} + \frac{q\vec{A}}{c}$ [Shankar, 1994, p. 84], and $\vec{A}$ does exist outside the solenoid (it circulates clockwise or counterclockwise for $\vec{B}$ into or out of the page, respectively). In this case, lattice gauge theory uses the 4-vector $(\Phi, \vec{A})$ on its 4D graphical lattice. Thus, *modified lattice gauge theory provides a probability amplitude for the distribution of spacetimesource elements in the Aharonov–Bohm experiment using the field $(\Phi, \vec{A})$ of lattice gauge theory with large links configured in accord with the AGC.* Again, there is no worldline traversing the space between the Source and a Detection Event, rather each spacetimesource element (Figure 5.4) represents the loss of mass m at the Source and the subsequent addition of mass

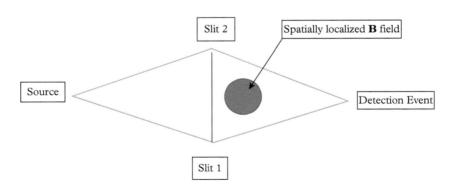

Figure 5.4 *The Aharonov–Bohm effect. The temporal dimension of Figure 4.7 is suppressed so the boxes are now lines connecting Source to Slits to Detection Event. A spatially localized magnetic field $\vec{B}$ oriented perpendicular to the page has been added behind the slit screen.*

m at a Detection Event. This is the adynamical explanation of the Aharonov–Bohm effect according to our version of God's-eye physics. So, the distribution of spacetimesource elements is contextual in a spatially holistic sense since it depends on the $\vec{B}$ *flux*, not the $\vec{B}$ location *per se* [Shankar, 1994, p. 499]. Wise used the holonomic[9] property of QFT in his graphical approach [Wise, 2006]. Healey notes something similiar: "There need be no action at a distance if the behavior both of the charged particles and of electromagnetism are non-separable processes" [Healey, 2016]. However, unlike Healey [2004], quantum non-separabilility and the Aharonov–Bohm effect have precisely the same origin in our approach: both result from the AGC so that detector clicks evidence the non-separable (contextual) nature of the devices in the experiment.

In summary, that **K** possesses a non-trivial null space is the graphical equivalent of gauge invariance and restricting $\vec{J}$ to the row space of **K** provides a natural gauge fixing, that is, restricting the integral of the transition amplitude $Z(J)$ to the row space of **K**. That **K** possesses a non-trivial null space also means the determinant of **K** is zero, so the set of vectors constituting the rows of **K** is not linearly independent. That some subset of these row vectors is determined by its complement follows from having the graphical set relationally constructed. That $\vec{J}$ reside in the row space of **K** means it is divergence-free. Thus, the AGC of RBW states simply that **K** be constructed with a non-trivial null space and $\vec{J}$ be restricted to the row space of **K** so as to represent a conserved exchange of energy–momentum. As with the graphical version of GR (Regge calculus, see chapter 6) and QM, RBW simply says the graphical version of QFT (lattice gauge theory) is fundamental to its continuum counterpart. Since RBW has direct action, these graphical counterparts end up with links connecting non-neighboring points on the corresponding spacetime manifold M of classical physics, so-called disordered locality. Thus, lattice gauge theory and Regge calculus are modified per disordered locality in RBW.

5.7 Regularization and renormalization

It should be apparent that in RBW the role of the field is very different than in QFT where it pervades otherwise empty, continuous space to mediate the exchange of energy–momentum between sources. In our view (and that of lattice gauge theory), a field is simply a map of scalars or vectors to elements of the graph. One obtains QFT results from lattice gauge theory by letting the lattice spacing go to zero. In fact, one can understand QFT renormalization through this process of lattice regularization ['t Hooft, 2007]. As it turns out, however, this limit does not always exist, so calculated values are necessarily obtained from small, but non-zero, lattice spacing [Rothe, 1992, p. 40]. More specifically, in a

---

[9] Holonomy is a topological property dealing with closed curves. In this case, the closed path is shown by the spacetimesource element in Figure 5.4.

standard particle physics calculation, a problem arises with the momentum space version of $Z(J)$ in that an integral over all field configurations entails all momentum configurations and $Z(J)$ is typically divergent for infinite momentum [Zee, 2003, pp. 145–51] (called an "ultraviolet (UV) divergence").[10] Thus, the integral must be "cut off" at a certain value of momentum in a procedure called "regularization." Most physicists view this ad hoc cut-off as the result of QFT being only a low energy approximation to some more comprehensive, underlying theory (sometimes called "theory X" [Wallace, 2006]). Thus, QFT is sometimes called "an effective theory." In our view, the need for regularization results from spatiotemporal ontological contextuality, that is, the classical context of the experimental setup and procedure. Most experimentalists would agree that the accelerator beam does not contain any particles with infinite momentum, so there is no reason to include such possibilities in the computation of $Z(J)$.[11]

5.8   Poincaré invariance, inequivalent representations, and Haag's theorem

A common complaint with graphical approaches is the (apparent) loss of Poincaré invariance: the invariance of physics under spatial translations, spatial rotations, temporal translations, and boosts from one frame to another in relative motion. Such symmetries are considered to be of fundamental importance for many physicists and we agree. So, it is worth mentioning that our graphical approach automatically satisfies Poincaré invariance, since the properties for our spacetimesource elements are obtained from their classical context and that context satisfies Poincaré invariance.

We conclude by explaining how RBW's modified lattice gauge theory deals with the inequivalent representations which exist in QFT due to the failure of the Stone–von Neumann theorem to apply to the (indenumerably) infinite degrees of freedom generated by the underlying (associated) spacetime manifold. As Wallace notes, there are two ways in which inequivalent representations can occur in QFT: one associated with the short distance and high-energy (ultraviolet) degrees of freedom (UV inequivalence) and one associated with the long-distance (infrared) degrees of freedom (IR inequivalence) [Wallace, 2006, p. 55]. As Wallace points out, discrete QFT "has only finitely many degrees of freedom per space-time point, and hence no UV inequivalent representations" [Wallace, 2006, p. 55]. Obviously, this applies to modified lattice gauge theory even though it differs from lattice gauge theory proper. Of course, one can dismiss this problem nearly as neatly if one already subscribes to QFT as an effective theory, as described earlier [Zee, 2003, p. 437].

---

[10] One also finds so-called infrared (IR) divergences associated with integrals over infinite spacetime volumes.

[11] Likewise, the experimental outcomes need only consider finite spacetime regions, so IR divergences are also avoided by a "cut off."

The "global" or IR inequivalences might arise if one attempted to apply RBW to a finite number of sources with infinite spatiotemporal separations. However, as per RBW, quantum physics requires Sources and sinks be coupled and reside in a classical context, so infinite separation is not feasible, even if one wants to study situations in cosmology. Indeed, if outcomes in some finite spatial region could be affected by sources at infinite spatial separation, spurious experimental outcomes could always be blamed on events outside the experiment's causal horizon. This view resonates with Jackiw's sentiment that "the consensus is that infrared divergences do not arise from any intrinsic defect of the theory, but rather from illegitimate attempts at forcing the theory to address unphysical questions" [Jackiw, 1999, p. 149]. And, it is also in accord with Wallace's belief that, within its domain of applicability, QFT has no foundational problems with IR or UV inequivalences.

Closely related to representational inequivalances is Haag's theorem [Earman and Fraser, 2006]. To summarize the problem:

> Everyone must agree that as a piece of mathematics Haag's theorem is a valid result that at least appears to call into question the mathematical foundation of interacting quantum field theory, and agree that at the same time the theory has proved astonishingly successful in application to experimental results.
>
> [Teller, 1997, p. 115]

Essentially, Haag proved that the Hilbert space representation of free fields, which is used to describe incoming particles prior to interaction and outgoing particles after interaction, cannot be housed with the Hilbert space representation of the interaction form for those same fields, that is, they are inequivalent representations. So, how does one toggle between the two spaces as necessary to connect incoming states to interaction states to outgoing states? In QFT, one simply expands $Z(J)$ with free field states in powers of the coupling constant from the interaction Hamiltonian. Given Haag's theorem, why does this work?

> [T]he interaction picture presupposes all of the assumptions needed to prove [Haag's] theorem; but this theorem shows that the interaction picture cannot be used to represent a non-trivial interaction. And yet the interaction picture and perturbation theory work. Some explanation of why they work is called for.
>
> [Earman and Fraser, 2006, p. 322]

Ideally, one would like to negate an offending assumption in the proof of Haag's theorem in order to explain the effectiveness of the interaction picture. One assumption of Haag's theorem is Poincaré invariance, but clearly for us Poincaré invariance is not the offending assumption, as we described earlier. Rather, we believe the offending assumption is not articulated in the proof of Haag's theorem but is a tacit assumption of QFT in general—the fundamental nature of the field quanta, that is, the normal modes of the classical fields quantized by QFT which supply a basis for Fock space (Hilbert space of free field states). As we stated earlier, QFT is an approximation needed to solve problems with large numbers of detector clicks (again, no one in their right mind would try to use RBW to compute probabilities for the distributions of 100,000 detector clicks). In RBW, there are only spacetimesource elements connecting Sources and sinks. Contrast this with QFT whereby one has a set of free fields for incoming and outgoing particles and a set of interaction fields with the interaction fields on the same ontological footing as the free fields.

But, per RBW, both sets are ontologically specious, that is, the fields of lattice gauge theory are simply computational devices for computing the probability amplitudes for the distribution of spacetimesource elements representing the distribution of energy–momentum exchange. Thus, QFT is simply providing a computational algorithm for producing the probability amplitudes using the approximation of continuous trajectories. Exactly why this algorithm works is an interesting mathematical question, probably related to the fact that field theory in general can be used to model all sorts of phenomena, for example, "finding the optimal connectivities in an audio-visual cortex" [Kardar and Zee, 2002, p. 15894] and "finding possible topologies of pseudoknots in single-stranded RNA molecules" [Pillsbury et al., 2005, p. 1]. But, at writing, as pertains to RBW, the "unreasonable effectiveness" of perturbative QFT in light of Haag's theorem is an interesting mathematical curiosity without ontological implications.

Foundational Physics for Chapter 5

In this foundational physics thread, we show how classical field theory relates to QFT and briefly explain gauge invariance. As it turns out, classical field theory is satisfied by the extremum field configurations of QFT in complete analogy to the spacetime paths of classical mechanics being extremum solutions in the path integral formulation of QM. As we pointed out in the main thread per [Lancaster and Blundell, 2014, p. 19], QM could be thought of as treating particles like waves while QFT could be thought of as treating waves like particles.

5.9   Classical field theory and QFT

We begin with the transition amplitude

$$Z = \int D\phi \exp\left[\frac{i}{\hbar}\int L\left(\dot{\phi},\phi\right)d^4x\right] \qquad (5.3)$$

where $L\left(\dot{\phi},\phi\right) = \frac{1}{2}(d\phi)^2 - V(\phi) + J(x,t)\phi(x,t)$ and the integral is over all field configurations $\phi(x,t)$ with source locations $J(x,t)$. Typically the Lagrangian field density $L\left(\dot{\phi},\phi\right)$ is written $L\left(\dot{\phi},\phi\right) = \frac{1}{2}\phi\mathbf{D}\phi + J(x,t)\phi(x,t)$ where $\mathbf{D}$ is a differential operator. The discrete version of Z is

$$Z(J) = \int \ldots \int dQ_1 \ldots dQ_N \exp\frac{i}{\hbar}\left[\frac{1}{2}\vec{Q}\cdot\mathbf{K}\cdot\vec{Q} + \vec{J}\cdot\vec{Q}\right] \qquad (5.4)$$

where $\mathbf{K}$ is the difference matrix (also known as "relations matrix") corresponding to the differential operator $\mathbf{D}$, $\vec{J}$ is the source vector on the graph corresponding to $J(x,t)$, and $\vec{Q}$ is the vector whose components correspond to the field values $\phi(x,t)$ on the graph. One uses $Z(J)$ to compute probability amplitudes for various distributions of spacetimesource elements, as we showed in the twin-slit experiment in Foundational Physics for Chapter 4. The adynamical global constraint (AGC) of RBW then dictates the construct of $\mathbf{K}$ and $\vec{J}$, that is, $\mathbf{K}$ is constructed with a non-trivial null space and $\vec{J}$ resides in the row space of $\mathbf{K}$ so as to represent a conserved exchange of energy–momentum in its classical context. The solution to Eq. 5.4 involves $\mathbf{K}^{-1}$,

$$Z(J) = \left(\frac{(2\pi)^N}{\det(\mathbf{K})}\right)^{1/2} \exp\left[\frac{1}{2}\vec{J}\cdot\mathbf{K}^{-1}\cdot\vec{J}\right] \qquad (5.5)$$

but since $\mathbf{K}$ contains a non-trivial null space, it doesn't have an inverse. This is the graphical counterpart to gauge invariance [Zee, 2003, p. 168–70] and reflects a redundancy in the mathematical representation created by the relational/self-referential construct of $\mathbf{K}$, as we explained previously and as noted by Rovelli [2013]. In order to solve Eq. 5.3, one "factors out infinities" in a process called "gauge fixing." The graphical counterpart to gauge fixing is to restrict the integral of Eq. 5.4 to the row space of $\mathbf{K}$. This requires $\vec{J}$ reside in the row space of $\mathbf{K}$, which means its components must sum to zero, that is, $\vec{J}$ is a "divergence-free" source. Recall, this same constraint applies to the stress–energy tensor in general relativity where it represents the local conservation of energy–momentum on the spacetime manifold M. Likewise here, divergence-free $\vec{J}$ is a necessary requirement for the conservation of energy–momentum, although disordered locality (large links of our modified lattice gauge theory) allows for non-local exchanges in the context of M. Thus, we see that gauge invariance and gauge fixing are necessarily germane to our spatiotemporal ontological contextuality.

In the eigenbasis of $\mathbf{K}$ with eigenvalues a_j ($a_N = 0$), $\vec{Q} \to \tilde{Q}$, and $\vec{J} \to \tilde{J}$ we have

$$Z(J) = \int_{-\infty}^{\infty} \cdots \int_{-\infty}^{\infty} d\tilde{Q}_1 \ldots d\tilde{Q}_{N-1} \exp\left[\sum_{j=1}^{N-1} \left(-\frac{1}{2}\tilde{Q}_j^2 a_j + \tilde{J}_j \tilde{Q}_j \right) \right] \qquad (5.6)$$

which has the solution

$$Z(J) = \left(\frac{(2\pi)^{N-1}}{\prod_{j=1}^{N-1} a_j} \right)^{1/2} \prod_{j=1}^{N-1} \exp\left[\frac{\tilde{J}_j^2}{2a_j} \right] \qquad (5.7)$$

Notice that this does not involve the field $\tilde{Q}$. In order to recover $\tilde{Q}$, so as to see what role it plays, we have to look for the extremum configuration. The probability of configuration $\tilde{Q}_j = \tilde{Q}_o$ is given by Lisi, 2006[12]:

$$P\left(\tilde{Q}_j = \tilde{Q}_o\right) = \frac{Z\left(\tilde{Q}_j = \tilde{Q}_o\right)}{Z} = \sqrt{\frac{a_j}{2\pi}} \exp\left[-\frac{1}{2}\tilde{Q}_o^2 a_j + \tilde{J}_j \tilde{Q}_o - \frac{\tilde{J}_j^2}{2a_j} \right] \qquad (5.8)$$

The extremum configuration is then given by

$$\delta P\left(\tilde{Q}_j = \tilde{Q}_o\right) = 0 \Rightarrow \delta\left[-\frac{1}{2}\tilde{Q}_o^2 a_j + \tilde{J}_j \tilde{Q}_o - \frac{\tilde{J}_j^2}{2a_j} \right] = 0 \Rightarrow a_j \tilde{Q}_o = \tilde{J}_j \qquad (5.9)$$

which is just $\mathbf{K} \cdot \vec{Q}_o = \vec{J}$, that is, the classical field theory solution (which can also be obtained by the stationary phase method [Zee, 2003, p. 15]). Thus, we understand classical field theory gives the extremum distribution of field values on the graph of lattice gauge theory just as the classical particle path in spacetime is an

---

[12] We are using the Euclidean or Wick-rotated version of $Z(J)$ here. That is why the probability has this statistical mechanics form.

extremum of the Feynman path integral for QM. In electromagnetism, for example, the electric field $\vec{E}$ is given by the spatiotemporal gradient of the graphical 4-field $A^\alpha = \left(\Phi, \vec{A}\right)$, that is,

$$\vec{E} = -\nabla \Phi - \frac{\partial \vec{A}}{\partial t} \tag{5.10}$$

and the average (classical) energy density (which goes as the number of energy–momenta quanta (clicks) per second per unit volume) is given by the square of the electric field. This is how the fields of lattice gauge theory, which give the distribution of spacetimesource elements, relate to a measurement of energy–momentum transfer from Source to sink (detector).

5.10 Gauge invariance

This example is from Zee [2003, pp. 226–7]. Suppose we transform the N-component scalar field ϕ in the Lagrangian density $\partial \phi^\dagger \partial \phi - V(\phi^\dagger \phi)$ according to $\phi(x) \rightarrow U\phi(x)$ where U is an element of SU(N) and can vary from place to place, that is, $U(x)$. Since $\phi^\dagger \rightarrow \phi^\dagger U^\dagger$ the quantity $\phi^\dagger \phi$ is invariant under this transformation (since $U^\dagger U = 1$), so $V(\phi^\dagger \phi)$ is also invariant under this transformation as long as its a polynomial in $\phi^\dagger \phi$. However, the derivative term $\partial \phi^\dagger \partial \phi$ will not be invariant. In order to make the derivative term invariant we introduce the "covariant derivative":

$$D_\mu \phi(x) = \partial_\mu \phi(x) - iA_\mu(x)\phi(x) \tag{5.11}$$

The field $A_\mu(x)$ is called a "gauge potential" and our Lagrangian density is now $(D_\mu \phi)^\dagger (D_\mu \phi) - V(\phi^\dagger, \phi)$. The $A_\mu(x)$ are N by N matrices and there are as many of them as there are generators of S(N). For example, there are three of them for SU(2) and eight of them for SU(3). In order for the field $A_\mu(x)$ to have dynamics, we construct a "field strength" $F_{\mu\nu} = \partial_\mu A_\nu - \partial_\nu A_\mu + [A_\mu, A_\nu]$ and add a term proportional to $F_{\mu\nu}F^{\mu\nu}$ to the Lagrangian density. These gauge fields represent new particles called "gauge bosons" and mediate a "gauge force" between components of the field $\phi(x)$ at different values x.

In lattice gauge theory, the derivative of a field in QFT is represented by a difference between the values of that field at adjacent graphical locations. Differencing vectors in two different vector spaces is not defined in general, so the gauge field can be understood to define this differencing process. This is called "parallel transport" in curved spaces, that is, a means of transporting $A_\mu(x)$ to the vector space of $A_\mu(x + \Delta x)$, so a difference can be computed. (In the case of U(1), the ambiguity in differencing a 1-component field is simply a phase.) Thus, the existence of gauge bosons, that is, gauge forces, can be understood to result necessarily from the ambiguous nature of differencing a field at different locations in spacetime.

6

Relational Blockworld Approach to Unification and Quantum Gravity

Not only do we not know what dark energy might be, that would be making the universe expand faster and faster, we don't even know whether really the answer will turn out to be a new energy in the universe. It's possible that we've just discovered an extra wrinkle in Einstein's Theory of Relativity, and that that would be the real final result.

Nobel Laureate Saul Perlmutter [Chodos, 2011, p. 1]

6.1 Introduction

In chapter 1, we explained the motivation behind our Lagrangian Schema Universe (LSU) approach by painting a picture of the current impasse of theoretical physics and philosophy of physics caused by the moribund nature of the Newtonian Schema Universe (NSU) approach across the board. We then sketched an alternative methodological and explanatory LSU approach and its ontological implications, which are instantiated by our Relational Blockworld (RBW) model. In chapter 2, we introduced special relativity (SR) and argued that it alone strongly suggests a block universe, which is the ontological basis of our LSU approach. In chapter 3, we introduced general relativity (GR), argued that it also strongly suggests a block universe, and then showed how several mysteries related to GR were caused by NSU thinking and could be resolved with an LSU approach. In chapter 4, we introduced quantum mechanics (QM) and its many mysteries caused by NSU thinking, then showed how our particular RBW model of the LSU approach could resolve these QM mysteries. To that point, RBW didn't require any new physics, just an adynamical approach to current physics. In chapter 5, we showed how the history of particle physics evolved per the dynamical universe culminating in quantum field theory (QFT) with its NSU characterization. At this point, we reached the impasse of particle physics with respect to unification and quantum gravity (QG). Again, we attributed this impasse to NSU thinking and suggested a new adynamical way to view particle physics per RBW. That means new physics, so in this chapter we will outline our speculative, RBW approach to

Beyond the Dynamical Universe. Michael Silberstein, W.M. Stuckey and Timothy McDevitt, Oxford University Press (2018). © Michael Silberstein, W.M. Stuckey and Timothy McDevitt. DOI 10.1093/oso/9780198807087.001.0001

unification, quantum gravity (QG), dark matter, dark energy, and the black hole firewall paradox.

In the main thread and in the philosophy of physics thread we will motivate the need for QG (the reconciliation of GR and QFT) having already raised the worry for unification in chapter 5. We will also outline the RBW approach to QG in the main thread. RBW's contextual approach to QG is very different than popular approaches whereby one is forced to assume that high-energy phenomena are by definition more fundamental than low energy phenomena. Consequently, the resolution of the dark energy problem is squarely in the purview of QG, as in RBW. Thus, in the main thread, we will use our account of QG to explain dark energy data via modified GR (as first introduced in chapter 3, details in the following). That is, disordered locality as per RBW, rather than some inexplicable new form of energy, will be the explanation for the so-called the dark energy-related phenomenon. Dark matter phenomena are handled via the contextuality already inherent in GR. The computational details for those data fits are provided in Foundational Physics for Chapter 6. Finally, in the main thread we will provide a resolution of the black hole firewall paradox, a paradigm case where GR and QFT come into conflict when using dynamical explanation in the mechanical universe. While we have striven to keep the main thread of each chapter accessible to advanced undergraduates, the nature of the subject (QG) profoundly constrains our ability to do so here, for the many reasons that will be articulated in Philosophy of Physics for Chapter 6.

In Philosophy of Physics for Chapter 6, we will situate RBW with respect to some other discrete accounts of QG and explicate our RBW approach to QG in particular by arguing the current impasse is once again caused by the moribund nature of the NSU approach. That is, we will suggest why, perhaps, current accounts flounder as a result of making certain assumptions endemic to the dynamical/mechanical paradigm. Finally, we argue that an adynamical global constraint as the basis for QG in the block universe provides a self-vindicating unification of physics.

6.2   Explaining the need for quantum gravity

Let us start with a reminder about why we need a theory of QG in the first place. The standard ways of describing and explaining the need for QG are as follows:

> [General relativity and quantum field theory] become meaningless in the regimes where relativistic quantum gravitational effects are expected to become relevant. These effects are not currently observed; they are negligible at currently accessible scales and are expected to become relevant only in extreme physical regimes. For instance, they should govern the end of the evaporation of black holes, the beginning of the life of the universe near the Big Bang, and any measurement involving an extremely short length scale (the "Planck scale") or a very high energy. "Quantum gravity" is the name given to the theory-to-be-found that should describe these regime.

[Rovelli, 2007, p. 1287]

There is a paradox at the heart of our understanding of the physical world. General relativity and quantum mechanics, the two jewels that the twentieth century has left us, have been prolific in gifts—for comprehending the world and for today's technology. From the first of these, cosmology has been developed, as well as astrophysics, the study of gravitational waves, and of black holes. The second has provided the foundations for atomic physics, nuclear physics, the physics of elementary particles and of condensed matter, and of much else besides. And yet between the two theories, there is something that grates. They cannot both be true, at least not in their present forms, because they appear to contradict each other. The gravitational field is described without taking quantum mechanics into account, without accounting for the fact that fields are quantum fields—and quantum mechanics is formulated without taking into account the fact that spacetime curves and is described by Einstein's equations.

[Rovelli, 2017, p. 147]

What are the main motivations for developing a quantum theory of gravity? Since there are currently no experimental hints, the main reasons are conceptual. Within general relativity, one can prove singularity theorems which show that the theory is incomplete: under very general conditions, singularities are unavoidable, cf. Hawking and Penrose (1996). Singularities are borders of spacetime beyond which geodesics [paths of extremal length] cannot be extended; the proper time of any observer comes to an end. . . . Singularities . . . can have diverging curvatures and energy densities. In fact, the latter situation seems to be realized in two important physical cases: the Big Bang and black holes. . . . One needs a more comprehensive theory to understand these situations, and the general expectation is that a quantum theory of gravity is needed, in analogy to quantum mechanics in which the classical instability of atoms has disappeared. The origin of our universe cannot be described without such a theory, so cosmology remains incomplete without quantum gravity.

[Kiefer, 2011, p. 665]

GR, which is the field theory that describes gravity when we can disregard its quantum properties, has changed our understanding of space and time. QM, which has replaced classical mechanics as our general theory of motion, has modified the notions of matter, field and causality. At present, we haven't yet found a consistent conceptual frame in which these modifications make sense together. Thus, our understanding of the physical world is currently badly fragmented. In spite of its empirical effectiveness, fundamental physics is in a phase of deep conceptual confusion. The problem of QG is to combine the insights of GR and QM into a conceptual scheme in which they can coexist.

[Rovelli, 2007, p. 1287]

6.3   RBW on quantum gravity and unification

While RBW is certainly applicable in principle to the phenomena and regimes mentioned by Rovelli, a couple of things need to be stated that will be reinforced in this chapter. First, as the reader has already seen in chapter 3, the RBW account of phenomena related to singularities such as in black holes and the Big Bang is going to be very different to any account hailing from the dynamical/mechanical paradigm. Second, again, because of our adynamical approach in RBW, QG-related phenomena are not restricted to short length scales and high-energy regimes, so empirical evidence for RBW's version of QG exists in the form of the phenomenon dubbed "dark energy."

We don't have a complete theory of QG, for example, we still need the action for modified GR (see 1) as justified by "disordered locality" per RBW, but we are assuming that this action will produce only a "small" correction to the un-modified GR solution so as to avoid its implications of exotic new (dark) energy. We have used this assumption to find the functional form of a metric perturbation ("small correction to GR solution") that results in a reasonable fit of the Supernova Cosmology Project (SCP) Union2/2.1 supernovae type Ia data, most commonly explained by a cosmological constant in GR and called "dark energy." That fit is shown graphically in this main thread and the calculations are summarized in Foundational Physics for Chapter 6. Our next goal is to use this empirically deduced perturbation as a clue in our search for the action proper. This fundamental action would be the basis for our account of QG.

To summarize, in this chapter, we will provide our adynamical-global, constraint-based account of QG and unification and show that it has the following consequences:

1. As a starting point for building a theory of QG, the assumption that the particles and forces of *high-energy* physics are the most fundamental onto-logical elements of objective reality, as one might infer from the Standard Model of particle physics and unification, is false.

2. General relativity doesn't need to be quantized to obtain quantum gravity, as in loop quantum gravity.

3. Galactic rotation curves, X-ray cluster mass profiles, gravitational lensing results, and the angular power spectrum of the cosmic microwave background do not necessarily indicate the existence of non-baryonic dark matter, despite popularized accounts otherwise.

4. The SCP Union2/2.1 supernovae type Ia data do not necessarily entail the accelerated expansion of the universe and a cosmological constant, dark energy, exactly as Perlmutter acknowledges [Chodos, 2011], despite the 2011 Nobel Prize in Physics citation to the contrary.

5. An adynamical resolution of the black hole firewall paradox is provided.

In proposing a radical new adynamical direction for future physics, we are not claiming that the current dynamical approach is wrongheaded or misguided. On the contrary, in chapters 3, 4, and 5 we showed how GR, QM, and QFT are all well-founded according to RBW. Just as GR, QM, and QFT did not replace Newtonian physics where it was already successful, RBW will not replace GR,

QM, or QFT where they are already successful. Given our dynamical experience and a dedicated, critical, and empirical approach to the study of nature, GR, QM, and QFT were probably inevitable and, like Newtonian physics, will continue to serve successfully in modeling a great number of phenomena. What we're claiming is that the *explanation* of those phenomena will simply have to be given the proper LSU context in order to further future physics and overcome the impasse of foundational physics and philosophy of physics.

So, in that spirit, let us begin what is admittedly a speculative description of that proposed new adynamical direction for future physics here in the main thread of chapter 6. Keep in mind that we are not proposing an *evolutionary* change to physics, for example, adding a new field to the Lagrangian density of particle physics. We are proposing a *revolutionary* change to physics via "disordered locality" which violates what it means to be "nearest neighbors" in the differential manifold sense (Figure 6.3), and the contextuality of mass, that is, that matter can simultaneously possess different values of mass by virture of different contexts. This form of contextuality is already inherent in GR, but to our knowledge it has not previously been employed to explain dark matter phenomena where Newtonian mechanistic thinking with mass as an intrinsic property of matter is the norm. Since we implement disordered locality graphically, we have to modify discretized graphical GR (called "Regge calculus") first introduced in chapter 3, and discretized graphical QFT (called "lattice gauge theory") which we briefly introduced in chapter 5. While both formalisms are well known to those who work on QG, neither are mainstream. We will say a little more about the background of these formalisms and how they relate to RBW in the philosophy of physics thread for this chapter. So, do not be worried if you don't fully grasp our modified Regge calculus and modified lattice gauge theory from this main thread explanation. We are simply trying to convey that buying into the LSU approach provides a new path to future physics.

First, as we will explain later in this chapter, as per RBW QG phenomena aren't restricted to extremely large energy densities or spacetime curvatures but are also associated with quantum exchanges over large spacetime scales. Thus, the dark energy phenomenon actually constitutes an experimental hint of QG, in our view. Second, we assume the graphical and discrete version of GR called "Regge calculus" [Regge, 1961] is fundamental to GR, not the converse as per conventional thinking. As noted, in Regge calculus, the metric and stress–energy tensor reside on the links of the graph, so there are no infinite energy densities or curvatures for any finite matter one might want to associate with a given node. This necessarily avoids the singularities of the Big Bang and black holes resulting from the continuum approach of GR. Third, when one obtains a Regge cosmology solution, they retro-time-evolve the spatial hypersurface of homogeneity. This is similar to the notion of extending geodesics into the past which leads to the singularity theorems of Penrose and Hawking in GR, but as we explained in Foundational Physics for Chapter 3, in Regge calculus the restrospective time evolution of the spatial hypersurface "freezes up" when adjacent nodes acquire a large relative velocity. This is called the "stop point problem" in Regge cosmology because the solution ceases to follow the GR counterpart. Of course, this is not a computational problem for Regge cosmology since one can regularly check the speed of adjacent nodes in the computational algorithm and refine the size of the lattice to ensure the speed between nodes remains small.

However, in RBW the graphical approach is fundamental, so lattice refinements are not mere mathematical adjustments but instead would constitute new "mean" configurations of matter–energy. Such refinements are certainly required in earlier cosmological eras, but one would expect that there exists a smallest spatial scale (associated with smallest nodal matter) so that eventually (evolving backward in time) the stop point could not be avoided and the minimum graph spacing would be reached. This justifies a choice for non-zero a(t) in the solution of Einstein's equations for cosmology.

At this point, one would have to avoid asking "why is **this** the initial state of the universe?" As we explained in chapter 3, no point on the spacetime manifold M is more or less explanatory than any other point in the block universe view. The stop point of the computational method is not of any ontological or epistemological import. Thus, Regge calculus cosmology coupled with the God's-eye view naturally avoids the initial singularity of its GR counterpart, precisely as expected of QG. So, what exactly is our RBW approach to QG?

In Stuckey et al. [2016c], we showed that the relations matrix **K** of our modified lattice gauge theory (MLGT) satisfies RBW's adynamical global constraint (AGC) for the Klein–Gordon, Dirac, Maxwell, and Einstein–Hilbert graphical actions and can be extended to the graphical action for the Standard Model of particle physics (as summarized in chapter 5). While this is the formal heart of QG and unification for RBW, there isn't space to reproduce this here, so we will provide only a summary.

To introduce our approach to QG briefly, recall in Foundational Physics for Chapter 4 we showed how the AGC applies to a spacetimesource element per the Schrödinger graphical action using the non-relativistic limit of the Klein–Gordon action for scalar fields on nodes. In Stuckey et al. [2016c], we showed similarly how the AGC applies to a spacetimesource element per the Einstein–Hilbert action (used for GR without matter–energy added). Therein, a first-order metric perturbation $h_{\alpha\beta}$ ("small correction to the GR solution") is a scalar field on the plaquettes (faces) of the (hypercubical) spacetimesource element and the eigenbasis representing exchanges between faces (plaquettes) of the hypercube mirrors the eigenbasis for exchanges between nodes of the rectangular spacetimesource element shown in Figures 4.21, 4.22, 4.23, and 4.24 for the twin-slit experiment. That is, instead of exchanges between nodes of a rectangular spacetimesource element, we have exchanges between plaquettes of a hypercubical spacetimesource element.

These exchanges represent small changes to the spacetime geometry and therefore constitute a different configuration for the graphical action of modified Regge calculus (the classical context). Modified lattice gauge theory would then be used to compute the probability amplitudes for the different possible graphical configurations. The requirement of a classical context for quantum physics avoids the "problem of time" introduced in chapter 1 for canonical approaches to QG, because the proper time of the GR spacetime context is used for establishing time between measurement events, so the GR proper time is "external" to the quantum system being measured just as Newtonian time is for QM and SR time is for QFT. That is, the quantum measurement events are events in the classical spacetime of GR, so their distribution in the classical spacetime of GR is just as unambiguous as any other events in the classical spacetime of GR. This in a nutshell is QG according to RBW. Now we provide a summary of our approach to unification that will return us to the RBW resolution of the dark matter and dark energy problems.

As we stated in chapter 5 concerning our graphical approach, the role of the field is different than in the continuum approach of QFT where it pervades otherwise empty, continuous space to mediate the exchange of matter–energy between sources. In MLGT, as in lattice gauge theory, the field is a scalar or vector associated with nodes, links, or plaquettes on the graph. Specifically, the Schrödinger and Klein-Gordon actions use scalars on nodes, the Dirac action uses vectors on nodes, the Maxwell action uses scalars on links, and the Einstein-Hilbert action uses scalars on plaquettes. As with lattice gauge theory, MLGT notes that the graphical action involves differences between vectors at different nodes on the graph. In general, taking the difference between vectors residing in distinct vector spaces requires a rule for aligning the vector spaces, i.e., parallel transport (see Foundational Physics for Chapter 5). The SU(N) gauge field on links of the Standard Model graph fills this role when there are N vectors at each node. While we do not subscribe to the existence of "click-causing entities" moving through space, we agree that clicks trace classical paths, as we noted in the main thread of chapter 5. Indeed, this is the basis for our approach and consequently the results and analyses of particle physics experiments are very important. Perhaps we should quickly review the relationship between quantum and classical phenomena to emphasize this point.

As we explained in chapters 4 and 5, respectively, classical mechanics gives the classical particle path in spacetime as an extremum of the Feynman path integral for QM and classical field theory gives the extremum distribution of field values on the graph of lattice gauge theory. The difference between classical phenomena and quantum phenomena resides in the probability of their occurrence, so that individual detector clicks are classical when they occur with probability ~ 1 and quantum otherwise. Thus, classical physics is understood probabilistically from quantum physics as would be expected. This classical-quantum ontology follows from the classical-quantum probability. That is, quantum probability allows for the cancellation of possibilities per the Born rule (add amplitudes then square) while classical probability, which assumes classical objects and therefore already assumes quantum probability ~ 1 for each object, does not allow for the cancellation of possibilities (square amplitudes then add). Consequently, when dealing with "approximately 100,000 individual measurements of either energy or spatial information" [Frisch, 1993] (Figure 5.3) in a particle physics "event," it is best to assign individual detector clicks to classical particle tracks (called "pattern recognition") and let the individual tracks constitute the fundamental constituents. This can be done easily since the location of the second detector click for a given trajectory already has quantum probability ~ 1, as shown theoretically by [Mott, 1929] and verified empirically in countless experiments. Thus, only the collection of first clicks (one for each trajectory) needs to be determined by QFT, i.e., QFT is determining the probability for the number and type of particles in the event. And, what we mean by "type of particle" is obtained when the tracks (worldlines) are parameterized per classical physics to obtain their dynamical characteristics (called "geometrical fitting") [Fernow, 1986, §1.7.1, 1.7.2, & 1.7.3]. The properties (mass, charge, momentum, energy, etc.) for a given worldline are defined relationally in the context of the accelerator Source and particle detector. This is the process by which worldlines are inferred from the individual detector clicks in particle physics detectors and the worldlines constitute what we mean by time-evolved "classical objects." Thus, as with QM, QFT depends on the spatiotemporal context of classical physics even while its ontology of spacetimesource elements underwrites the classical ontology, that is, we have spatiotemporal ontological contextuality.

With this understanding of the Standard Model and particle physics we see that the next logical addition to our collection of fundamental spacetimesource elements would be those constructed from the gradient (differencing) of vector fields on links. The scalar field on plaquettes (basis for graviton, i.e., the particle carrying the gravitational force in standard lattice gauge theory) would define parallel transport for the vector fields on links in the manner scalar fields on links define parallel transport for the vector fields on nodes. This is the standard approach to QG in the particle physics community. The problem with this is, of course, that we simply have gravitons in flat spacetime; we still need spacetime curvature as in GR. In our view, since MLGT is necessarily contextual, that is accomplished by understanding the context of the properties in question as in the simplices of Regge calculus (Figure 6.1).

According to Regge calculus, gravity is a scalar field on the plaquettes of its simplices, that is, Newtonian gravity in flat spacetime, and spacetime curvature (variable spacetime geometry) is accounted for via a "deficit angle" between simplices in the global structure (Figure 6.2). This spacetime curvature is a function of the matter–energy content of spacetime to include all forms, not just the graviton. Thus, if we were able to construct experiments with individual gravitons, a quantum gravity experiment could be the simple twin-slit experiment with gravitons in flat spacetime, that is, in a single simplex. On the other hand, we could view graviton interference patterns generated between a Source and detector that don't occupy the same simplex, in which case the spacetimesource element for the exchange of gravitational energy would have simplex-to-simplex segments between Source and detection event. To compute the amplitude for the exchange of gravitational energy associated with that spacetimesource element, one would pick up phase factors associated with the deficit angles between simplices, just

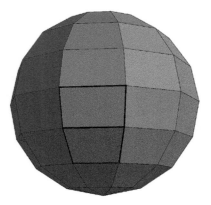

Figure 6.1 *In Regge calculus, small flat sections called "simplices" are pieced together to approximate the curved manifold. Typically, simplices are "triangular" rather than "rectangular," as shown here.*

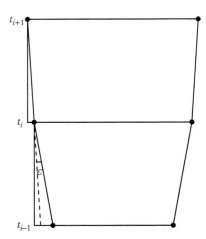

Figure 6.2 *Curvature manifests itself as a "deficit angle" ε between adjacent simplices. Unlike this figure, 2D simplices are typically triangular in Regge calculus.*

as a photon amplitude picks up phase factors associated with reflection from mirrors and beam splitters as computed between Source and detection event in an interferometer.

While we have finally succeeded in measuring gravitational waves [Abbott et al., 2016], we don't yet have the technology to manipulate individual gravitons. Thus,

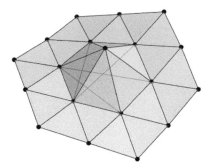

Figure 6.3 *A direct connection between non-neighboring nodes (in the context of the continuum manifold approximation) on the graph is called "disordered locality."*

we might instead explore the effect of variable geometry on spacetimesource elements connecting different simplices for the exchange of photons (as on astrophysical scales). The direct connection between non-neighboring points on the corresponding classical spacetime manifold is referred to as "disordered locality," as we introduced in Philosophy of Physics for Chapter 5 [Caravelli and Markopoulou, 2012] (Figure 6.3).[1] Since different energy distributions mean different spacetime geometries per Regge calculus, an exchange of energy, quantum or otherwise, has geometric implications.[2] And, since geometry affects the value of the action in Regge calculus, one should expect that energy lost at the Source will not only reappear at the detection event but will also appear in the spacetime geometry, as compared to the geometry where the photon energy exchange is not taken into account. That quantum mechanical phenomena might have non-negligible impact on the spacetime geometry of GR is not a new idea. Maldacena writes:

> Although we identified the connection between wormholes and entangled states using black holes, it is tempting to speculate that the link is more general—that whenever we have entanglement we have a kind of geometric connection. This expectation should hold true even in the simplest case, in which we have only two entangled particles. In such situations, however, the spatial connection could involve tiny quantum structures that would not follow our usual notion of geometry. We still do not know how to describe these microscopic geometries, but the entanglement of these structures might somehow give rise to spacetime itself.
>
> [Maldacena, 2016, p. 31]

---

[1] Of course, direct action doesn't entail disordered locality in the globally flat spacetime of special relativity, since every location is local to every other location in a globally flat spacetime per the differential geometry sense.

[2] While GR is typically viewed as providing a geometrization of gravity, Weinberg and others view gravity as a tensor field, hence the talk about gravitons. Notice that in RBW that dualism evaporates.

Scalar field on nodes	One vector each node	One vector each link
Scalar field on links	Two vectors each node	Two vectors each link
Scalar field on plaquettes	Three vectors each node	Three vectors each link

Figure 6.4 *Unification scheme per RBW.*

Of course, we are not "building" spacetime in "Lego" fashion with our space-timesource elements; rather, these "quantum structures" that form a "geometric connection" in "even the simplest case" require the context of a GR spacetime manifold per our spatiotemporal ontological contextuality (as described in Philosophy of Physics for Chapter 4).

At this point the reader should appreciate that underwriting interacting classical objects via spacetimesource elements leads to a relatively simple picture of unification (Figure 6.4) compared to that based on fundamental particles (Standard Model Lagrangian density; credit T. D. Gutierrez as transcribed from the appendices of *Diagrammatica* [Veltman, 1994]; two missing parentheses were added):

$$-\tfrac{1}{2}\partial_\nu g^a_\mu \partial_\nu g^a_\mu - g_s f^{abc}\partial_\mu g^a_\nu g^b_\mu g^c_\nu - \tfrac{1}{4}g^2_s f^{abc}f^{ade}g^b_\mu g^c_\nu g^d_\mu g^e_\nu + \tfrac{1}{2}ig^2_s(\bar{q}^\sigma_i \gamma^\mu q^\sigma_j)g^a_\mu + \bar{G}^a\partial^2 G^a +$$
$$g_s f^{abc}\partial_\mu \bar{G}^a G^b g^c_\mu - \partial_\nu W^+_\mu \partial_\nu W^-_\mu - M^2 W^+_\mu W^-_\mu - \tfrac{1}{2}\partial_\nu Z^0_\mu \partial_\nu Z^0_\mu - \tfrac{1}{2c^2_w}M^2 Z^0_\mu Z^0_\mu - \tfrac{1}{2}\partial_\mu A_\nu \partial_\mu A_\nu -$$
$$\tfrac{1}{2}\partial_\mu H \partial_\mu H - \tfrac{1}{2}m^2_h H^2 - \partial_\mu \phi^+ \partial_\mu \phi^- - M^2\phi^+\phi^- - \tfrac{1}{2}\partial_\mu \phi^0 \partial_\mu \phi^0 - \tfrac{1}{2c^2_w}M\phi^0\phi^0 - \beta_h[\tfrac{2M^2}{g^2} + \tfrac{2M}{g}H +$$
$$\tfrac{1}{2}(H^2 + \phi^0\phi^0 + 2\phi^+\phi^-)] + \tfrac{2M^4}{g^2}\alpha_h - igc_w[\partial_\nu Z^0_\mu(W^+_\mu W^-_\nu - W^+_\nu W^-_\mu) - Z^0_\nu(W^+_\mu \partial_\nu W^-_\mu -$$
$$W^-_\mu \partial_\nu W^+_\mu) + Z^0_\mu(W^+_\nu \partial_\nu W^-_\mu - W^-_\nu \partial_\nu W^+_\mu)] - igs_w[\partial_\nu A_\mu(W^+_\mu W^-_\nu - W^+_\nu W^-_\mu) - A_\nu(W^+_\mu \partial_\nu W^-_\mu -$$
$$W^-_\mu \partial_\nu W^+_\mu) + A_\mu(W^+_\nu \partial_\nu W^-_\mu - W^-_\nu \partial_\nu W^+_\mu)] - \tfrac{1}{2}g^2 W^+_\mu W^-_\mu W^+_\nu W^-_\nu + \tfrac{1}{2}g^2 W^+_\mu W^-_\nu W^+_\mu W^-_\nu +$$
$$g^2 c^2_w(Z^0_\mu W^+_\mu Z^0_\nu W^-_\nu - Z^0_\mu Z^0_\mu W^+_\nu W^-_\nu) + g^2 s^2_w(A_\mu W^+_\mu A_\nu W^-_\nu - A_\mu A_\mu W^+_\nu W^-_\nu) +$$
$$g^2 s_w c_w[A_\mu Z^0_\nu(W^+_\mu W^-_\nu - W^+_\nu W^-_\mu) - 2A_\mu Z^0_\mu W^+_\nu W^-_\nu] - g\alpha[H^3 + H\phi^0\phi^0 + 2H\phi^+\phi^-] -$$
$$\tfrac{1}{8}g^2\alpha_h[H^4 + (\phi^0)^4 + 4(\phi^+\phi^-)^2 + 4(\phi^0)^2\phi^+\phi^- + 4H^2\phi^+\phi^- + 2(\phi^0)^2H^2] - gMW^+_\mu W^-_\mu H -$$
$$\tfrac{1}{2}g\tfrac{M}{c^2_w}Z^0_\mu Z^0_\mu H - \tfrac{1}{2}ig[W^+_\mu(\phi^0\partial_\mu \phi^- - \phi^-\partial_\mu \phi^0) - W^-_\mu(\phi^0\partial_\mu \phi^+ - \phi^+\partial_\mu \phi^0)] + \tfrac{1}{2}g[W^+_\mu(H\partial_\mu \phi^- -$$
$$\phi^-\partial_\mu H) - W^-_\mu(H\partial_\mu \phi^+ - \phi^+\partial_\mu H)] + \tfrac{1}{2}g\tfrac{1}{c_w}(Z^0_\mu(H\partial_\mu \phi^0 - \phi^0\partial_\mu H) - ig\tfrac{s^2_w}{c_w}MZ^0_\mu(W^+_\mu \phi^- -$$
$$W^-_\mu \phi^+) + igs_w MA_\mu(W^+_\mu \phi^- - W^-_\mu \phi^+) - ig\tfrac{1-2c^2_w}{2c_w}Z^0_\mu(\phi^+\partial_\mu \phi^- - \phi^-\partial_\mu \phi^+) + igs_w A_\mu(\phi^+\partial_\mu \phi^- -$$
$$\phi^-\partial_\mu \phi^+) - \tfrac{1}{4}g^2 W^+_\mu W^-_\mu[H^2 + (\phi^0)^2 + 2\phi^+\phi^-] - \tfrac{1}{4}g^2\tfrac{1}{c^2_w}Z^0_\mu Z^0_\mu[H^2 + (\phi^0)^2 + 2(2s^2_w - 1)^2\phi^+\phi^-] -$$
$$\tfrac{1}{2}g^2\tfrac{s^2_w}{c_w}Z^0_\mu \phi^0(W^+_\mu \phi^- + W^-_\mu \phi^+) - \tfrac{1}{2}ig^2\tfrac{s^2_w}{c_w}Z^0_\mu H(W^+_\mu \phi^- - W^-_\mu \phi^+) + \tfrac{1}{2}g^2 s_w A_\mu \phi^0(W^+_\mu \phi^- + W^-_\mu \phi^+) +$$
$$\tfrac{1}{2}ig^2 s_w A_\mu H(W^+_\mu \phi^- - W^-_\mu \phi^+) - g^2\tfrac{s_w}{c_w}(2c^2_w - 1)Z^0_\mu A_\mu \phi^+\phi^- - g^1 s^2_w A_\mu A_\mu \phi^+\phi^- - \bar{e}^\lambda(\gamma\partial + m^\lambda_e)e^\lambda -$$
$$\bar{\nu}^\lambda \gamma\partial \nu^\lambda - \bar{u}^\lambda_j(\gamma\partial + m^\lambda_u)u^\lambda_j - \bar{d}^\lambda_j(\gamma\partial + m^\lambda_d)d^\lambda_j + igs_w A_\mu[-(\bar{e}^\lambda \gamma^\mu e^\lambda) + \tfrac{2}{3}(\bar{u}^\lambda_j \gamma^\mu u^\lambda_j) - \tfrac{1}{3}(\bar{d}^\lambda_j \gamma^\mu d^\lambda_j)] +$$
$$\tfrac{ig}{4c_w}Z^0_\mu[(\bar{\nu}^\lambda \gamma^\mu(1 + \gamma^5)\nu^\lambda) + (\bar{e}^\lambda \gamma^\mu(4s^2_w - 1 - \gamma^5)e^\lambda) + (\bar{u}^\lambda_j \gamma^\mu(\tfrac{4}{3}s^2_w - 1 - \gamma^5)u^\lambda_j) + (\bar{d}^\lambda_j \gamma^\mu(1 -$$

$$\tfrac{8}{3}s_w^2 - \gamma^5)d_j^\lambda)] + \tfrac{ig}{2\sqrt{2}}W_\mu^+[(\bar{v}^\lambda\gamma^\mu(1+\gamma^5)e^\lambda) + (\bar{u}_j^\lambda\gamma^\mu(1+\gamma^5)C_{\lambda\kappa}d_j^\kappa)] + \tfrac{ig}{2\sqrt{2}}W_\mu^-[(\bar{e}^\lambda\gamma^\mu(1+$$

$$\gamma^5)v^\lambda) + (\bar{d}_j^\kappa C_{\lambda\kappa}^\dagger\gamma^\mu(1+\gamma^5)u_j^\lambda)] + \tfrac{ig}{2\sqrt{2}}\tfrac{m_e^\lambda}{M}[-\phi^+(\bar{v}^\lambda(1-\gamma^5)e^\lambda) + \phi^-(\bar{e}^\lambda(1+\gamma^5)v^\lambda)] -$$

$$\tfrac{g}{2}\tfrac{m_e^\lambda}{M}[H(\bar{e}^\lambda e^\lambda) + i\phi^0(\bar{e}^\lambda\gamma^5 e^\lambda)] + \tfrac{ig}{2M\sqrt{2}}\phi^+[-m_d^\kappa(\bar{u}_j^\lambda C_{\lambda\kappa}(1-\gamma^5)d_j^\kappa) + m_u^\lambda(\bar{u}_j^\kappa C_{\lambda\kappa}(1+\gamma^5)d_j^\kappa)] +$$

$$\tfrac{ig}{2M\sqrt{2}}\phi^-[m_d^\lambda(\bar{d}_j^\lambda C_{\lambda\kappa}^\dagger(1+\gamma^5)u_j^\kappa) - m_u^\kappa(\bar{d}_j^\lambda C_{\lambda\kappa}^\dagger(1-\gamma^5)u_j^\kappa)] - \tfrac{g}{2}\tfrac{m_u^\lambda}{M}H(\bar{u}_j^\lambda u_j^\lambda) - \tfrac{g}{2}\tfrac{m_d^\lambda}{M}H(\bar{d}_j^\lambda d_j^\lambda) +$$

$$\tfrac{ig}{2}\tfrac{m_u^\lambda}{M}\phi^0(\bar{u}_j^\lambda\gamma^5 u_j^\lambda) - \tfrac{ig}{2}\tfrac{m_d^\lambda}{M}\phi^0(\bar{d}_j^\lambda\gamma^5 d_j^\lambda) + \bar{X}^+(\partial^2 - M^2)X^+ + \bar{X}^-(\partial^2 - M^2)X^- + \bar{X}^0(\partial^2 - \tfrac{M^2}{c_w^2})X^0 +$$

$$\bar{Y}\partial^2 Y + igc_w W_\mu^+(\partial_\mu\bar{X}^0 X^- - \partial_\mu\bar{X}^+ X^0) + igs_w W_\mu^+(\partial_\mu\bar{Y}X^- - \partial_\mu\bar{X}^+ Y) + igc_w W_\mu^-(\partial_\mu\bar{X}^- X^0 -$$

$$\partial_\mu\bar{X}^0 X^+) + igs_w W_\mu^-(\partial_\mu\bar{X}^- Y - \partial_\mu\bar{Y}X^+) + igc_w Z_\mu^0(\partial_\mu\bar{X}^+ X^+ - \partial_\mu\bar{X}^- X^-) + igs_w A_\mu(\partial_\mu\bar{X}^+ X^+ -$$

$$\partial_\mu\bar{X}^- X^-) - \tfrac{1}{2}gM[\bar{X}^+ X^+ H + \bar{X}^- X^- H + \tfrac{1}{c_w^2}\bar{X}^0 X^0 H] + \tfrac{1-2c_w^2}{2c_w}igM[\bar{X}^+ X^0\phi^+ - \bar{X}^- X^0\phi^-] +$$

$$\tfrac{1}{2c_w}igM[\bar{X}^0 X^-\phi^+ - \bar{X}^0 X^+\phi^-] + igMs_w[\bar{X}^0 X^-\phi^+ - \bar{X}^0 X^+\phi^-] + \tfrac{1}{2}igM[\bar{X}^+ X^+\phi^0 - \bar{X}^- X^-\phi^0]$$

However, while we do not view particle physics as the study of what is ultimately fundamental in nature it will no doubt prove essential to understanding how the fundamental elements of spacetimesource are to be constructed and combined in accord with how the fields contribute to the Lagrangian density in the action of the transition amplitude. For example, the conserved properties of particles such as baryon number, electron lepton number, strangeness, and charge, must ultimately find counterparts in the graphical properties of MLGT. Thus, the results of the decades long, systematic investigation of particle physics will almost certainly provide information necessary to complete the adynamical unification picture of RBW (or whatever adynamical approach ultimately succeeds). As for QG in the context of the infinite energy densities of (continuum) classical physics where is it typically assumed to be needed, there is no such concern in a graphical approach to gravity, as we showed in chapter 3. Again, this is trivially true since matter can be placed at a point (node) without producing a singularity.

6.4 Black hole firewall paradox

To help understand our approach to QG, consider the black hole firewall paradox [Ouellette, 2012; Almheiri *et al.*, 2013]. Here is the paradox in a nutshell, as it follows from dynamical thinking.[3] According to the Hamiltonian description of particle physics (the description employing fields in space), virtual particle pairs are created from the vacuum (ground) state throughout space all the time. Normally, the pair recombines and annihilates in very short order without consequence, but near the event horizon of a black hole it is possible that one of the pair falls into the black hole and the other escapes to infinity. The particle

---

[3] There is widespread disagreement as to whether or not all of the assumptions leading to this paradox are reasonable, but we are not here to argue for or against those assumptions. We are simply explaining the paradox as it follows from a particular dynamical perspective and resolving it with our adynamical approach.

that falls into the hole carries negative mass causing the black hole to shrink a bit while its partner, who escaped to infinity, carries the same amount of positive mass. Thus, the net effect is that a particle tunneled out of the black hole. Of course, it is not supposed to be possible for anything to escape a black hole but in QM particles can pass through potential energy barriers even though they don't have enough kinetic energy. Classical mechanics says this can't happen but quantum mechanics says there is a small probability of it happening and experiments vindicate the QM prediction of "quantum tunneling." Indeed, there are even circuit elements called tunnel diodes that function on this principle. The radiation that is emitted by a black hole via this process is called "Hawking radiation" and in order to avoid the "black hole information paradox,"[4] it must be entangled with the other particles emitted before it, at least in this particular dynamical rendering of the paradox. But, this can't happen because the outgoing particle is still entangled with the infalling particle and a particle cannot become entangled with two independent systems at the same time (infalling particle and Hawking radiation emitted previously), a principle called "monogamy of entanglement."

Therefore, it must be the case that the entanglement between the infalling particle and its outgoing partner is broken immediately after the pair is created. If so, this would release an enormous amount of energy just outside the event horizon. This region of very high energy is called a "black hole firewall." Unfortunately, this solution to the information paradox violates the equivalence principle of GR. According to the equivalence principle, an object in free fall through the event horizon should not encounter anything different than being in a zero gravitational field far from any mass, so an infalling object should experience a smooth journey.[5] But, the firewall is a very extreme deviation from such a smooth journey indeed! As reported in *Nature*:

> Steve Giddings, a quantum physicist at the UCSB, describes the situation as "a crisis in the foundations of physics that may need a revolution to resolve" . . . The firewall idea "shakes the foundations of what most of us believed about black holes," said Raphael Bousso, a string theorist at the University of California, Berkeley, as he opened his talk at the meeting. "It essentially pits quantum mechanics against general relativity, without giving us any clues as to which direction to go next."
>
> [Merali, 2013]

So, how do we avoid the black hole firewall paradox?

Recall Feynman's quotation from chapter 4: "From the overall space-time point of view of the least action principle, the field disappears as nothing but bookkeeping variables insisted on by the Hamiltonian method" (Feynman in Brown,

---

[4] Information, as quantified by entropy for example, cannot simply disappear without violating the second law of thermodynamics and unitary evolution in quantum physics. So, when black holes evaporate the information they swallowed up must be recovered in some fashion.

[5] Here the assumption is that the black hole is massive enough that tidal forces are negligible.

2005b, p. xv). Thus, in the case of Hawking radiation, the infalling particle and outgoing particle don't exist; there is only a Source emission event inside the event horizon and a sink (detection) event somewhere outside the event horizon. The Hamiltonian bookkeeping method requires two time-evolved entangled particles proceeding from just outside the event horizon, but in the Lagrangian Schema there is only a single emission event and a single detection event, that is, "one particle." This one particle is clearly entangled with (carries information about) the Source, so no information is lost and the paradox is resolved. Does this mean that QFT and GR escape unscathed? No. This situation clearly requires disordered locality, which entails a correction to the action of Regge calculus as we described earlier. While we don't expect to acquire observational data from Hawking radiation anytime soon which will offer guidance as to the construction of a Regge calculus action modified as per disordered locality, there are other phenomena that may qualify.

As we explain in section 6.5, dark energy may be an example of disordered locality. There the energy lost by the photon per its cosmological redshift between Source (supernova) and detector (telescope) in the Einstein–deSitter (EdS) cosmology model based on matter alone (no electromagnetic contribution to the spacetime geometry, as shown in chapter 3) will be used to justify a first-order correction to the proper distance between Source and detector in that dust-filled cosmology model. That correction is small (scaled by a factor of $(8.38 \text{ Gcy})^{-1}$), but we show that it allows for a fit of distance modulus versus redshift for the SCP Union2.1 SN Ia supernovae data without accelerating expansion and, therefore, without dark energy. Thus, dark energy is explained via our approach to QG as in disordered locality. So what is important in our graphical approach with direct action is disordered locality and it becomes a concern for relatively small energy exchanges over relatively large spatiotemporal regions. We encounter that situation in astrophysics, so that is where we expect disordered locality according to our version of QG to become important.

6.5   RBW on dark matter and dark energy

So, according to the graphical approach of RBW, an exploration of QG doesn't require enormous energy densities (quite the contrary) and that means QG phenomena are empirically accessible as per RBW. Furthermore, the infinite density in continuum approaches that obtains when finite energy is squeezed into zero volume is automatically avoided in a graphical approach, since one can associate finite matter with a point (node) without creating an infinity. Even if one wants to create a spacetime solution by evolving a spatial hypersurface backward in time, Regge calculus indicates that the solution will asymptotically terminate at the fundamental lattice spacing (determined by whatever physics obtains at that era), thereby avoiding the singularity of its continuum counterpart. Therefore, our adynamical graphical approach to physics puts QG in the realm of empirical science and out of its current metaphysical and formal stalemate; more on this in Philosophy of Physics for Chapter 6. Specifically, in our approach to QG, we believe fits using spacetime metric perturbations of astronomical data associated with disordered locality, such as type Ia supernova data, should guide our search for modified Regge calculus. Regarding dark matter, we use the contextuality already inherent in GR to motivate a metric perturbation of mass for fits of galactic rotation curves, X-ray cluster mass profiles, and the cosmic microwave background (CMB) angular power spectrum.

Thus, our spatiotemporal ontological contextuality allows us to begin with empiricism and work toward producing a corresponding mathematical formalism. This is in stark contrast to most approaches including string theory, loop quantum gravity, causets, and causal dynamical triangulation, whereby the formalism is developed first—as prompted by ontological and methodological reductionism—and the prospect of producing a nexus to empirical data must be left as an end goal (there are no data to work with from their perspective). Accordingly, in this section we will explore our proposed resolution of two enormous problems with the standard cosmology model, viz., dark matter and dark energy, via QG and contextuality per RBW. We start with a review of these problems.

Since the early 1930s, galactic rotation curves (RCs) and galactic cluster masses have been known to deviate from Newtonian expectations based on luminous matter and known mass-to-luminosity ratios [Oort, 1932; Zwicky, 1933, 1937; Rubin and Ford Jr, 1970]. Astronomers compute the mass of a star in relation to its luminosity using assumptions about the physics inside the star, such as hydrostatic equilibrium between radiation pressure outward and gravitational attraction inward. The value of mass obtained by measuring a star's luminosity agrees with the value needed to explain binary stellar orbits, that is, two stars in close orbit around each other. We will refer to this "locally determined" value of mass for the matter of the star as its "dynamic mass." When you use the dynamic mass for many stars in a galaxy or galaxies in a cluster of galaxies, the value is well short of what is needed to explain the galactic or cluster dynamics. We will refer to this "globally determined" value of mass for the matter of the star as its "proper mass."[6] There is also a vast amount of interstellar and intracluster medium (ICM) gas with a locally determined value of mass, but adding this to the stellar dynamic mass is still well short of what is needed. It is as if there is (a lot of) matter in the galaxies and galactic clusters that doesn't give off electromagnetic radiation, hence the term "dark matter."

This is at least the third episode of dark matter in the history of astronomy. Irregularities in the orbit of Uranus caused astronomers to posit the existence of dark matter in the form of a new planet leading to the discovery of Neptune in 1846. And a slight increase (43 arcseconds per century) in Mercury's orbital precession rate above that predicted by Newtonian physics led astronomers to posit the existence of dark matter inside the orbit of Mercury. In this case,

---

[6] As we will see shortly, this term actually applies to the value of mass determined by observers inside the source of the Schwarzschild metric via the interior solution. But, as we will argue, it is this proper mass that then also surrounds the Schwarzschild vacuum in a nested solution giving rise to dark matter phenomena.

however, no such dark matter was found but rather a new theory of gravity (general relativity in 1915) ultimately accounted for the discrepancy. So, the two main approaches to the current dark matter problem involve finding some mysterious new form of matter or revising GR (a new theory of gravity). A seldom considered possibility is that GR already contains a mechanism for explaining dark matter phenomena without having to introduce exotic new matter or modifications to GR proper, and that is the avenue we are exploring via GR's contextuality.

Galactic RCs and galactic cluster masses are just two aspects of the dark matter problem [Garrett and Duda, 2011], there are others we will describe below, but given its unresolved history spanning 80+ years this is one of the most persistent problems in physics. No one disputes the existence of ordinary (so-called baryonic) dark matter such as brown dwarfs, black holes, and molecular hydrogen, but there is wide agreement that baryonic dark matter does not exist in large enough supply to resolve the dark matter problem [Bergstrom, 2000]. Thus, astronomers have posited the existence of non-baryonic dark matter (DM). The galactic and galactic cluster aspects of the dark matter phenomena were joined more recently by a discovery in the angular power spectrum of the cosmic microwave background (CMB).

Specifically, the second and third so-called acoustic peaks of the CMB angular power spectrum are about equal in height (Figure 6.9), which is taken to be strong evidence for the existence of dark matter [Carroll, 2011], as we will explain shortly. As a consequence, many approaches have been brought to bear on the dark matter problem, typically by way of new particles [Munoz, 2004; Feng, 2010], but also by way of modifications to existing theories of gravity, for example, modified Newtonian dynamics (MOND) [Milgrom, 1983, 2015; Sanders and McGaugh, 2002] and relativistic counterparts [Bekenstein, 2004; Sanders, 2005; Zlosnik et al., 2007; Zhao and Li, 2010; Blanchet and Le Tiec, 2008], Moffat's modified gravity (MOG), metric skew-tensor gravity (MSTG) and scalar-tensor-vector gravity (STVG) [Moffat, 2005; Brownstein and Moffat, 2006a,b; Moffat, 2006, 2008; Moffat and Rahvar, 2013, 2014; Israel and Moffat, 2016], renormalization group corrected general relativity [Rodrigues et al., 2012], and non-local general relativity [Mashhoon, 2015]. While the assumption of non-baryonic dark matter in ΛCDM cosmology[7] works well for explaining cosmological features (scales greater than 1 Mpc),[8] there is still no independent verification of DM [Young, 2016], galactic RCs do not conform to the theoretical predictions of ΛCDM for the distribution of DM on galactic scales [Gentile et al., 2005; de Blok, W., 2010; Clifton et al., 2012; McGaugh, 2015b; Wyse, 2017], and there are five seemingly incompatible properties that must be satisfied by DM: dark, cold, abundant, stable, and dissipationless [Carroll, 2015]. Let us briefly review Carroll's argument here.

---

[7] Λ is the cosmological constant explained in chapter 3. "CDM" stands for "cold dark matter."
[8] A parsec (pc) is 3.26 light years (cy). It is the distance to an object that exhibits 1 arcsecond of parallax as Earth orbits the Sun.

We have many particles that are "dark," for example, neutrons; they are simply particles that don't have electric charge and therefore do not interact via the electromagnetic force. DM must be dark simply because we don't find electromagnetic radiation emissions for it. DM must be "cold" because it must settle down to form gravitationally bound stable structures like galaxies and clusters. These first two characteristics tell us that our DM should be a heavy, electrically uncharged particle. To have the abundance of DM shortly after the Big Bang needed to explain the CMB acoustic peaks and the abundance of DM needed to form galaxies and clusters billions of years later, we need DM to interact weakly with massive anti-particles so that the DM and massive anti-particles don't mutually annihilate each other (disappear). Weakly interacting massive particles (WIMPs) fit that bill nicely, but LUX (the Large Underground Xenon dark-matter experiment) announced in July 2016 that its 20-month search had produced no evidence of WIMPs [Young, 2016].

Further complicating this situation, we need this heavy, uncharged, abundant particle to be stable, again so that it persists over billions of years to do its gravitational work. All of the stable particles we know are light, but supposing that there is a heavy stable particle? We need it to carry some sort of conserved property (properties define a particle after all). Conserved quantities result from symmetries of one's action typically associated with long-range forces. So, this new particle would probably have to introduce a new force in order to be stable. This presents considerable difficulties because the final characteristic of DM is that it has to be dissipationless. That is, ordinary matter loses energy through radiation and collisions which allows it to form galaxies while DM stays in large halos around galaxies. Therefore, DM must retain its energy longer than baryonic matter. But, long-range forces dissipate energy. So, we would need an entire "dark matter sector" like "dark photons" and "dark atoms" to interact with each other through its "dark force" while only interacting with baryonic matter via gravity. While Ackerman and colleagues did precisely that, that is, they created a "dark sector physics" [Ackerman et al., 2009], DM still can't lose too much energy via dark photons or it ends up in the galaxies with the baryonic matter instead of in DM halos. These problems prompted Sean Carroll to conclude [Carroll, 2015], "So should we be surprised that we live in a universe full of dark matter? I'm going to say: yes."

In addition to dark matter phenomena, astronomers reported "the discovery of the accelerating expansion of the Universe through observations of distant supernovae [NobelPrize, 2011]" in the late 1990s, ushering in the phenomenon of dark energy. The problem of cosmological "dark energy" is well established [Perlmutter, 2003; Garfinkle, 2006; Paranjape and Singh, 2006; Tanimoto and Nambu, 2007; Bianchi et al., 2010; Clarkson and Maartens, 2010]. Essentially, type Ia supernovae are farther away at large redshifts than our otherwise best cosmology model[9] predicts [Riess et al., 1998; Perlmutter et al., 1999; Suzuki et al., 2012].

---

[9] The flat, matter-dominated GR cosmology model called the Einstein–deSitter (EdS) cosmology model.

The easiest way to compensate for the discrepancy in distance modulus versus redshift for these type Ia supernovae is to add a cosmological constant to Einstein's equations and that causes the universe to change from deceleration to acceleration at about $z = 0.752$ [Suzuki et al., 2012]. The new "concordance model," that with the most robust fit to all observational data (ΛCDM), simply adds a cosmological constant Λ to the Einstein–deSitter cosmology model $(\Omega_M + \Omega_\Lambda = 1)$[10] and Λ then provides the mechanism for accelerated expansion, that is, it provides the dark energy. The "problem" is that our best theories of quantum physics tell us the cosmological constant should be exactly zero[Carroll, 2001] or something incredibly large [Weinberg, 2000], and neither of these two cases holds in ΛCDM. Thus, one of the most pressing problems in cosmology today is to account for the unexpectedly large distance moduli at larger redshifts observed for type Ia supernovae [Bianchi et al., 2010]. As stated by Padmanabhan:

> The late time evolution of the universe is characterized by the cosmological constant Λ, and the four constants (Λ, G, $\hbar$, c) describing nature thus lead to the dimensionless combination
>
> $$\Lambda \left(\frac{G\hbar}{c^3} \right) = 2.8 x 10^{-122}$$
>
> which is probably the smallest non-zero number relevant to physics! This issue is well known and has often been thought of as *the* most fundamental problem in theoretical physics today.
>
> [Padmanabhan, 2016, p. 10]

The most popular attempts to explain the apparent accelerating expansion of the universe include quintessence [Zlatev et al., 1999; Wang et al., 2000; Weinberg, 2000] and inhomogeneous spacetime [Garfinkle, 2006; Paranjape and Singh, 2006; Tanimoto and Nambu, 2007; Clarkson and Maartens, 2010; Ellis, 2011; Marra and Notari, 2011] (there are even combinations of the two such as Roos, 2000; Buchert *et al.*, 2006). Although these solutions have their critics [Zibin et al., 2008], they are certainly interesting approaches. For example, Ellis's work [2011] with inhomogeneous spacetime discards the "cosmological principle," that is, the assumption that matter-energy is distributed uniformly (homogeneously and isotropically) on the spatial hypersurfaces of spacetime. The cosmological principle was used to produce the FLRW (Friedmann–Lemaître–Robertson–Walker) cosmology models as described in chapter 3. And, as we stated in that chapter, we cannot observationally prove this principle. So, it is not unreasonable to try new cosmology models based on a "very different spacetime" than that assumed in the concordance model (ΛCDM). Another popular

---

[10] Ω_X is the contribution from X to the total matter–energy density relative to that needed to make space flat. Thus, in the spatially flat model $\Omega_{total} = 1$ by definition.

attempt to explain the SCP Union2.1 SN Ia data is the modification of GR. These approaches, such as f(R) gravity [Capozziello, 2002; Kleinert and Schmidt, 2002; Capozziello and Francaviglia, 2008; Bernal et al., 2011; Nojiri and Odintsov, 2011; Olmo, 2011], have stimulated much debate [Flanagan, 2004; Vollick, 2004; Barausse et al., 2008], which is a healthy situation in science.

Both DM and a cosmological constant Λ (dark energy) play important roles in the concordance model of cosmology ΛCDM [Planck Collaboration, 2016] where baryonic matter accounts for ∼ 4% of all the energy density in the universe, DM accounts for ∼ 23%, and dark energy accounts for ∼ 73%. The sum of these contributions results in a spatially flat radiation-dominated universe transitioning to a spatially flat matter-dominated universe which does a good job accounting for cosmological observations, for example, the angular power spectrum of the CMB [Hu, 1995] and the matter power spectrum (large-scale galactic distributions) [Eisenstein et al., 2007]. In other words, while DM and Λ serve us well in ΛCDM, they are also problematic. "As Tom Shanks once said, there are only two things wrong with ΛCDM: Λ and CDM" [Bull et al., 2016, p. 68]. For these reasons and others, there are efforts to explain gravitational phenomena on astrophysical scales without DM or Λ [Milgrom, 1983, 2015; Sanders and McGaugh, 2002; Bekenstein, 2004; Sanders, 2005; Brownstein and Moffat, 2006a,b; Garfinkle, 2006; Paranjape and Singh, 2006; Tanimoto and Nambu, 2007; Zlosnik et al., 2007; Blanchet and Le Tiec, 2008; Clarkson and Maartens, 2010; Zhao and Li, 2010]. However, as far as we know, there is no attempt to get rid of both DM and Λ, which is what we propose.

As we alluded to earlier, reality is fundamentally discrete according to RBW, so although the lattice geometry of Regge calculus [Regge, 1961; Misner et al., 1973; Barrett, 1987; Williams and Tuckey, 1992] is typically viewed as an approximation to the continuous spacetime manifold of GR, it could be that discrete spacetime is fundamental while "the usual continuum theory is very likely only an approximation" [Feinberg et al., 1984, pp. 345] and that is what we assume. Further, the links of a Regge calculus graph can connect non-neighboring points of the corresponding GR spacetime manifold leading to small corrections to the corresponding GR spacetime geometry as per disordered locality [Caravelli and Markopoulou, 2012] (Figure 6.3), and this has been used on astrophysical scales to provide a mechanism for dark energy [Smolin, 2009].[11] Our views deviate from the standard use of Regge calculus so we refer to our approach as "modified Regge calculus" (MORC).

As alluded to earlier, RBW's disordered locality results from a variation on the old idea of direct particle interaction [Wheeler and Feynman, 1949; Hawking, 1965; Davies, 1971, 1972; Hoyle and Narlikar, 1995; Narlikar, 2003] whereby the naïve notion of a mediating quantum field between sources is eliminated. A discrete graphical spacetime is obviously at odds with the differentiable spacetime manifold of GR, but particularly so when quantum matter–energy exchange

---

[11] In contrast, we are using disordered locality to avoid a need for dark energy in the first place.

occurs between sources at distances exceeding the validity of a flat spacetime approximation; for example, a Source and its sink don't reside inside the same simplex of Figure 6.1. Mathematically speaking, disordered locality violates the local character[12] of our divergence-free stress–energy tensor, that is, the local conservation of energy–momentum.

Indeed, this conflict with GR is largely responsible for researchers abandoning the idea of direct action. The difference here is that we are proposing a spatiotemporally holistic, adynamical global constraint approach to physics, so disordered locality resulting from direct action in curved spacetime is not a problem in our approach, one simply finds the required correction to the metric quantities. When all link lengths are small, that is, in the absence of disordered locality, the Regge calculus solutions would correspond to GR solutions and we have standard Regge calculus [Miller, 1995; Brewin, 2000; Brewin and Gentle, 2001]. Accordingly, MORC can explain the dark energy phenomenon without Λ. A conceptual introduction is provided here and the corresponding formal details can be found in Foundational Physics for Chapter 6.

If one favors an exotic new kind of matter to account for dark matter phenomena then it is important to note that the exotic new matter is far more prevalent than ordinary matter. But, if one favors a GR correction to dynamic mass to account for dark matter phenomena, then it is important to note that the required corrections represent small deviations from the uncorrected spacetime geometry. It is customary to expect modifications to GR for strong gravitational fields with their large spacetime curvature, but attempting to account for dark matter phenomena via a GR correction of dynamic mass actually requires just the opposite. Thus, contrary to conventional thinking, what we're advocating is a geometric view of even weak gravitational fields. Indeed, Cooperstock and colleagues used GR instead of Newtonian gravity in fitting galactic RCs and found that the nonluminous matter in galaxies per GR is much less than that found using Newtonian gravity [Cooperstock and Tieu, 2007; Carrick and Cooperstock, 2012; Magalhaes and Cooperstock, 2015]."[13] And, Moffat and Rahvar used the weak field approximation of MOG as a perturbation of "the metric and the fields around Minkowski spacetime" in fitting galactic RCs [Moffat and Rahvar, 2013, p. 1] and X-ray cluster mass profiles [Moffat and Rahvar, 2014] without DM. Certainly MOND can be viewed in this fashion, since MOND advocates an extremely small modification to acceleration on astronomical scales in the context of flat spacetime (Newtonian gravity), and acceleration due to gravity in flat spacetime is replaced by curved spacetime in GR as we explained in chapter 3.

So, MOND could be viewed as claiming that a small change in spacetime curvature (equivalent to a small change to acceleration in Newtonian gravity) replaces the need for a greatly increased mass in accounting for galactic dynamics.

---

[12] Local in the sense of a "small" region about the point in question.
[13] There are dissenting opinions on this result; see [Korzynski, 2005; Cross, 2006; Fuchs and Phleps, 2006; Menzies and Mathews, 2006; Martila, 2015].

In fact, McGaugh and team found "a strong relation between the observed radial acceleration g_{obs} and that due to the baryons, g_{bar}" which "suggests that the baryons are the source of the gravitational potential" [McGaugh et al., 2016, p.5]. Concerning this result, David Merritt stated:

> Galaxy rotation curves have traditionally been explained via an ad hoc hypothesis: that galaxies are surrounded by dark matter. The relation discovered by McGaugh et al. is a serious, and possibly fatal, challenge to this hypothesis, since it shows that rotation curves are precisely determined by the distribution of the normal matter alone. Nothing in the standard cosmological model predicts this, and it is almost impossible to imagine how that model could be modified to explain it, without discarding the dark matter hypothesis completely.
>
> [Merritt, 2016]

Thus, MOND could be interpreted as saying precisely what we are claiming with RBW: small spacetime curvature perturbations on astronomical scales can account for dark matter phenomena without DM.

That we are discussing small differences in spacetime geometry even though we have relatively large differences between M and M_p ($M_p \sim 10M$) can be seen in the Schwarzschild metric, for example. The deviation from flat spacetime on typical galactic scales per $\dfrac{2GM_p}{c^2 r}$ of the Schwarzschild metric is $\sim 10^{-6}$, so we're talking about an empirically small metric correction of flat spacetime even when taking into account the much larger value M_p.

Another way to quantify the difference geometrically between M and M_p is to use $^{(3)}R = \dfrac{16\pi G\rho}{c^2}$ on galactic scales (where ρ is the mass density) [Wong, 1971]. Assuming constant mass density (typical in a galactic bulge) out to radius r where we have orbital velocity v gives $^{(3)}R \sim 10^{-45}/m^2$ (details in Foundational Physics for Chapter 6). So, a factor of 10 one way or the other in ρ is geometrically inconsequential in this context. Another way to state this in terms of Regge calculus is that the variation in the deficit angles that accounts for the variation of ρ we are proposing on galactic scales is minuscule. ICM gas is even more rarefied and the potentials used for FLRW metric perturbations leading to CMB anisotropies are already $\ll 1$ to include DM. Therefore, the large difference between M and M_P is contained in a small metric perturbation of a background spacetime.

The metric modification for mass to account for dark matter phenomena that we propose has to do with contextuality as motivated by GR. That is to say, it is well known that the proper mass of the matter interior to the Schwarzschild solution can differ from the mass of that same matter per the exterior Schwarzschild metric [Wald, 1984, p. 126].[14] The exterior value of mass for the Schwarzschild

---

[14] The difference between M and M_p pointed out by Wald is the "binding energy" and goes as $\dfrac{2GM}{c^2 r}$, which is very small as noted earlier. The difference we will allude to can be much larger, as we will show.

metric M could be determined dynamically by geodetic orbits, as explained in Foundational Physics for Chapter 3, that is, M is the "locally determined" value mentioned earlier. That this Schwarzschild vacuum is static even though it surrounds (or is surrounded by) a time-dependent matter distribution follows from Birkhoff's theorem [Wald, 1984, p. 125], so we can ignore that complication and talk unambiguously about "space." Now suppose a Schwarzschild vacuum surrounds a sphere of FLRW dust (recall the matter-dominated cosmology solutions from chapter 3). The dynamic mass M of the surrounding Schwarzschild metric is not necessarily equal to the proper mass M_p of the FLRW dust [Stuckey, 1994]; $M = M_p$ in the flat model, $M \geq M_p$ for the open model, and $M \leq M_p$ for the closed model. Let us relate this difference in mass to a difference in spatial curvature, since $^{(3)}R$ is zero for both the flat FLRW dust and the Schwarzshild vacuum where $M = M_p$. Now a simplistic model for a collapsing star is a ball of FLRW dust inside a Schwarzschild vacuum [Misner et al., 1973, pp. 851–3]. Any of the three FLRW models will work, depending on initial conditions, but we will consider the closed model, since $M \leq M_p$ therein. Depending on where the connection between the FLRW and Schwarzschild metrics is made, the ratio $\frac{M_p}{M}$ can be large enough to account for dark matter phenomena (Figure 6.5), as we now argue.

Suppose that the Schwarzschild vacuum is surrounded by the remaining FLRW dust, that is, the ball of FLRW dust has collapsed out of its FLRW cosmological context and is now separated from that cosmological context by the Schwarzschild vacuum. The spacetime geometry of the surrounding FLRW dust will be unaffected by the intervening Schwarzschild vacuum, so observers

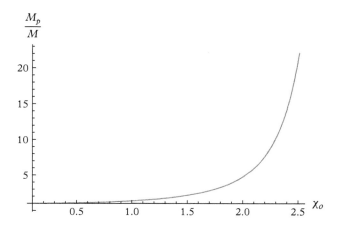

Figure 6.5 *This graph shows the ratio of proper mass to dynamic mass $\frac{M_p}{M}$ as a function of χ_o, the radial coordinate location where the FLRW dust ball is connected to the surrounding Schwarzschild vacuum. At $\chi_o = \frac{\pi}{2}$ the ratio is already 2.36 and at $\chi_o = 0.8\pi$ the ratio is 22.1.*

in the surrounding FLRW dust (global context) will obtain the "globally determined" proper mass M_p for the mass of the collapsed dust while observers in the Schwarzchild vacuum (local context) will obtain the "locally determined" dynamic mass for the collapsed dust M. Of course, this is an idealization and the actual situation in a galaxy would be far more complicated since such a nested solution would have to marry up with other such nested solutions. Indeed, the surrounding FLRW dust may not even remain, having coalesced into other bodies and clouds. Given the complexity of GR, no such solution can be expected, so some approximation method for obtaining the proper mass from the dynamic mass must be motivated and checked for efficacy against astrometric data.

To obtain a first guess for this approximation method, imagine that little if any of the matter in the original uncoalesced state remains. In that case, the global dynamics, for example, of galactic rotation curves, is determined in (large) part by the curvature of empty space remaining between the many regions of Schwarzshild vacuum surrounding coalesced matter. In other words, as the matter coalesces to the centers of Schwarzschild vacuum regions it leaves a footprint of its proper mass in the curvature of the spatial boundaries between those Schwarzschild regions. This proper mass footprint is in large part responsible for the global dynamics, since proper mass M_p is much larger than the Schwarzschild dynamic mass M.[15] Essentially, we imagine that coalescing matter sweeps up and concentrates spatial curvature, leaving flatter empty space in its wake and residual spatial curvature in regions between coalescent centers. So, there is no DM responsible for forming baryonic matter concentrations; rather, it is the baryonic matter that is responsible for forming spatial curvature concentrations which change the value of mass in the spacetime metric, $M \rightarrow M_p$. This is consistent with the findings of McGaugh and team [2016] explained earlier where kinematic differences typically attributed to DM trace the ordinary baryonic matter in galaxies. Unfortunately, there is no realistic way to find an exact GR solution dictating M_p given the "locally determined" M obtained from mass-to-luminosity ratios. Thus, we will have to propose and test an ansatz for obtaining M_p from M based on experience from GR. The ansatz will then be tested to search for trends and subsequently refined in an iterative process to find the most robust form in accounting for all dark matter phenomena. This is a relatively common practice when the phenomenon under investigation is too complex to allow a methodologically reductive account from first principles, for example, in chemistry where molecular energy levels are much too complex to allow for an exact solution of the Schrödinger equation.

Accordingly, in the first step of this iteration we used a simple geometric modification of dynamic mass M to obtain proper mass M_p for fits [Stuckey et al., 2016a] of 12 high-resolution galactic RCs from the HI Nearby Galaxy Survey [Walter et al., 2008] used by Gentile and colleagues to explore

---

[15] This is a simple heuristic since, of course, we advocate direct action and are not spacetime substantivalists. Indeed, we apply our correction of dynamic mass to the matter components proper.

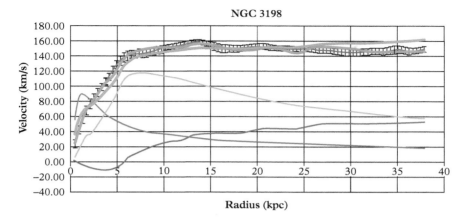

Figure 6.6 *This graph shows test (turquoise) and MOND (burnt orange) fits of a THINGS galactic RC (with error bars). The disk curve is green, the gas curve is red and the bulge curve is blue. The mean square error (MSE) for this fit is* 10.1 *(km/s)²* $(km/s)^2$ *for our test fit and* 74.2 *(km/s)²* $(km/s)^2$ *for MOND. Eleven more graphs are available in [Stuckey et al., 2016a].*

modified Newtonian dynamics (MOND) fits [Gentile et al., 2011] (Figure 6.6). We also used this same technique to fit the mass profiles of the 11 X-ray clusters found in Brownstein [Brownstein, 2009] as obtained from Reiprich and Böhringer [Reiprich, 2001; Reiprich and Böhringer, 2002] using combined ROSAT (ROentgen SATellite) and ASCA (Advanced Satellite for Cosmology and Astrophysics) data (Figure 6.7). As can be seen from the results, our test fits compare favorably with the MOND and MSTG fits of these data. (Computational details are in Foundational Physics for Chapter 6.)

What about the gravitational lensing data of the Bullet Cluster (1E0657-558) originally touted as "direct empirical proof of the existence of dark matter" [Clowe et al., 2006]? As it turns out, this can also be explained without DM [Brownstein and Moffat, 2006a]. What happened in this case is a small galactic subcluster ("bullet cluster") collided with the larger main cluster. The galaxies of both clusters passed through the collision region relatively unaffected, but the intracluster medium (ICM) gas of the two clusters was left behind in the collision region due to the ram pressure. The result was four lobes of baryonic matter aligned as follows: the galaxies of the main cluster, the gas of the main cluster, the gas of the subcluster, and the galaxies of the subcluster (Figure 6.8). If one accepts that the mass of the cluster galaxies is only 10% of the baryonic mass then in the absence of DM one would expect gravitational lensing maps of this region (blue lobes in Figure 6.8) to overlap X-ray images of the gas lobes (red lobes in Figure 6.8), since the ICM gas possesses 90% of the baryonic mass. What Clowe and colleagues rather found [Clowe et al., 2006] was that the lensing peaks were located in the galaxy lobes, so the galaxies are inside the blue lobes of Figure 6.8. Their conclusion was that there exists large quantities of DM which

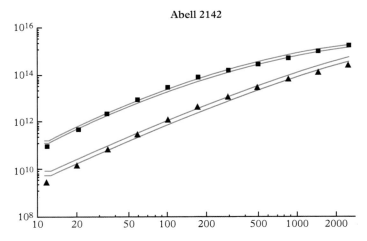

Figure 6.7 *This log–log plot shows test and MSTG fits of an X-ray cluster mass profile (compiled from ROSAT and ASCA data). Vertical scale is in solar masses and horizontal scale is in kpc. The test fit is increasing the dynamic gas mass (triangles) to fit the proper mass (squares). MSTG is decreasing the proper mass to fit the dynamic gas mass. The sizes of the objects are approximately equal to their errors. The test fit is the upper pair of lines (connecting fit points) over the squares where line separation corresponds to error. The MSTG fit is the lower pair of lines (connecting fit points) over the triangles where line separation corresponds to error. MSE is $(\Delta Log(M))^2$ and for this fit the test MSE = 0.00786 and MSTG MSE = 0.0302. Ten more graphs are available in [Stuckey et al., 2016a].*

Figure 6.8 *Bullet Cluster X-ray and lensing composite (false color) image from NASA Release 06-297. Blue lobes are lensing data in vicinity of the galaxies. Red lobes are X-ray images of ICM gas left behind after the subcluster (right side) passed through the main cluster (left side). Reproduced by permission of image author Maxim Markevitch.*

passed through the collision with the galaxies. Brownstein and Moffat [2006a] explained the offset lensing peaks using MSTG because $G(r)$ associated with the galaxies increases more than $G(r)$ associated with the gas, since the galaxies are farther removed from the center of the Newtonian gravitational potential. The explanation per our heuristic is that the ram pressure that swept the gas out of the galactic clusters did not displace the small spatial curvature originally swept into the clusters via gravitationally collapsing matter. Thus, the spatial curvature resides in the galactic lobes rather than the gas lobes and that produces more lensing about the galactic lobes than the gas lobes. This same effect is responsible for anisotropies in the angular power spectrum of the CMB, as we will explain shortly. This does not mean that the ICM gas *cannot* carry spatial curvature in such collision processes. Indeed, Jee et al. found a significant "dark core" in the X-ray gas region of a cluster collision that was not associated with any bright cluster galaxies [Jee et al., 2012]. And, Bradac et al. used both weak and strong lensing to find a non-negligible mass concentration coincident with the main X-ray peak in the Bullet Cluster [Bradac et al., 2006].

Of course, this is just a heuristic, the bottom line in GR is that one must find a metric and stress-energy tensor that satisfy Einstein's equations everywhere on the spacetime manifold to have an allowable configuration. Without a complete spatiotemporal solution (allowable configuration), there is no way to tell a definitive corresponding dynamical story. In order to model gas having collapsed into dynamical collections of stars, galaxies, clusters, super clusters, filaments, etc., with differing values of mass as shown above, we would need multiply connected spacetime regions well beyond the simple FLRW-Schwarzschild combination. Unfortunately, there is likely no such exact solution forthcoming, so a GR-motivated ansatz may be the best we can do in exploring this proposition. In any event, a fit of the Bullet lensing offset using our GR ansatz at this point is of no interest, since the number of fitting parameters would equal the number of data points. Thus, fitting these lensing data will have to wait until the fitting parameters of our ansatz are constrained by theory or observations otherwise.

After the Bullet Cluster, the most compelling evidence for DM is perhaps the CMB angular power spectrum [Carroll, 2011]. In this situation, small primordial inhomogeneities in the otherwise homogeneous energy density ρ of the the background FLRW cosmology model lead to anisotropies in the CMB angular power spectrum C_ℓ versus ℓ (typically plotted D_ℓ versus ℓ where $D_\ell = \frac{\ell(\ell+1)C_\ell}{2\pi}$). The various components Δ_i typically considered are photons, baryons (this term includes electrons), neutrinos, and DM. Before recombination, the photons are coupled to the baryons and together their perturbations oscillate, gravity pulling them in while the photon gas pressure pushes them out. The neutrino perturbations simply stream free and the DM perturbations stay put, since DM is not affected by the photon gas pressure; that is, DM does not interact via the electromagnetic force. As a result, the DM acts to enhance the contraction phase of the photon–baryon Δ_i oscillations while suppressing the rarefaction phase which manifests itself as enhanced odd peaks relative to even peaks in D_ℓ versus ℓ. Thus, the size

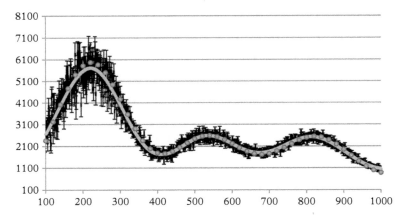

Figure 6.9 *This is a plot of D_ℓ in $(\mu K)^2$ versus ℓ in the range $100 \le \ell \le 1000$ for the Planck 2015 CMB data [Collaboration, 2016b] (black error bars), the Planck consortium's best ΛCDM fit [Collaboration, 2016a] (solid line), and our best HuS sCDM fit (dots). MORC and STVG trivially reproduce the HuS fit without DM. Since MORC does not have Λ, its best fit to these data would equal the HuS sCDM best fit. The root-mean-square error (RMSE) for the HuS sCDM fit points shown is 225 $(\mu K)^2$. STVG can also trivially replace DM in ΛCDM and STVG keeps Λ, so the STVG best fit to these data would equal the ΛCDM best fit. The RMSE for the ΛCDM fit shown corresponding to the HuS fit points shown is 240 $(\mu K)^2$, although this fit is for all ℓ in the range $30 \le \ell \le 2508$.*

of the third peak relative to the second peak is taken to be an indication of DM and the Planck 2015 CMB data clearly show that the second and third peaks have very nearly the same amplitude (Figure 6.9) in accord with the existence of DM.

The relativistic version of MOND, known as the Tensor-Vector-Scalar (TeVeS) theory, does not show such an enhanced third peak [Skordis et al., 2006; Skordis, 2009], so the Planck 2015 CMB data would appear to present a serious challenge for MOND [McGaugh, 2015a], although Skordis is working on a new approach and expects results in 2017 [Skordis, 2016]. Angus [2009] has argued that non-relativistic MOND would have no cosmological affect and one can obtain the enhanced third peak using a sterile neutrino with mass 11eV. This also allows MOND to account for "the dark matter of galaxy clusters without influencing individual galaxies, . . . and potentially fit the matter power spectrum" [Angus, 2009, p. 1]. Whether this constitutes a true extension of MOND or a "cheat" is subject to debate, as cosmology certainly resides in the relativistic domain, so some relativistic version of MOND that accounts for the CMB anisotropies is needed [McGaugh, 2016].

The situation with MOG is less contentious as its practitioners seem to agree that it has addressed astronomical data at all scales with some degree of success. Recently, Moffat and Toth [2013] used STVG to successfully account for the matter power spectrum per data from the Sloan Digital Sky Survey, as well as the CMB angular power spectrum per data from WMAP and the Boomerang experiment for $\ell \leq 1000$, which includes the third peak. In both cases, they used an enhanced gravitational constant G in a fashion similar to the MOG account of the Bullet Cluster. Since STVG is the best modified gravity fit of the CMB angular power spectrum to date, we compared our test fitting procedure to that of STVG. As it turns out, both can reproduce the standard cold dark matter (sCDM) fit of the first three acoustic peaks in a rather trivial fashion (Figure 6.9).

Having completed the first step in this iterative process, we must next explore the possibility that the fitting constants in our ansatz have a basis in astrometric quantities since the results show interesting trends in the fitting factors of our dynamic mass correction (see Foundational Physics for Chapter 6 for details). Qualitatively speaking, for our galactic rotation curve fits we see no radial correction in the six bulge contributions and only a relatively small radial correction in 5 of 12 disk contributions, while there were relatively large radial corrections to all 12 gas contributions. The constant fitting factor was small to nonexistent in all 12 cases (0.55 ± 0.81). Moving to our X-ray cluster mass profile fits we see again large radial corrections to the ICM gas mass with an increased contribution from the constant fitting factor although still one deviation within zero (600 ± 1100). For the primordial density perturbations, the radial variation disappears and the correction is entirely in the constant fitting factor. Also, the X-ray cluster results don't seem to follow a simple scaling with cluster mass. Clearly, other geometrical factors are at work here according to the ansatz. Therefore, it would appear that the functional form for our ansatz is relevant and it is not

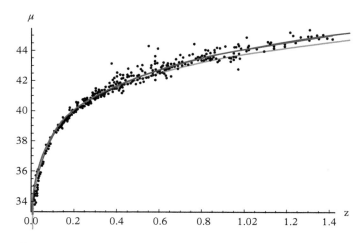

Figure 6.10 *Plot of SCP Union2.1 SN Ia data (distance modulus versus redshift) along with the best fits for EdS (green), ΛCDM (blue), and MORC (red). The MORC curve is terminated at z = 1.4 in this figure so that the ΛCDM curve is visible underneath.*

unreasonable to search for astrometric hints in these fits in order to revise the ansatz for the second iteration. For a more detailed conclusion see [Stuckey et al., 2016a].

As for the dark energy associated with the SCP Union2.1 SN Ia supernova data, we simply corrected proper distance in EdS cosmology per standard GR perturbation theory as justified by disordered locality per MORC [Stuckey et al., 2012a,b]. This MORC fit was done before the test fits of THINGS, ROSAT/ASCA, and Planck 2015 data, so the test ansatz was chosen to have the same functional form as the perturbation theory fit of MORC. The type Ia supernova data to be fit are distance modulus (μ) versus redshift (z) leading to the fit shown in Figure 6.10. As with the dark matter phenomena, we see that the MORC fit compares favorably with the ΛCDM fit. (Computational details are in Foundational Physics for Chapter 6.)

6.6   Summary

In summary, quantum physics is necessarily relational and contextual, that is, the properties of a quantum system are obtained relationally from its classical context. This deviation from methodological and ontological reductionism avoids the pitfalls of conventional programs along those lines, as we will explain in Philosophy of Physics for Chapter 6. Thus, our spatiotemporal ontological contextuality leads to a straightforward understanding of QM and QFT, and a relatively simple empirical approach to QG. If the classical context is that of flat spacetime, then QFT (which is based on SR) suffices to provide the fundamental formalism to model objective reality. Modified lattice gauge theory (MLGT), which assumes reality is fundamentally discrete, resolves the conundrums of QM, and justifies otherwise questionable techniques of QFT. In other words, *while adynamical explanation in the block universe is a conceptually radical deviation from dynamical explanation in the mechanical universe, the current formal structures of QM and QFT*—Lagrangian or Newtonian, on the lattice or the differentiable manifold—*are well justified per RBW*. The technical difficulties facing modern physics—unification, QG, dark matter, and dark energy—are the result of not accepting the discrete, spatiotemporally contextual and relational nature of physics.

To deal successfully with the challenge of unification, we have to accept that reality is fundamentally adynamical, so it doesn't rest on a foundation of fundamental particles governed by fundamental forces. Rather, objective reality is fundamentally about scalars and vectors on the nodes, links, and plaquettes of graphical spacetimesource elements distributed in a classical context per RBW's adynamical global constraint (AGC). While the fundamental spacetimesource element is not itself a diachronic entity, the AGC distribution of spacetimesource elements is in statistical accord with the trans-temporal identification necessary for the diachronic objects of classical physics. Thus, objective reality per physics is necessarily self-referential, just as Einstein's equations constitute a self-consistency constraint between the spacetime metric and stress–energy tensor (recall chapter 3). Classical physics provides a spacetime metric and stress–energy tensor context for quantum physics and quantum physics provides a distribution of spacetimesource elements in some subset of that classical context, that is, an ontological decomposition of that subset. As we explained in chapters 4 and 5, the diachronic objects of classical physics are the relata for the relations of spatiotemporal ontological contextuality. Accordingly, one can only use quantum physics to relationally and contextually decompose a proper subset of classical objective reality, since relations need relata and the properties of relations are contextual.

With this understanding of objective reality, we see that the truly fundamental experimental data are individual detector clicks, not the collection of clicks that comprise particle tracks of a high-energy physics experiment. The most fundamental (unifying) collection of ontological objects according to RBW is the small collection of scalars and vectors on nodes, links, and plaquettes for a graphical, spatiotemporal decomposition of the individual clicks in the context of a subset of classical reality (Figure 6.4) rather than the baroque Lagrangian density of the Standard Model of particle physics (see T. D. Gutierrez earlier in this chapter).

A typical question posed concerning unification is "what am I to do when the quantum fields are very dense and require a curved spacetime?" According to MLGT, this is a mistaken belief carried over from lattice gauge theory that there are actual matter–energy fields on a spacetime lattice in the region in question. Recall that in MLGT the fields are merely calculational devices to find the distribution of spacetimesource elements in a classical context. The spacetimesource elements represent a loss of energy–momentum at the Source(s) and a gain at the sink(s), but do not represent the existence of matter–energy in the regions between, that is, direct action. If the exchange of matter–energy is substantial enough to warrant a deviation in the Regge calculus graph as compared to the situation whereby the exchange didn't occur, then the geometry of the graph under the exchange must be taken into account as a final boundary condition as part of the classical context, just as the locations and settings of the detectors must be taken into account. Thus, the question is a non-starter for RBW.

Similarly, in order to deal with the other three challenges (QG, dark matter, and dark energy), our classical context is the curved spacetime of GR. So, again, we note the necessity of a classical context for quantum physics. If one ignores a classical context they are led to theories like the Wheeler–DeWitt equation with its "problem of time." Attempting to model objective reality using quantized GR with no classical context means reifying configuration space and that leaves us with the problem of recovering a single spacetime structure for objective reality, that is, a similar problem faced by the Many-Worlds interpretation of QM. A reified configuration space is nonsensical, empirically speaking.

Our proposed method for dealing with the most fundamental objective reality (individual clicks in a classical context) is modified lattice gauge theory in modified Regge calculus per disordered locality, that is, large simplices that deviate from the curved manifold approximation (Figure 6.3). Concerning the distribution function for individual clicks in a GR spacetime context, we have small energy–momentum exchanges over large distances (by assumption), so we need to modify the action of Regge calculus or GR per disordered locality. We then used this modification to find a small correction to the metric of Einstein–deSitter cosmology to deal with the data from type Ia supernova without having to introduce a cosmological constant, that is, no dark energy and no accelerated expansion. Thus, RBW presents quite a different approach to QG with astrophysical consequences for dark energy.

Concerning dark matter phenomena, we used contextuality already inherent in GR to justify two different values of mass for one and the same gravitationally bound matter in galaxies, galactic clusters, and primordial inhomogeneities in the early universe. Since all of these scenarios represent small perturbations to the surrounding spatial curvature, we used the same functional form for the test fitting ansatz in these phenomena that we used for the dark energy fit. Thus, we were able to explain galactic rotation curves, X-ray cluster mass profiles, gravitational lensing, and the angular power spectrum of the CMB without recourse to non-baryonic dark matter.

In conclusion, our adynamical explanation in the block universe has resolved the puzzles, problems, paradoxes, and conundrums of GR, QM, and QFT, presented a new path to unification and QG, and dismissed the need for dark energy and dark matter. While there is still much to be done, we believe this body of work justifies interest in adynamical explanation.

But, before we can claim that the Sun has truly set on the dynamical universe, we must address the main complaint against adynamical explanation in a block universe, its seeming incompatibility with time as experienced. That will be addressed in the next two chapters. The reader interested only in foundational physics may choose to skim chapters 7 and 8 after finishing Philosophy of Physics for Chapter 6 and Foundational Physics for Chapter 6 and proceed to the Coda for Ants.

Philosophy of Physics for Chapter 6

6.7   Introduction

We begin by reminding the reader why we need a theory of QG. But, the primary goal of this thread is to situate RBW with respect to some other, possibly related, discretized accounts of QG. We make those comparisons with respect to the fundamentality of time and its nature, including the "problem of time," empiricism, explanatory goals, nature of beables, and the like. We will not provide an exhaustive description of these accounts, formal or otherwise. What we want to suggest is that these programs founder perhaps because they cannot fully transcend the dynamical/mechanical paradigm. We want to state up front that we appreciate the promise of many of these programs and the brilliance of the researchers involved. However, we admit to being puzzled by how long (decades in some cases) some researchers have remained dedicated to various approaches given the lack of results. As Rovelli states:

> A theory begins to be credible only when its original predictions are reasonably unique and are confirmed by new experiments. Neither loop quantum gravity nor string theory—nor any other tentative theory of QG—are yet credible in this sense. Furthermore, in spite of much effort, both theories are still badly incomplete and far from being clearly understood. The problem of QG must therefore be considered still fully open.
>
> [Rovelli, 2007, p. 1302]

We realize, of course, that the RBW approach to QG is in all probability false, demonstrably so for all we know. We humbly submit that whatever the fate of RBW's approach to QG, it is perhaps time to backtrack and reconsider some of the axiomatic assumptions at work in the field of QG that derive from the dynamical paradigm.

6.8   Motivations for quantum gravity revisited

Weinstein and Rickles describe the need for and the project of QG as follows:

> Quantum Gravity, broadly construed, is a physical theory (still 'under construction') incorporating both the principles of general relativity and quantum theory. Such a theory is expected to be able to provide a satisfactory description of the microstructure of spacetime at the so-called Planck scale, at which all fundamental constants of the ingredient theories, c (the velocity of light in vacuo), $\hbar$ (the reduced Planck's constant), and G (Newton's constant), come together to form units of mass, length, and time. This scale is so remote from current experimental capabilities that the empirical testing of quantum gravity proposals along standard lines is rendered near impossible.

In most, though not all, theories of quantum gravity, the gravitational field itself is also quantized. Since the contemporary theory of gravity, general relativity, describes gravitation as the curvature of spacetime by matter and energy, a quantization of gravity seemingly implies some sort of quantization of spacetime geometry: quantum spacetime. Insofar as all extant physical theories rely on a classical (non-quantum) spacetime background, this presents not only extreme technical difficulties, but also profound methodological and ontological challenges for the philosopher and the physicist. Though quantum gravity has been the subject of investigation by physicists for almost a century, philosophers have only just begun to investigate its philosophical implications.

The second reason for the absence of consensus is that there are no experiments in quantum gravity, and little in the way of observations that might qualify as direct or indirect data or empirical evidence. This stems in part from the lack of theoretical predictions, since it is difficult to design an observational test of a theory if one does not know where to look or what to look at. But it also stems from the fact that most theories of quantum gravity appear to predict departures from classical relativity only at energy scales on the order of 1019 GeV. (By way of comparison, the proton-proton collisions at Fermilab have an energy on the order of 103 GeV.)

However, it should be noted, finally, that to date neither of the main research programs has been shown to properly reproduce the world we see at low energies. Indeed, it is a major challenge of loop quantum gravity to show that it indeed has general relativity as a low-energy limit, and a major challenge of string theory to show that it has the standard model of particle physics plus general relativity as a low-energy limit.

[Weinstein and Rickles, 2016]

We quoted this excellent overview at length to note again that most researchers assume that unifying QFT and GR is largely going to come into play for high energy and small-length phenomena such as allegedly to be found at or below the Planck scale. As Weinstein and Rickles make clear, it is this high-energy assumption that is partially responsible for the frozen state of progress in QG both with respect to constructing theories and making correspondence with observables. And as we noted earlier, the primary motivations for that high-energy assumption, in addition to a general kind of methodological and ontological "compositional" reduction assumption that high-energy and small-length scale phenomna are by definition more fundamental, are unification of forces at high energies (the fields merge as one goes back toward the Big Bang) and cases where high-energy densities such as black holes demand that QFT is generally performed with a curved spacetime.

So it is important to remind the reader that in the main thread to this chapter and in other chapters (3, 4, and 5) we have already discharged all three of those motivations. We will return to these concerns with specific discrete accounts shortly. Just keep in mind that the only thing demanded of a theory of QG is to remove any "tension" between QFT and GR, and to have a formalism that can in principle be applied to any reasonably tractable situation. In any other requirement, such a theory is going to show a pre-reflective bias about the nature of reality. Also keep in mind that how one evaluates various QG programs is going to be determined by not only what they manage to derive but by what one thinks it is reasonable to give yourself at bottom. For example, is starting with discretized spacetime cheating? Must one begin with something pregeometric or more essentially quantum mechanical in nature?

6.9 RBW and its ancestors

Let us attempt to situate RBW's version of QG among its predecessors and discrete cousins. While there are obviously many essential differences, RBW bears some resemblance to other nascent theories of quantum physics and QG such as spin-foam models and quantum graphity. This isn't surprising because, as Crowther notes:

> There are various approaches to QG which might be called 'discrete.' The first of these grew out of Regge calculus (Regge 1961), which is a means of modelling spacetime in GR (i.e., solutions to the Einstein field equations), using discrete elements called *simplices*. One of the several approaches which stemmed from Regge calculus is dynamical triangulations, problems with which led to the development of another theory, called *causal dynamical triangulations*. Most of the approaches that utilize, or stem from, Regge calculus involve taking a sum-over-histories in order to make the geometry quantum mechanical.
>
> [Crowther, 2016, p. 147]

Of course, while our modified Regge calculus in RBW is discrete, and formally uses the Lagrangian method, as we noted in the main thread it differs from the standard conception of Regge calculus in a number of ways. We will re-emphasize that point for the reader shortly.

Obviously, any discretized account of QG has something discrete at bottom, in our case, as Crowther noted, simplices. There are many possible motivations, both formal and ontological, for assuming that QG must be discrete [Crowther, 2016, pp. 129–30], perhaps most obviously the belief that QM is fundamental to relativity in some sense. However, merely being discrete certainly isn't sufficient for being quantum. In RBW discreteness is both an ontological commitment and a calculational tool. Our ultimate justification is simply that it allows us to solve a lot of problems across the board.

Let us now characterize some of the other discrete approaches to QG that bear some resemblance to our own. Discrete theories of QG fall into two categories, the geometric and the pregeometric. Both want to recover approximately or statistically the continuum of spacetime. In most discretized approaches spacetime is conceived of as an effective, low-energy manifestation of some very different high-energy degrees of freedom. Typically, in these approaches spacetime in the continuum sense emerges statistically under some dynamical operation or phase transition of the fundamentally discrete whatever-it-is, allegedly at higher energy and representing some fundamental length scale. The idea is to think of spacetime as in thermodynamics, at high energy it is composed of discrete elements, interactions result in effective spacetime at low energy. Obviously pregeometric approaches have a harder job in some sense compared to an account that gives itself a discrete spacetime.

Pregeometric approaches assume a physical phase transition in which space-time geometry comes into being. As Crowther states, pregeometric approaches don't calculate QM spacetime using a path integral over classical histories but attempt to "build in" the quantum nature of spacetime at the outset, encoding it in the dynamics of the theory [Crowther, 2016, p. 175]. Pregeometric approaches hold that "time" is fundamental and not geometric. One might ask, how does one define time in the absence of geometry? What does "time" mean in such cases? The answer is that "time" in such cases is just the evolution of whatever fundamental pregeomtric dynamics presumably leads to recovering spacetime, for example, dynamics just means changes in states of the vertices. The pregeometric theory has its own time tied to the pregeomtric dynamics wherein geometry appears in some phase transition. Therefore the pregeometric dynamics is responsible for the emergence of spacetime and diffeomorphism invariance in GR at low energies.

The obvious question is why "go" pregeometric when it is so much harder to recover spacetime? The answer is that pregeometric approaches hope to avoid problems regarding dynamics that befall other discrete approaches. Crowther describes the situation:

> Most of the 'traditional' discrete/background independent theories, which utilize a sum-over-histories-type approach, encounter significant difficulties with their main aim of recovering relativistic spacetime as a low energy limit. . . . In response some physicists have suggested that perhaps, rather than utilizing a sum-over-histories-type approach, which presupposes some sort of quantum geometry, we need to implement a stronger sense of background, where there is no underlying micro-geometry. These approaches are thus known as *pre-geometric* approaches to quantum gravity. . . . The promise of these approaches is 'truly emergent' geometry, as well as gravity. One if the core principles of these approaches is that the geometry be defined *intrinsically* via the interactions of the micro-elements described by the theory, and is dependent on the dynamics of these micro-elements; because of this, these approaches are set to recover geometry that is dynamical [in the GR sense], as well as being emergent.
>
> [Crowther, 2016, p. 168]

> A main advantage touted by the pre-geometric approaches compared with the 'traditional' background-independent theories . . . is its ability to deal naturally with *dynamics*. It seems that the dynamics of the micro-theory is one aspect that does manifest itself in the effective theory.
>
> [Crowther, 2016, pp. 169–70]

So if any readers are still wondering why some researchers are wedded to pregeo-metry, then we are of one mind. Let us try and imagine the motivations for doing just this. The first possibility, as suggested earlier, is that blanket reductionism assumes that the fundamental theory must be something pregeometric or otherwise essentially quantum-like, that the idea being that reductive base must have different essential theoretical concepts to count as a reduction; that is, giving oneself something like spacetime is cheating; for example, Crowther says "pregeometric approaches claim to go some way toward *explaining* gravity, rather than just quantising it" [Crowther, 2016, p. 172]. And if we understand Smolin, who is especially concerned about background independence, when he says "time" is fundamental, he has something like the pregeometric notion of time in mind. Consider the following:

> So long as we cling to the absolute view of time and space, regarding them as entities separate from the phenomena that they would somehow house, rather than as orderings of events, we cannot do justice to the inclusive reality of time.
>
> [Unger and Smolin, 2015, p. 251]

> Non-emergent, global, irreversible, and continuous time resembles, in may respects, time as described in Newton's famous scholium. It differs from Newton's time, however, because from the standpoint of the temporal and relational naturalism that we espouse, it is not a thing, or stage, or backdrop separate from the phenomena. It is a fundamental and primitive feature of nature: the susceptibility of all nature to differential change. . . . The subsequent developments of physics, whether in the direction of general and special relativity or in the direction of quantum mechanics, has on the whole narrowed rather than widened the space for the reception of a view of time as non-emergent, global, irreversible and continuous.
>
> [Unger and Smolin, 2015, p. 237]

> The universe cannot have a history unless time is real. For time to be fully real, it must be non-emergent, global, irreversible and continuous. For time to be inclusive as well as real, there must be nothing outside it, nothing safe from its ravages and surprises, not even the laws of nature.
>
> [Unger and Smolin, 2015, p. 239]

So the point is that someone like Smolin, if not Smolin himself, can hope for some pregeometric dynamical process that constitutes a fundamental arrow of time, that is, that process that yields an "ordering of events." And that arrow of time must somehow explain or correspond with time as experienced (Passage, Presence, and Direction):

> What then is permanent? Only one aspect of reality: time. The condition of the radical and inclusive reality of time is the impermanence of everything else. Time is internally related to change and causal connection: equally fundamental and primitive feature of nature.
>
> [Unger and Smolin, 2015, p. 243]

Smolin's reasoning is to the dynamical/mechanical universe what postmodernism is to modernism: "modernism with a gun to its head." The reasoning is that if change is fundamental then even the dynamical laws of nature must be susceptible to change, all driven by that pregeometric process.

But to be fair, many defenders of pregeometric approaches would say they are motivated by failures of most discrete theories (which typically employ sum-over-histories) to obtain spacetime or the metric tensor of GR. Such a defence often assumes that these failures stem from those theories being insufficiently background-independent. Rather than employ sum-over-histories in a discrete model, pregeometric discrete models such as "quantum graphity" hope to build quantum features into the fundamental pregeometric basis. As Crowther describes it:

> Quantum graphity is another example of the pre-geometric approaches, and also draws explicitly from techniques and theorems of condensed matter physics. What distinguishes this approach is its dynamics: the dynamics is not movement or 'birthing' of points, but rather a change in the connections between points. The connections, represented by the edges of the graph, are able to be in two states 'on' or 'off', and, being quantum-mechanical, are able to exist in superpositions of both 'on' and 'off' states.
>
> [Crowther, 2016, p. 170]

Even though this account is often touted as being compatible with a block universe because it is not about the growth or movement of points and the graph is supposed to be interpreted as the "entire universe," there is still the fact that the states of the links evolve in "time" via a Hamiltonian and the pregeometric arrow of time given by that process is not equivalent to the geometric notion of time, let alone time as experienced.

As ever, the question is why should we believe that some obscure pregeometric dynamical evolution explains or corresponds with time and change as experienced? This is a hard case for even a particular physical/geometric arrow of time to make as we will see in chapters 7 and 8. The first goal of any pregeometric account is to recover spacetime or the metric tensor of GR and currently they all fail to do that [Crowther, 2016, chapter 6], let alone recover time as experienced. This isn't surprising since there is no micro-geometry on these accounts; geometry must be truly emergent on this view. Geometry must be defined intrinsically via the interactions/evolution of of whatever discrete micro-elements are at bottom in the formalism. So geometry must be both truly dynamical and emergent which perhaps is why these accounts hope to underwrite the dynamical nature of GR, but as things stand they underwrite nothing.

The pregeometric view is an instance of mechanical paradigm thinking par excellence. The reductionist idea is that there is some funky pregeometric dynamical process that underwrites spacetime. The thought is that what is fundamental must be dynamical or process-like, so if we can't find a good perch for that idea in relativity or QM, then let's build into the pregeomtric base. This raises a host of formal and conceptual questions especially if one is a realist about Minkowski spacetime and therefore the block universe. How and in what sense can spacetime—the block universe—be explained by some pregeometric process? Conceptually it makes no sense to say that a block universe is *generated* dynamically by a more fundamental process, even a pregeometric process. Of course, recalling chapters 1 and 2 we know Smolin will say "dump the block universe," but most QG researchers accept it. However, even if it did make sense, the idea violates background independence because that pregeometric process constitutes a meta-time (a background time).

Let us now consider other non-pregeometric discrete accounts of QG. Again, most of these approaches employ sum-over-histories, but standard "causal set theory" (CST) is entirely classical and yet still can't recover relativistic spacetime. CST gives itself a partial ordering that corresponds to "causality," that is, a microscopic conception of before and after. Causal sets have dynamics that grow by creation of additional elements. The dynamics are a discretized and stochastic Markov process. The only basic elements are the causets. The birthing process of these causets is the arrow of time in this model. More generally, CST gives itself "Bell-type causality" and discrete general covariance. Bell causality says that a spacelike separated birth in one region of the causet can't be influenced by any subsequent birth in a region spacelike to it. Obviously, there is no continuum conception of time and distance intervals in this model (nothing corresponding to length, duration, spacetime intervals, or spacelike hypersurface) and the goal is to explain how relativistic spacetime emerges at large scales.

Again, CST fails to give rise to spacetime manifolds and can't recover relativistic spacetime. As Crowther notes, this failure is symptomatic of more general problem in discrete theories of QG, which is that such models are always dominated numerically by non-manifold-like discreta. And a uniform distribution over sample space will render these undesirable ones much more likely. The assumption is that this could only be corrected by implementing a dynamics that constrained peaks on manifold-like discreta [Crowther, 2016, chapter 6]. While CST is not strictly speaking categorized as a pregeometric account, it suffers all the same problems and conceptual concerns as quantum graphity, and not only does it fail to recover spacetime, unless one adds a sum-over-histories element, it has no way to model quantum behavior.

A truly non-pregeometric account of discrete QG is causal dynamical triangulation (CDT). This account does employ the Feynman machinery but sums over virtual paths rather than physical trajectories. The fundamental discrete elements are 4-simplices. Ultimately, the fundamental elements are the superposition of all virtual "paths" or "spacetime histories" the universe can follow in "time." CDT provides a rule for gluing the 4-simplices such that it leads to a well-defined path integral and a four-dimensional spacetime. To get four-dimensional spacetime CDT also gives itself "causality" and a global integer-valued proper time t. As Crowther notes: "the building blocks" (4-simplices) are not considered real— although they are the basic elements described by the theory, they are removed from the theory before any claims about spacetime are made. In other words, they are mathematical tools, used simply to approximate the spacetime integral. "This is in contrast with causal set theory, where the elements of the causal set are taken as real and fundamental" [Crowther, 2016, p. 166].

This is why in the CDT account, spacetime is not considered a low-energy manifestation of something more fundamental at high-energy; the high-energy degrees of freedom are still spatiotemporal, though two-dimensional and not four-dimensional as in the low-energy case. This is a fancy way of saying that while CDT has much more success recovering spacetime than CST, it doesn't have matter or any idea how to bring matter into the picture, no beables whatsoever, in fact. Even though CDT is supposed to be a model of quantum systems at the Planck scale, the model has no fundamental physical elements. This is a fancy way of saying that CDT doesn't really start with some quantum beables and derive spacetime. As Crowther write, "spacetime is not thought to physically "emerge" from these elements" [Crowther, 2016, p. 167].

Furthermore, given that CDT gives itself 4-simplices, "causality," and a proper time t, it is an open question as to whether CDT could fail to recover some notion of spacetime. Aside from all the other concerns about CDT, even assuming it was clear what the ontology is and the beables are, the idea of gluing 4-simplices together via some dynamics that constitute some pregeometric temporal process to dynamically generate the block universe is incoherent. In other words, CDT suffers the same problems as the previous accounts.

This brings us to loop quantum gravity (LQG), which as we know is the leading competitor to string theory. LQG began with the idea the the Yang–Mills theory is really a theory of loops. As Crowther states, "in LQG the holonomy is a quantum operator that creates a 'loop state'" [Crowther, 2016, p. 180]. Think of loops in terms of the holonomy and gauge potential discussed in Philosophy of Physics for Chapter 5. A loop state is such that the field vanishes everywhere except along a single Faraday line. However, LQG starts out straightforwardly with a Hamiltonian formulation of GR, which then undergoes a canonical quantization procedure. This procedure has not been fully completed to date. In LQG we abstract graphs called spin networks that represent networks of loop interactions. Spin networks model networks of interwoven loops with "spin" representations sitting on the nodes and edges of the network. These spin representations quantify the discretely valued quantum "volume" to which each node corresponds. The discretely valued quantum "area" of the edge correspond to the surface of adjacency of the connected "volumes" [Crowther, 2016, pp. 182–3].

The actual dynamics of the spin networks are not agreed on at present, but the idea is that some Hamiltonian operator will act upon them as explained by Huggett and Wüthrich:

> In general terms, the action of the Hamiltonian on a node of the spin network will either be identity, i.e. the node simply persists through 'time,' or it splits the node into several nodes, or it fuses the node and some other nodes into a single node. The resulting structure is taken to be the quantum analog of a four-dimensional spacetime and is called 'spin foam.'

The spin networks bear a superficial resemblance to the discrete lattices introduced above, but there are two relevant differences. First, the actually existing and physically fundamental structure is supposed to be a quantum superposition of something like these spin networks, and not just a single spin network. Since all the different structures in the superposition will have a different connectivity (and perhaps different cardinality), and in this mathematical sense be different structures altogether, what is local in one term of the superposition will in general not be local in others. Except perhaps for very special states, local beables can thus not be part of the fundamental reality, but must instead emerge in some limit – presumably the same as that in which locality emerges. How such local, i.e. topological, structures like relativistic spacetimes emerge from spin networks is at present little understood.

Secondly, not only does the quantum superposition frustrate the applicability of locality criteria, but there is a sense in which even a spin network corresponding to a single term in the superposition is not amenable to the kind of localization that may be required to ensure empirical coherence. The problem is that any natural notion of locality in LQG—one explicated in terms of the adjacency relations encoded in the fundamental structure—is at odds with locality in the emerging spacetime. In general, two fundamentally adjacent nodes will not map to the same neighbourhood of the emerging spacetime. Hence the empirically relevant kind of locality cannot be had directly from the fundamental level.

[Huggett and Wüthrich, 2012, p. 6]

Once again the message is that LQG has yet to recover relativistic spacetime, in part because it has no local beables. Rovelli, one of the fathers of LQG writes, "even disregarding the incompleteness of the theory; the conceptual price for this result is heavy: the theory gives up unitarity, time evolution, Poincaré invariance at the fundamental level, and the very notion that physical objects are localized in space and evolve in time" [Rovelli, 2007, p. 1302]. As Huggett and Wüthrich explain:

The entire programme of LQG stands and falls with the success of this endeavour: if we fail to recover relativistic spacetimes from the spin foams and spin networks of LQG, and hence fail to understand the relationship between general relativity and LQG, there is no prospect of explaining why general relativity was as empirically successful as in fact it was while at the same time false, and therefore in need of replacement. However, it should be noted that every theory hoping to dislocate a fundamental theory which is 'incumbent' at the time must discharge this explanatory debt. We see no other way for LQG of doing so unless it is understood how spacetimes emerge from spin networks. Once this is understood, however, the threat to LQG's empirical coherence is ipso facto averted.

[Huggett and Wüthrich, 2012, p. 20]

6.10   The problem of time

Here is how Weinstein and Rickles describe the problem of time in quantum gravity:

It is not surprising, then, that a theory of quantum spacetime would have a problem of time, because there is no classical time against which to evolve the 'state.' The problem is not so much that the spacetime is dynamical; there is no problem of time in classical general relativity (in the sense that a time variable is present). Rather, the problem is roughly that in quantizing the structure of spacetime itself, the notion of a quantum state, representing the structure of spacetime at some instant, and the notion of the evolution of the state, do not get any traction, since there are no real 'instants.'

[Weinstein and Rickles, 2016]

Nonetheless, if one had not just read the quotation from Rovelli, one might ask if the defenders of LQG also claim that time is fundamental because they associate time with the most basic dynamical evolution of the spin networks; whatever those dynamics turn out to be. One can easily imagine Smolin making such a claim and there is nothing in the formalism that prevents such an interpretation. But the most ardent defender of LQG, Rovelli, states just the opposite:

In a certain sense, space no longer exists in fundamental theory; the quanta of the gravitational field are not *in* space. In the same sense, time no longer exists in the fundamental theory: the quanta of gravity do not evolve *in* time. Time just counts their interactions. As evidenced with the Wheeler–DeWitt equation, the fundamental equations no longer contain the time variable. Time emerges, like space, from the quantum gravitational field.

[Rovelli, 2017, p. 176]

It is well known that LQG allows for many exact solutions (too many) to the Wheeler–DeWitt equation and that Rovelli accepts that some equation like that is the master equation for the universe as a whole. So even though Rovelli embraces the block universe and even more so the Wheeler–DeWitt equation, we still have a QG program where one is attempting to recover spacetime and the block universe from a quantum process that, as Crowther notes, "would involve both the micro/macro transition as well as the quantum/classical transition, where the latter perhaps will need to be understood before the former can be implemented" [Crowther, 2016, p. 198]. Her point is that LQG as of yet has mastered neither transition and that LQG must first resolve all the standard interpretational problems in QM before it can progress. Perhaps Rovelli has already married his relational interpretation to LQG? If so, we are unaware of this. Clearly when Rovelli denies the fundamentality of time he means both spacetime and time as a variable.

6.11   Local beables

Does RBW have local beables? There has been a great deal of hand-wringing lately in the foundation literature on QG as to whether the most fundamental unifying theory from which spacetime emerges must have local beables to be empirically coherent and correspondfully with higher-level physical theories and the experienced world. Maudlin notes that [Maudlin, 2007, p. 3157], "local beables do not merely exist: they exist somewhere," or, as Bell explains [Bell, 1987, p. 234], beables are "definitely associated with particular space-time regions." Of course there is less consensus about the necessary and sufficient conditions for being a local beable. So where does this leave the status of spacetimesource elements?

Huggett and Wüthrich describe "local beables":

> The bottom line is that empirical science as we understand it presupposes the existence of what John Bell (1987, 234) called 'local beables.' 'Beables' are things that we take to be real, from fundamental objects of the theory to more familiar objects of experience (in the context of interpreting quantum mechanics they are candidates for being-hence they are 'be-able'). 'Local' is a term with a number of meanings, some technical, but Bell has in mind locality in the sense of being "definitely associated with particular space-time regions"—so 'local' here does not carry any direct consequences for interactions or their propagation. What he has in mind is closely related to the principle of 'separability' proposed by Albert Einstein (1948, §2). For our purposes it is generally adequate to take a beable to be local if the degrees of freedom describing it are associated with an open region of spacetime. It should be noted that this condition for locality is very weak as it stands – entities spreading across different galaxy clusters still qualify as 'local.' Hence, we think of this locality condition as necessary, but likely not sufficient, for observables.
>
> [Huggett and Wüthrich, 2012, p. 5]

To return to the main question about the status of spacetimesources, local beables are thought of as being separate from but located somewhere in spacetime, whereas, again, spacetimesources are *of* space, time, and sources. That said, the various source values (observables) of our fundamental elements are certainly localized on the graphs.

Concerning the locality of beables, Einstein writes:

> if one asks what is characteristic of the realm of physical ideas independently of the quantum theory, then above all the following attracts our attention: the concepts of physics refer to a real external world, i.e., ideas are posited of things that claim a 'real existence' independent of the perceiving subject (bodies, fields, etc.), and these ideas are, on the other hand, brought into as secure a relationship as possible with sense impressions. Moreover, it is characteristic of these physical things that they are conceived of as being arranged in a spacetime continuum. Further, it appears to be essential for this arrangement of the things introduced in physics that, at a specific time, these things claim an existence independent of one another, insofar as these things 'lie in different parts of space'. Without such an assumption of mutually independent existence (the 'being-thus') of spatially distant things, an assumption which originates in everyday thought, physical thought in the sense familiar to us would not be possible.
>
> [Einstein, 1948, pp. 321–2]

Einstein appears to conflate (or at least highlight) several different notions of "local" in this passage, including,

1. Local as localized in spacetime.
2. Local as possessing primitive thisness with intrinsic properties.
3. Local as in no superluminal interactions.
4. Local as in being otherwise independent (e.g., statistically) of entities at other points in spacetime.

Our beables are local in the first and third sense. Is this sufficient for beables being local? Huggett and Wüthrich say the following:

> 'Normal' theories, which postulate space and time, allow for local beables in their fundamental ontology, so the observable local beables can potentially be understood as composed, spatiotemporally, of the fundamental ones. A theory that does not postulate familiar space and time in its fundamental ontology precludes fundamental local beables, and there is then no obvious strategy for identifying observable local beables—and the question of the theory's empirical significance becomes acute. Indeed, the problem is not merely one of not knowing how to test such theories."
>
> [Huggett and Wüthrich, 2012, p. 12]

Per our modified Regge calculus, RBW has discretized spacetime at bottom. And as Huggett and Wüthrich note "mere discreteness is not a great conceptual leap from ordinary spacetime." Discretized GR is neither radically non-temporal nor radically non-spatial:

> One step removed from relativistic spacetimes, then, we find discrete lattices consisting of a set of basal events exemplifying a structure spanned by the spatiotemporal relations obtaining among pairs of such events. At this first step, it is important that the relevant relations be interpreted spatiotemporally, in such a way that the only relevant difference is that what was a continuum before is now a lattice. To facilitate the discussion, let us take the image of a lattice somewhat seriously and imagine, without much loss of relevant generality, that all spatiotemporal relations among the lattice points supervene on a basis of elementary spatiotemporal relations capturing spatiotemporal adjacency. We can thus think of these elementary relations as 'spacing' the basal events—the nodes of the lattice—equidistantly in spatiotemporal terms. The structure thus possesses a minimum spatiotemporal distance. The discreteness of the structure is captured by the demand that any two basal events can be connected by elementary relations with only finitely many intermediate events.
>
> [Huggett and Wüthrich, 2012, p. 7]

So RBW, as opposed to, say, configuration space fundamentalists, wave function realists, and LQG, seems safe in terms of positing local beables. However, in RBW the observable local beables are not spatiotemporally composed of the fundamental ones—spacetimesource elements. Given spatiotemporal contextuality, the absence of autonomous entities with intrinsic properties, and disordered locality, the relationship between spacetimesource elements and observables is not one of composition in any ordinary sense. Though again, properties that live on the links of the graph are certainly localized in spacetime. So we see no reason to worry that RBW is "empirically incoherent" or epistemically self-undercutting in terms of justification, either with respect to time, space, or mereology. Nor, given RBW, do we see any necessary reason why QG must go to more radical options that do threaten local beables and empirical coherence such as LQG and certain pregeometric accounts.

6.12   Stuck in the dynamical paradigm: RBW to the rescue

Let us take stock. On one side we have those "time lovers" who want to save our everyday experience of time based on some pregeometric process that is somehow supposed to explain time as experienced and underwrite the dynamical nature of spacetime in GR, most of whom nonetheless admit the existence of a block universe. On the other side, we have those "time haters" who embrace Wheeler–DeWitt and yet still advocate for some funky quantum process at bottom that recovers the timeless block universe. It seems to us that both sides, in spite of their apparent differences and rhetoric about the place of time, are stuck in the mechanical paradigm where explanation is still fundamentally dynamical, pregeometric or otherwise, and there is some basic entity that must be evolved dynamically such as loops, however counter intuitive they may be.

In short, we don't see that much difference between the time lovers and the time haters. It seems to us that the compositional/levels reductionism and the assumption that explanation is fundamentally dynamical reinforce each, forcing researchers of all stripes further and further away from each other's arguments, and further and further from reality at the level of experience and as described by classical physics. We hope the reader sees that RBW provides another alternative all together.

In other words, despite decades of effort by some of the most brilliant and inventive mathematical minds of our time, no one has managed to derive GR's spacetime structure from something that doesn't already posit essential features of spacetime. We believe that this reason alone warrants an entirely new attitude toward QG, although, as we argued in the main threads of chapters 4, 5, and 6, we were led to our novel attitude toward QG for independent reasons. Again, per our 4D spatiotemporal ontological contextuality, the quantum and classical are co-dependent upon each other, neither "emerges" from the other so there is no pretense of recovery or emergence of spacetime from some "thing" or "things" without spatiality or temporality. Rather, diachronic objecthood, space, and time of classical spacetime are co-fundamental per the AGC, as we explained in Foundational Physics for Chapter 4.

That is, the relations matrix $\mathbf{K}$ of our modified lattice gauge theory approach can be constructed from boundary operators on the graph, $\mathbf{K} = \partial_1 \partial_1^T$ where $\partial_1 \partial_2 = 0$ is the graphical counterpart to $\partial\partial = 0$. $\partial_1 \partial_2 = 0$ is merely saying that if an oriented link goes from node 1 to node 2 as part of the boundary of an oriented plaguette, then "out of node 1" cancels "into node 2" in a global sum. Demanding that the source vector $\vec{J}$ reside in the row space of this $\mathbf{K}$ per the AGC entails divergence-free sources which, in conjunction with the AGC's demand for conserved exchanges, mathematically characterizes an ontology for quantum statistics in the context of the worldlines/worldtubes of classical spacetime as per the Born rule (as explained in the main thread of chapter 4). So, we propose that this topological tautology $\partial_1 \partial_2 = 0$ underwrites the transtemporal identity employed tacitly in all dynamical theories. Accordingly, our discrete (graphical) starting point provides a topological basis for sources $\vec{J}$, space, and time. For example, the process by which the worldlines for two sources in spacetime is modeled using $\vec{J}$ on the graph of Figure 4.19 is an organization of the set $\vec{J} = (J_1(t_1), J_1(t_2), J_1(t_3), J_2(t_1), J_2(t_2), J_2(t_3))$ on two levels: there is the split of the set into two subsets $(J_1(t_1), J_1(t_2), J_1(t_3))$ and $(J_2(t_1), J_2(t_2), J_2(t_3))$, one for each source, and there is the ordering t_i over each subset. The split represents space, the ordering represents time, and the result is relational, diachronic objecthood. In this sense, space, time, and sources are relationally and inextricably co-constructed in our formalism. Consequently, we believe the articulation of the otherwise tacit construct of dynamical entities, that is, "the idea that individual entities exist at all" [Siegfried, 2008, p. 28], has a mathematical counterpart fundamental to the action, viz., the boundary of a boundary principle, $\partial\partial = 0$, that encodes "the central role of quasi-separability" [Albrecht and Iglesias, 2012, p. 67] at the fundamental level.

We hope RBW so described is the realization of a dream anticipated by Wheeler as follows from his list of "five clues" about QG:

> *First clue:* The boundary of a boundary is zero. This central principle of algebraic topology, triviality, tautology, though it is, is also the unifying theme of Maxwell electrodynamics, Einstein geometrodynamics, and almost every version of modern field theory. That one can get so much from so little, almost everything from almost nothing, inspires hope that we will someday complete the mathematization of physics and derive everything from nothing, all laws from no law."
>
> [Wheeler, 1994, p. 302]

This topological characterization of sources, space, and time (in a graphical sense) would constitute a reductive approach to QG if the spatiotemporal properties of the graphical spacetimesource element were independent of the classical spacetime metric. But, this would introduce a totally superfluous ontological aspect, since we don't have "quantum entities" mediating quantum exchanges, as argued in the main thread of chapter 4. Thus, again, it makes far more sense to close the explanatory system upon itself per spatiotemporal ontological contextuality—contextuality writ large for the explanatory "process" as a whole. As we explained in the main thread of chapter 4, in the language of quantum information theory, we are simply saying that no counterfactual definiteness means that information availability has bottomed out. So, if there is no information available for that spacetime region and any such screened-off quantum entity is equivalent to direct action, then why bother positing a screened-off quantum entity at all? Consequently, QM motivates direct action which motivates our spatiotemporal ontological contextuality approach to all fundamental physics, which means the intractable problem of recovering GR spacetime from pre-geometric QG is a non-starter for us. This point about information and contextuality was Wheeler's second clue [Wheeler, 1994, p. 303].

From the perspective of dynamism, a unifying enterprise with something like ontological contextuality at bottom might seem an Ouroboros-like cheat. We want to know why classical reality is the way it is. As Weinberg wrote, concerning his "grand reductionism," "you have to ask, why is it the way it is? And it isn't always in terms of something smaller, but in terms of something deeper" [Weinberg, 1999, p. 384]. Since our AGC is based on the same topological maxim ($\partial\partial = 0$) as classical field theory (as we explained in Foundational Physics for Chapter 4 and reviewed earlier), in what sense is it explaining the appearance of classical reality, rather than simply positing this appearance? And since the AGC isn't dynamical, how can it be a "deeper explanation" per "grand reductionism?" In response, we can only ask those of dynamical persuasion to temporarily set aside this prejudice in order to consider an alternative account of grand reductionism, one in which an adynamical global constraint for the block universe provides contextuality writ large as a deeper *self-vindicating unification of physics*. Keep in mind that even Weinberg cautions against what he calls "petty reductionism." We only ask that you do the same:

> We first of all ought to distinguish between what . . . I like to call grand and petty reductionism. Grand reductionism is what I have been talking about so far—the view that all of nature is the way it is (with certain qualifications about initial conditions and historical accidents) because of simple universal laws, to which all scientific laws may in some sense be reduced. Petty reductionism is the much less interesting doctrine that things behave the way they do because of the properties of their constituents: for instance, a diamond is hard because the carbon atoms of which it is composed can fit together neatly. Grand and petty reductionism are often confused because much of the reductive progress in science has been in answering what things are made of, but one is very different from the other. Petty reductionism is not worth a fierce defense. Sometimes things can be explained by studying their constituents— sometimes not . . . In fact, petty reductionism in physics has probably run its course . . . It is also not possible to give a precise meaning to statements about particles being composed of other particles. We do speak loosely of a proton as being composed of three quarks, but if you look very closely at a quark you will find it surrounded with a cloud of quarks and antiquarks and other particles, occasionally bound into protons . . . It is grand reductionism rather than petty reductionism that continues to be worth arguing about.

> [Weinberg, 1999, p. 384]

In summary, the so-called quantum and classical in our view are ontologically co-dependent. Properly understood, they are both necessary and sufficient for one another, as we explained in chapters 4 and 5. The approach introduced here avoids the dilemma posed by Smolin and Redhead in chapter 1 concerning reductionism because it admits *a priori* that the entire enterprise is one of self-consistency via contextuality. Thus, the ultimate expression/explanation is not "at bottom" some thing or dynamical entity or even a stand-alone ontic modal structure as in ontological structural realism (OSR), conceived at higher energies and smaller spatiotemporal scales, begging for justification from some thing at some yet "deeper" scale, but contextuality writ large for the explanatory "process" as a whole. Contextuality writ large is just extremum thinking writ large, which truly transcends the dynamical perspective; it is the heptapod's Theory of Everything.

So, while some might complain that our spatiotemporal ontological contextuality is a cop-out, we would respond that they are just assuming some broadly "atomistic" picture of reality and we think that is precisely what QM and relativity are telling us we must abandon. Also lost is the complete openness of the future, given the reality of the future in the block universe, and we know many find this distressing on a number of fronts.[16] Further, for many all this will be too high a price to pay. This is why, again, it is important to understand that RBW is not just another interpretation of quantum physics. RBW is an approach to QG and unification that provides a novel alternative to the long-standing impasse of fundamental physics and foundations of physics as explained in this book.

6.13 Summary

The differences between our approach to QG and the others are profound:

1. RBW is a rejection of the mechanical universe with its dynamical bias.
2. RBW puts space, time, and matter on the same footing in terms of fundamentality (spacetimesource elements).
3. Our spatiotemporal ontological contextuality is a rejection of the kind of constitutive entity-based reductionism that is assumed in almost all theories of QG; there are nothing like loops or strings at bottom for us, nor any pregeometric entities. There are no autonomous entities with intrinsic properties.

While quantum graphity [Prescod-Weinstein and Smolin, 2009; Caravelli and Markopoulou, 2012] also employs disordered locality, it and all other extant formal models of QG, even those attempting to recover spacetime [Oriti, 2013],

---

[16] We will address these concerns in chapters 7 and 8.

are dynamical according to Carroll [2012]. Rather than finding a rule for time-evolved entities per Carroll (e.g., causal dynamical triangulation [Ambjorn et al., 2013]), the AGC leads to the self-consistency of a graphical spacetime metric and its relationally defined sources. While we talk about "constructing" or "building" spatiotemporal objects we are not implying any sort of "evolving block universe" as in causet dynamics [Sorkin, 2007b]. Our use of this terminology is merely in the context of a computational algorithm. This is certainly something we share with ontological structural realism (OSR) [Ladyman, 2007; Ladyman and Ross, 2007]. However, while OSR merely changes the base of what is fundamental in the axiomatic structure of the universe to modal structures, RBW is a rejection of any kind of constitutive or compositional reductionism in which autonomous entities such as strings or loops at smaller length scales and higher energies determine everything else.

Rickles and Bloom start to capture this contextuality and how to think about it with regard to the world of experience:

> Clearly, eliminating things as the fundamental ontological category (out of which the world is composed) will not really alter our everyday interactions with the world: we will still drink tea out of teacups, and not some flimsy structural counterparts thereof. Moreover, at least in terms of our direct experimental dealings, it will be things that form the objects of discourse (results of observations and experiments) and thing-language that is used to speak about them. However, *when, e.g., the experimentalist speaks of a click in a counter or a spark on a screen or a number on a dial, he is really speaking elliptically about a component in a relation (that is, a relation in disguise).*
>
> [Rickles and Bloom, 2015, p. 105]

Foundational Physics for Chapter 6

This foundational physics thread contains mathematical details omitted from the main thread of chapter 6. Essentially, per modified Regge calculus (MORC) the Lagrangian density is modified by the addition of the following from Padmanabhan, 2004:

$$L = -\partial_\lambda h_{\alpha\beta}\partial^\lambda h^{\alpha\beta} + 2\partial_\lambda h_{\alpha\beta}\partial^\beta h^{\alpha\lambda} \tag{6.1}$$

so that the metric $g_{\alpha\beta}$ for the spacetime manifold M becomes $g_{\alpha\beta} + h_{\alpha\beta}$ in regions on M affected by disordered locality. This modification to L is for the classical field theory, but it can be written as $\frac{1}{2}\vec{Q}\cdot\mathbf{K}\cdot\vec{Q}$ on the graph for a quantum exchange of energy–momentum associated with the scalar field $h_{\alpha\beta}$ on plaquettes [Stuckey et al., 2016c]. To find the functional form of the classical field theory correction $h_{\alpha\beta}$ we solve

$$\Box h_{\alpha\beta} - \frac{1}{2}\eta_{\alpha\beta}\Box h = -16\pi G T_{\alpha\beta} \tag{6.2}$$

where h is the trace of $h_{\alpha\beta}$ and $\Box$ is the d'Alembertian given by $-\frac{1}{c^2}\frac{\partial^2}{\partial t^2} + \nabla^2$. This becomes simply $\Box h_{\alpha\beta} = 0$ in vacuum. This small correction to the metric can account for the SCP Union2.1 SN Ia data without Λ using Regge calculus EdS cosmology with a correction of the proper distance $D_p = \sqrt{D_p \cdot D_p} \to \sqrt{1 + \overline{h_{ii}}}\, D_p$ where $\nabla^2 h_{ii} = 0$ in the flat space of EdS cosmology. That is, $\frac{d^2}{dD_p^2}h_{ii} = 0$ where D_p is proper distance per the EdS metric. Thus, $D_p \to \sqrt{1 + \frac{D_p}{A}}\, D_p$. A is an arbitrary constant used as a fitting parameter, although ultimately we would expect A to be specified or constrained by an underlying graphical action for Regge calculus modified per disordered locality.

For dark matter phenomena, we must find an approximation method for $M \to M_p$. Since $dM_p = \sqrt{g_{rr}}\, dM$ for computing the difference in mass due to binding energy in a spherically symmetric spacetime [Wald, 1984, p. 126], we will assume $dM_p = \sqrt{1 + h}\, dM$ in analogy with our correction of proper distance above. Further, $\nabla^2 h = 0$ with spherical symmetry assumed for galactic rotation curves, X-ray cluster mass profiles, and the baryon–photon perturbations in pre-recombination FLRW cosmology. It does not have to be the case that $h \ll 1$, that is, h does not have to be a small change, as is obviously the case with dark matter phenomena where proper mass can be ~ 10 times as large as dynamic mass. However, as we will show, in these cases the corrected spacetime geometry *proper* is still perturbatively small in an absolute sense. In all three cases, we are modifying what is already a weak term in the Schwarzschild metric, so we are essentially correcting a Newtonian potential (first-order Schwarzschild metric). In the FLRW case, where the background FLRW metric being corrected is truly relativistic, we must specify the gauge. However, as we will see shortly, there is a Newtonian gauge for the perturbations that accommodates this view.

We should point out that even the relativistic EdS solution allows for a Newtonian-like characterization per Regge calculus. As we obtained previously [Stuckey et al., 2012b], the Regge equation for EdS cosmology with continuous time is

$$\frac{\pi - \cos^{-1}\left(\frac{v^2/c^2}{2(v^2/c^2+2)}\right) - 2\cos^{-1}\left(\frac{\sqrt{3v^2/c^2+4}}{2\sqrt{v^2/c^2+2}}\right)}{\sqrt{v^2/c^2+4}} = \frac{Gm}{2rc^2} \tag{6.3}$$

With $v^2/c^2 \ll 1$ an expansion of the LHS of Eq. 6.3 gives

$$\frac{v^2}{4c^2} + \mathcal{O}\left(\frac{v}{c}\right)^4 = \frac{Gm}{2rc^2} \tag{6.4}$$

Thus, to leading order (defining "small" simplices required to accurately approximate GR) we have $\dfrac{v^2}{2} = \dfrac{Gm}{r}$, which is just a Newtonian conservation of energy expression for a unit mass moving at escape velocity v at distance r from mass m. So, in the case of Regge EdS cosmology, we see that spacetime curvature enters as a deficit angle between adjoining Newtonian simplices. We are certainly not advocating a Newtonian approach to gravity, quite the contrary. As we explained in the main thread for chapter 6, we are advocating is a geometric view of even weak gravitational fields.

Returning to our ansatz applied to the dark matter phenomena, the solution $h = \frac{A}{r} + B$ of $\frac{1}{r^2}\frac{d}{dr}\left(r^2\frac{dh}{dr}\right) = 0$ is then used to obtain proper mass M_p from dynamic mass M per $dM_p = \sqrt{1+h}\,dM = \sqrt{\frac{A}{r}+B}\,dM$. As with dark energy, the arbitrary constants A and B are used as fitting parameters at this stage in the program. The values of these fitting parameters in galactic, galactic cluster, and cosmological contexts will then provide guides to the construction of the next guess in the iterative process. Indeed, as we will see, the best fit values of A and B show interesting trends across the three datasets.

Again, in order to fit the SCP Union2.1 SN Ia supernova data matching that of ΛCDM, we simply corrected proper distance D_p in EdS cosmology as outlined above. The type Ia supernova data to be fit are distance modulus (μ) versus redshift (z), that is, $\mu = 5\log\left(\frac{D_L}{10pc}\right)$ where the luminosity distance $D_L(z) = (1+z)D_p$ and D_p depends on the cosmology model being used. In MORC, we have

$$D_L = (1+z)D_p \rightarrow (1+z)D_p\sqrt{1+\frac{D_p}{A}} \tag{6.5}$$

where D_p is the proper distance obtained using the Regge calculus EdS solution and A is a fitting parameter from the solution for $h_{\alpha\beta}$. From our Regge calculus EdS solution we have

$$D_p = \int \left(\frac{F'(b)}{bF(b)} \sqrt{1 + \frac{b^2}{4}} \right) db \tag{6.6}$$

where

$$F(b) = \frac{\sqrt{4 + b^2}}{2[\pi - cos^{-1}(\frac{b^2}{4+2b^2}) - 2cos^{-1}(\frac{\sqrt{4+3b^2}}{2\sqrt{2+b^2}})]} \tag{6.7}$$

with $b = \frac{R\dot{a}}{c}$ and R is the lattice spacing. Recall from chapter 3 that $a(t)$ is the scaling factor for the spatial part of the metric in FLRW cosmology. We find the MORC sum of squares error (SSE) for $\frac{\mu}{5} - 8$ is robust against variation in R and nodal mass m. Specifically, SSE = (1.630 ± 0.002) for $A = (7.48 \leftrightarrow 10.6)$ Gcy with a current Hubble constant of $H_o = (69.9 \leftrightarrow 75.1)$ km/s/Mpc using the MORC values $R = (2.11 \leftrightarrow 8.39)$ Gcy and $m = (0.301 \leftrightarrow 17.5) \times 10^{51}$ kg. The best fit ΛCDM gave SSE = (1.639 ± 0.003) using $H_o = (68.9 \leftrightarrow 70.1)$ km/s/Mpc, $\Omega_M = (0.24 \leftrightarrow 0.28)$ and $\Omega_\Lambda = (0.72 \leftrightarrow 0.76)$. Both of these fits were superior to the EdS best fit with SSE = 2.67 and $H_o = 60.9$ km/s/Mpc (Figure 6.10). A recent study has found $H_o = 73.00 \pm 1.75$ km/s/Mpc [Riess et al., 1998].

As we explained in the main thread for chapter 6, the metric modification for dark matter phenomena we propose has to do with contextuality already inherent in GR. For example, suppose a Schwarzschild vacuum surrounds a sphere of FLRW dust connected at the instanteously null Schwarzschild radial coordinate. The dynamic mass M of the surrounding Schwarzschild metric is related to the proper mass M_p of the FLRW dust by the following, from Stuckey, 1994:

$$\frac{M_p}{M} = \begin{cases} 1 & \text{flat model} \\ \dfrac{3(\eta - \sin(\eta))}{4\sin^3(\eta/2)} \geq 1 & \text{closed model} \\ \dfrac{3(\sinh(\eta) - \eta)}{4\sinh^3(\eta/2)} \leq 1 & \text{open model} \end{cases} \tag{6.8}$$

That $M = M_p$ for the spatially flat FLRW dust (EdS cosmology) follows in part from the fact that the spatial curvature scalar equals zero $^{(3)}R = 0$ for the Schwarzschild metric. Accordingly, small differences in spacetime geometry can be characterized by relatively large differences in the values of mass. For example, while M_p can be $\sim 10M$, the deviation from flat spacetime on typical galactic scales per $\dfrac{2GM_p(r)}{c^2 r}$ in Eq. 3.12 is negligible given current technical limits on astronomical observations. Assuming circular orbits (which is common for fitting galactic rotation curves) we have $v^2 r = GM_p(r)$, where $M_p(r)$ is the proper mass inside the circular orbit at radius r and v is the orbital speed ("global determined" value for proper mass inside the orbit about the galactic center). This gives $\dfrac{2GM_p(r)}{c^2 r} = 2\dfrac{v^2}{c^2}$. The largest galactic rotation speeds are typically only $10^{-3}c$ so the metric deviation from flat spacetime is $\sim 10^{-6}$; thus, $M_p \sim 10M$ still constitutes an empirically small metric correction $h_{\alpha\beta}$.

The other way we quantified the geometric difference between M and M_p was to use $^{(3)}R = \dfrac{16\pi G\rho}{c^2}$ from Wong, 1971, on galactic scales (where ρ is the mass density). Assuming constant mass density (typical in a galactic bulge) out to radius r where we have orbital velocity v, we have $\dfrac{16\pi G\rho}{c^2} = \dfrac{12v^2}{r^2 c^2}$. Constant density means $M(r) \sim r^3$ so

$$v(r) = \sqrt{\frac{GM(r)}{r}} \tag{6.9}$$

increases linearly with distance from zero to its max value v_{max}. Assuming this happens at the radius of the bulge r_b (galactic rotation curves actually peak farther out, so this overestimates the effect), we have $^{(3)}R = \dfrac{12v_{max}^2}{r_b^2 c^2}$ for the spatial curvature scalar in the annulus at r rather than its Schwarzschild value of zero. $v_{max} \approx 300$ km/s and $r_b \approx 2000$pc gives $^{(3)}R \sim 10^{-45}/\text{m}^2$. So, a factor of 10 one way or the other in ρ is geometrically inconsequential in this context.

As we explained in the main thread and depicted in Figure 6.5, if we consider the ratio $\frac{M_p}{M}$ for a ball of FLRW closed-model dust surrounded by Schwarzschild vacuum as joined at FLRW radial coordinate χ_o we have, following Stuckey, 1994,

$$\frac{M_p}{M} = \frac{3(2\chi_o - \sin(2\chi_o))}{4\sin^3(\chi_o)} \tag{6.10}$$

For $\chi_o = \frac{\pi}{2}$ Eq. 6.10 gives $\frac{M_p}{M} = 2.36$ and for $\chi_o = 0.8\pi$ Eq. 6.10 gives $\frac{M_p}{M} = 22.1$. Although the extrinsic curvature of the interface changes sign for $\chi_o > \frac{\pi}{2}$ so this region is probably not physically realistic, the point is simply that GR allows for large differences between proper and dynamic mass. As explained in the main thread, we attribute this mass difference to a difference in spatial curvature, since $M = M_p$ for the spatially flat FLRW model surrounded by Schwarzschild vacuum and $^{(3)}R = 0$ for both the spatially flat FLRW dust and Schwarzschild vacuum.

Thus, as explained in the main thread, we assume that little if any of the matter in the original uncoalesced state remains. Therefore, the global dynamics, for example, of galactic rotation curves, is determined in (large) part by the curvature of empty space remaining between the many regions of Schwarzshild vacuum surrounding coalesced matter. In other words, as the matter coalesces to the centers of Schwarzschild vacuum regions, it leaves a footprint of its proper mass in the curvature of the spatial boundaries between those Schwarzschild regions. This proper mass footprint is in large part responsible for the global dynamics,[17] since proper mass M_p is much larger than the Schwarzschild dynamic mass M.

Here is a (very) short explanation of our geometric modification of dynamic mass M to obtain proper mass M_p for our test fits of 12 high-resolution galactic RCs from the HI Nearby Galaxy Survey [Walter et al., 2008] (THINGS) used by Gentile and colleagues to explore MOND fits [Gentile et al., 2011]. [For details see Stuckey et al., 2016a.] We modified the dynamic mass ΔM of each (discrete) annulus of galactic matter to obtain its proper mass ΔM_p (i^{th} component, where bulge, disk, and gas are the possible components) per

$$\Delta M_{p_i} = \sqrt{A + \frac{2B}{r_2 + r_1}}\,\Delta M_i \qquad (6.11)$$

with B and A (same for all components) fitting parameters. The THINGS data to be fit are rotation velocity versus orbital radius. Gentile and tea describe these data as "the most reliable for mass modelling, and they are the highest quality RCs currently available for a sample of galaxies spanning a wide range of luminosities." Our test fits rival MOND fits which were deemed "very successful" for these data [Gentile et al., 2011] (Figure 6.6).

We used this same technique to fit the mass profiles of the eleven X-ray clusters found in Brownstein [2009] as obtained from Reiprich and Böhringer [Reiprich, 2001; Reiprich and Böhringer, 2002] using combined ROSAT (ROentgen SATellite) and ASCA (Advanced Satellite for Cosmology and Astrophysics) data. Specifically, we used the continuum version of Eq. 6.11

$$M_p(r) = \int_0^r \sqrt{A + \frac{B}{r'}}\, dM = 4\pi \int_0^r \sqrt{A + \frac{B}{r'}}\, \rho(r')r'^2 dr' \qquad (6.12)$$

to modify the dynamic mass of each annulus of intracluster medium gas to obtain its proper mass. The X-ray cluster mass profile data to be fit are mass versus radius. Our test fits rival MSTG fits which were better than the MOND and STVG fits of these same data [Brownstein, 2009] (Figure 6.7).

Finally, we briefly explain our test and STVG fits of the CMB angular power spectrum per Planck 2015 data. Moffat and Toth [2013] modified G in Mukhanov's analytic approach[Mukhanov, 2005] such that "[w]hen a quantity containing G appears in an equation describing a gravitational interaction, G_{eff} must be used. However, when a quantity like Ω_b is used to describe a nongravitational effect, the Newtonian value of G must be retained." The modification is given by

$$G_{eff} = G\left[1 + \alpha\left[1 - \left(1 + \frac{\mu a}{k}\right)\exp\left(-\frac{\mu a}{k}\right)\right]\right] \qquad (6.13)$$

---

[17] Again, this is a simple heuristic since, of course, we advocate direct action and are not spacetime substantivalists. Indeed, we apply our correction of dynamic mass to the matter components proper.

Ω_o is the current total matter–energy density divided by the critical matter–energy density (required to make flat spatial hypersurfaces in FLRW cosmology). Ω_m is the current contribution to the total energy density due to matter. In the model here, we have $\Omega_m = \Omega_o$, that is, $\Omega_\Lambda = 0$, and the current photon and neutrino contributions are negligible. Ω_b is the current contribution to the total energy density due to baryons, so for our test fit and STVG, $\Omega_b = \Omega_m$, that is, $\Omega_{DM} = 0$. Since STVG is the best modified gravity fit of the CMB angular power spectrum to date, we compared our test fitting procedure to that of STVG. As it turns out, both can reproduce the standard cold dark matter (sCDM) fit of the first three acoustic peaks in a rather trivial fashion. To see this, we provide a review of an sCDM fitting procedure below. Before we start, we point out that STVG gets rid of DM but it retains Λ while RBW gets rid of both DM and Λ. Therefore, our comparison of STVG and RBW in this context will only involve the replacement of DM in sCDM.

The analytic method of Hu and Sugiyama [Hu and Sugiyama, 1995, 1996] (HuS) maps closely to the full numerical solution in the first three acoustic peaks and its physics is transparent, so we used HuS. Since the Planck 2015 data strongly suggest a spatially flat universe [Planck Collaboration, 2016], that is, $\Omega_o = 1$, we work in that context. Further, we have no Λ, so $\Omega_m = \Omega_o = 1$ while a proper STVG fit would have $\Omega_m \approx 0.3$ and $\Omega_\Lambda \approx 0.7$. Here is a brief outline of HuS in this context with an explanation of our test and STVG modifications thereto.

The spatially flat FLRW metric (we will use $c = 1$ in what follows), from Hu, 1995,

$$ds^2 = -dt^2 + a^2(t)\delta_{ij}dx^i dx^j \tag{6.14}$$

is modified in the conformal Newtonian gauge by the existence of the perturbations $\Delta_i = \frac{\delta\rho_i}{\rho_i}$ to read

$$ds^2 = -(1 + 2\Psi)\,dt^2 + (1 + 2\Phi)\,a^2(t)\delta_{ij}dx^i dx^j \tag{6.15}$$

where Ψ is called "the Newtonian potential" and Φ is a perturbation to the spatial curvature. When pressure is negligible, $\Psi = -\Phi$. Since the spatial part of the Schwarzschild metric can be written, as per Wong, 1971:

$$ds^2 = \left(1 + \frac{GM}{2r}\right)^4 \delta_{ij}dx^i dx^j \tag{6.16}$$

we have for small $\dfrac{GM}{2r}\left(= -\dfrac{\Phi}{2}\right)$

$$ds^2 \approx \left(1 + \frac{2GM}{r}\right)\delta_{ij}dx^i dx^j \tag{6.17}$$

which explains the perturbation Φ in Eq. 6.15 as a small Newtonian potential placed in a spatially flat FLRW background. Thus, as with galactic rotation curves and X-ray cluster mass profiles, we will be correcting what is already a small effect, that is, Ψ and Φ. STVG will be correcting G as shown above. Einstein's equations directly relate Ψ and Φ to the total energy density ρ, which only appears coupled to G. That is, Einstein's equations give

$$k^2 \Phi = 4\pi G\rho \left(\frac{a}{a_o} \right)^2 \Delta_T \tag{6.18}$$

and

$$\Phi + \Psi = -\frac{8\pi G\rho}{k^2} \left(\frac{a}{a_o} \right)^2 \Pi \tag{6.19}$$

in the total matter rest frame gauge in k space where Π is the anisotropic stress and Δ_T is the total perturbation. Thus, a multiplicative correction to Ψ and Φ on the LHS of Eqs. 6.18 and 6.19 per our test fit could equally be viewed as a multiplicative STVG correction to G on the RHS. Eqs. 6.18 and 6.19 are represented in HuS as

$$\bar{\Phi}(a,k) = \frac{9}{10} \left(\frac{k_{eq}}{k} \right)^{3/2} \frac{a+1}{a^2} \left[1 + \frac{2}{5} f_v \left(1 - \frac{1}{3} \frac{a}{a+1} \right) \right] U_G(a) \tag{6.20}$$

and

$$\bar{\Psi}(a,k) = -\frac{6}{5} \left(\frac{k_{eq}}{k} \right)^2 \frac{a+1}{a^2} f_v \frac{\bar{N}_2}{(a+1)} - \bar{\Phi}(a,k) \tag{6.21}$$

where f_v is the fraction of the radiation energy density contributed by the neutrinos (0.405),

$$U_G(a) = \frac{9a^3 + 2a^2 - 8a - 16 + 16\sqrt{a+1}}{9a(a+1)} \tag{6.22}$$

and

$$\bar{N}_2 = \frac{12}{45a} \frac{\sqrt{k}}{\sqrt{k_{eq}}} \left[8 + 4a - 3a^2 - 8\sqrt{1+a} + 4a\log \frac{a}{4} + 4a\log \left(\frac{\sqrt{1+a}+1}{\sqrt{1+a}-1} \right) \right] \tag{6.23}$$

is the neutrino quadrupole contribution to the anisotropic stess with

$$k_{eq} = 7.46 \times 10^{-2} \left(\frac{T_o}{2.7} \right)^{-2} h^2 \Omega_o \tag{6.24}$$

T_o is the current temperature of the CMB and h is the current Hubble constant divided by 100 km/s/Mpc. Again, Planck data strongly suggest $\Omega_o = 1$, so we used that model. We found h smaller than accepted (0.49(1) instead of ~ 0.7) and Ω_b too large (0.090(1) instead of ~ 0.05), but the combination $h^2\Omega_b$ is close to the accepted range (0.02222 ± 0.00023) [Planck Collaboration, 2016]. Again, the point of the test fit wasn't to find the fit parameters per se, although that will certainly be important once we are at the stage of building a cosmology model. Rather, we wanted to find out what change is required to bring the third peak up to the second peak without DM in the context of our ansatz and compare that with an established program in modified gravity, STVG.

In order to remove the DM from this sCDM fit, we look to the very first step since that is the representation of Einstein's equations in the metric perturbations Ψ and Φ. There, we see that the only place matter of any kind enters the equations is through k_{eq} by way of $h^2\Omega_m$. Of course, that is because $h^2\Omega_m \propto G\rho_m(a_o)$. Thus, eliminating DM from $\rho_m(a_o)$ means a loss of 91% of the critical density necessary for $\Omega_m = \Omega_o = 1$. The correction needed to restore the HuS fit, having removed the DM, is straightforward and exact. Since we have $\Omega_m \to \Omega_b$ in removing DM, in order to restore the HuS fit, we must multiply $h^2\Omega_m$ by Ω_b^{-1}, i.e., $\Omega_b^{-1}(h^2\Omega_m \to h^2\Omega_b) = h^2 = h^2\Omega_m$. In order to have that change to k_{eq} in the RBW view, Φ and Ψ are multiplied by $\Omega_b^{-3/2}$, since they are both proportional to $k_{eq}^{3/2}$. In STVG, we simply need $G \to \Omega_b^{-1}G$ which Eq. 6.13 gives for $\mu = 0$ and $\alpha = \Omega_b^{-1} - 1$.

While the RBW ansatz is couched in real space and this correction is in k space, the correction factor is a constant, so it is the same in both spaces. Thus, the RBW correction to Φ and Ψ is simply $A = 0$ and $B = \Omega_b^{-3}$. As with the corrected Newtonian potentials above, even though this correction is much larger than 1, the corrected Φ and Ψ are still much less than 1 (they were already small to include DM).

Again, in STVG, we simply need[18] $G \to \Omega_b^{-1}G$ which Eq. 6.13 gives for $\mu = 0$ and $\alpha = \Omega_b^{-1} - 1$. Again, wherever you have $\Omega_m h^2$ (you have $G\rho_m(a_o)$) in your fitting procedure, $G \to \Omega_b^{-1}G$ exactly restores the loss of Ω_{DM} from Ω_m. In the standard best fit for Planck [2015], that is, ΛCDM, $\Omega_m \approx 0.3$, and $\Omega_b \approx 0.05$, so $G \to 6G$ allows baryons alone to do the work of DM plus baryons in the ΛCDM fit. That is what Moffat and Toth found [2013] using $\mu^{-1} \approx$ particle horizon, which is $\mu \approx 0$, so $G \to (1 + \alpha)G$. Since $G \to G_{eff}$ per STVG regardless of the bound baryon-photon perturbations, the primordial perturbations aren't necessary to produce the apparent enhanced mass of the baryons as is the case with RBW.

We must next explore the possibility that the fitting constants in our ansatz have a basis in astrometric quantities, since the results show interesting trends in the fitting factors A and B of our dynamic mass correction. That is, for our galactic rotation curve fits we see no correction in the six bulge contributions A_{bulge} and only a relatively small correction in five of twelve disk contributions A_{disk}, while there were relatively large corrections to all twelve gas contributions A_{gas}. The

---

[18] The k dependence is necessary for the matter power spectrum, so nonzero μ is required there.

constant B was small to non-existent in all 12 cases (0.55 ± 0.81). Moving to our X-ray cluster mass profile fits we see again large corrections to the ICM gas mass A with an increased contribution from the constant B although still one deviation within zero (600 ± 1100). For the primordial density perturbations, $A = 0$ and the correction is entirely in B. Also, the X-ray cluster results don't seem to follow a simple scaling of A with cluster mass. For example, we have $A = 826$ kpc for the bullet cluster with a proper mass $\approx 10^{15} M_\odot$, $A = 27800$ kpc for the Perseus cluster with a proper mass $\approx 10^{15} M_\odot$, and $A = 20900$ kpc for the Fornax cluster with a proper mass $\approx 10^{13.5} M_\odot$. Clearly, other geometrical factors are at work here according to the ansatz. Therefore, it would appear that the functional form for our ansatz is relevant and it is not unreasonable to search for astrometric hints in these fits.

Part III

Adynamical Explanation:
Time as Experienced

Part III Overview

Part III deals specifically with time as experienced and with the hard/generation problem of conscious experience more generally, as the the former is a subset of the latter. In particular we will discuss why these problems are made even harder in a block universe. This is important to do in its own right because RBW enables a kind of neutral monism that can simultaneously deflate the generation problem and explain the experience of time in a block universe. But it is also important because one of the most damning arguments against the block universe is allegedly the argument from the experience of time. As we discussed in chapter 1, there seems to be a conflict between time as experienced and time as characterized in foundational physics. At the very least, physics doesn't seem to have the resources to explain time as experienced. The block universe model is among the worst offenders in this regard, along with the time-symmetric nature of dynamical laws and certain accounts of quantum gravity (QG).

In Part II we used our version of adynamical explanation in the block universe to:

1. Resolve mysteries associated with the Big Bang, closed timelike curves, the flatness problem, the horizon problem, the low entropy problem, the black hole firewall paradox, and quantum mechanics.
2. Ameliorate the technical concerns of quantum field theory.
3. Provide a new path to unification and quantum gravity.
4. Dismiss the need for (non-baryonic) dark matter and dark energy (cosmological constant).

But despite the explanatory power of God's-eye physics, it does not have wide appeal in the foundations of physics community because, in part, it lacks a formal counterpart to the time-evolved Now, that is, "the ant's-eye view of human consciousness, which senses a succession of events in time." Thus, we perceive only the present instant (Now) and experience this present moving through time into the unknown experiences called the future while leaving a path of memories called the past. We will call the specialness of the present-as-experienced "Presence." How does physics couched in a block universe account for the Now? We agree with Einstein "that this important difference does not and cannot occur within physics."

Presence isn't the only feature of time as experienced that people worry the block universe in particular has a hard time explaining. The other two features are Passage (the experience of ceaseless change) and Direction (that time appears to move in only one direction from past to future). Of course, there is a more general worry about how to explain conscious experience of any sort if matter is fundamental, what we now call the "hard problem" of consciousness. We argue the block universe only makes this generation problem worse.

The good news is that RBW suggests a resolution to both the generation problem and the problem of time as experienced. That is, we will perform a double elimination of the hard/generation problem of consciousness. The generation problem is deflated because:

1. In RBW, even in physics, fundamental explanation isn't about causation or generation, so since we have given up the ghost of the dynam-ical/mechanical worldview for physical phenomena, why not give up it for conscious experience as well? That is, we discard the idea that matter *generates* conscious experience; matter doesn't even *generate* matter, except from the ants-eye view.

2. We are advocating for neutral monism and therefore dumping the idea that matter and mind are essentially different and distinct phenomena to begin with. So there is no profound mystery about how subjectivity "pops" out of insensate brains.

Fortunately, neutral monism also explains every feature of time as experienced and shows that time is neither a projection of the brain or something that must be added to fundamental physics.

In chapter 7 we will explain why many people claim that the "argument from time as experienced" is the most damning argument against the block universe. We will show that as typically presented, the argument from experience is easily refuted. We will then argue that all the "A-series" alternatives such as dynamical presentism, frozen presentism, and growing–glowing block universe are, at the very least, no better off than the block universe itself at accounting for time as experienced. We will further argue that the standard defense of the block universe in terms of the "argument from the equivalence of experience" is a good one. But, the proponent of the block universe, having eliminated the competitors, still has to explain time as experienced. Here we argue that the standard moves in this regard, such as invoking some physical arrow of time or mental representations/neural mechanisms, fail. Indeed, we argue that the block universe hypothesis makes it impossible to resolve the generation problem of conscious experience, including time as experienced. This will pave the way for chapter 8 where we reject the dynamical model for explaining conscious experience.

The block universe model makes the generation problem unsolvable. Once again, the dynamical paradigm is the culprit because it assumes that insensate matter is fundamental and must somehow *generate* conscious experience. We think it is high time to rethink the nature of both conscious experience and matter, and their relationship to one another. After all, if one is a realist about conscious experience and believes that matter is fundamental, at the end of the day the hard problem of consciousness is really this: how do neural processes *cause* or *generate* conscious experience? Fortunately, RBW gives us a way to deflate this problem. In chapter 8 we show that given how RBW reconceives matter, rejigs the adynamical block universe and fundamental explanation, RBW naturally admits of a kind of neutral monism that simultaneously deflates the generation problem and explains time as experienced. And we will then have refuted the idea that the block universe is incompatible with time as experienced.

In sections two and three of chapter 8 we will explain how RBW allows for a type of neutral monism that deflates the generation problem and explains time as experienced. These sections will focus on the Passage and Presence of time in particular and show in detail how the neutral monism of RBW explains Passage without making it either a mental projection or some fundamental arrow of time in foundational physics. The fourth section is about the Direction of time and argues that while there is no Objective Direction in time, explaining Passage is sufficient for explaining Direction. The fifth section argues that embodied, embedded, and extended cognitive science and phenomology support the neutral monism of RBW and its explanation for the experience of time. The sixth section focuses on freedom, spontaneity, and creativity in the relational block universe. Many people object to the block universe picture because they think it negates free will in human action, and negates creativity and spontaneity in the universe writ large. We argue that the reality of the future entails none of those things, and that RBW has all the freedom, creativity, and spontaneity anyone could reasonably desire. Finally, in section 7 we characterize Presence in detail and its relation to time as experienced.

7

Conscious Experience and the Block Universe

> *Once Einstein said that the problem of the Now worried him seriously. He explained that the experience of the Now means something special for man, something essentially different from the past and the future, but that this important difference does not and cannot occur within physics. That this experience cannot be grasped by science seemed to him a matter of painful but inevitable resignation.*
>
> [Carnap, 1963, p. 37]

7.1 Introduction

Today many philosophers and scientists think that matter is fundamental and so must somehow explain consciousness, but historically Einstein wasn't alone. Many other giants in physics also shared Einstein's view. Here, for example, is a quotation from Erwin Schrödinger in an interview by J.W.N. Sullivan that appeared in the Observer on January 11, 1931:

> Although I think that life may be the result of an accident, I do not think that of consciousness. Consciousness cannot be accounted for in physical terms. For consciousness is absolutely fundamental. It cannot be accounted for in terms of anything else.
>
> [Sullivan, 1931a, p. 16]

And here is an excerpt from an interview of Max Planck by J.W.N. Sullivan that appeared in the Observer on January 25, 1931:

> Sullivan: Do you think that consciousness can be explained in terms of matter and its laws?

> Planck: No. I regard consciousness as fundamental. I regard matter as derivative from consciousness. We cannot get behind consciousness. Everything that we talk about, everything that we regard as existing, postulates consciousness.
>
> [Sullivan, 1931b, p. 17]

Beyond the Dynamical Universe. Michael Silberstein, W.M. Stuckey and Timothy McDevitt,
Oxford University Press (2018). © Michael Silberstein, W.M. Stuckey and Timothy McDevitt.
DOI 10.1093/oso/9780198807087.001.0001

The subject matter of this chapter is the relationship between conscious experience (in particular all aspects of temporal experience) and the physical world. In this chapter we will first describe the alleged problems of accounting for time as experienced in a block universe. The features of time as experienced in question are:

1. Presence (the experience that there is an objective present moment).
2. Passage (the experience of ceaseless change—a moving present).
3. Direction (the experience that time moves in only one direction from past to future).

Many people claim that the "argument from time as experienced" is the most damning argument against the block universe, so for that reason alone we need to address this issue directly. We will show that, as typically presented, the argument from experience is easily refuted. We will then argue that all the "A-series" alternatives such as dynamical presentism, frozen presentism, and growing–glowing block universe are, at the very least, no better off than the block universe itself at accounting for time as experienced. We will further argue that the standard defence of the block universe in terms of the "argument from the equivalence of experience" is a good one. But, the proponent of the block universe, having eliminated the competitors, still has to explain time as experienced. Here we argue that the standard arguments in this regard, such as invoking some physical arrow of time or mental representations/neural mechanisms, fail. Indeed, we argue that the block universe hypothesis makes it impossible to resolve the generation problem of conscious experience, including time as experienced. This will pave the way for chapter 8 where we will provide a totally different account of conscious experience such as the experience of time and how it relates to the physical world.

7.2 On the incompatibility of experience with the Block Universe

Given our block universe physics, what remains to be explained is the experience of time and change (phenomenal time or time as experienced) that motivated humans to adopt the dynamical/mechanical universe paradigm from the start. The question is why do we experience Presence, Passage, and Direction when none of these things is given to us in fundamental physics? That is, "how can we humans amid a fundamentally changeless universe have so vivid perceptions as of a fleeting Passage of time? As long as this explanatory debt is not discharged, any approach entailing a fundamentally frozen world comes with a promissory note" [Huggett et al., 2013, p. 250]. Hugget has in mind here explaining time as experienced in the Wheeler–DeWitt universe as interpreted by Barbour. While we discharged the "problem of time" in chapter 6, the concern remains even for a block universe such as ours. This is an especially poignant question considering that these features of conscious experience are arguably the most essential and fundamental aspects of daily conscious experience, and the features of experience that foundational physics is now deeply concerned about how to explain.

There are two sets or types questions that we can focus on and it is important to begin by distinguishing them, at least for now. There are metaphysical questions and there are phenomenological or experiential questions. Traditionally, starting with Husserl, in a neo-Kantian transcendental spirit, the discipline of phenomenology has sought to "bracket" questions of experience from metaphysical questions. One may attempt to dismiss or deflate the metaphysical questions but everyone is obliged to respond to the phenomenological ones. And as we will see, there are those who believe that the metaphysical and physical arrows of time explain the phenomenological ones.

Following Price, the metaphysical questions are as follows [Price, 2011, p. 277]:

1. Is the Present moment objectively distinguished such that it is a frame- or perspective-independent fact about which events are present as opposed to past or future?

2. Does time have an objective Direction such that it is a frame- or perspective-independent fact about which two non-simultaneous events one is the earlier and one the later (is there an objective fact of the matter about which Direction is toward the past, such as allegedly the Big Bang, and which toward the future, such as allegedly the heat death of the universe)?

3. Irrespective of conscious observers, their frame of reference, or perspectives in the universe, is it an objective fact that there is a Passage or flow of time as suggested, for example, by dynamical presentism? [Price, 2011, p. 277]

On the phenomenological side, what we want to explain in this chapter is the experiential arrow of time which has the following features:

- Passage: the world is in constant flux such that the future becomes the present and the present becomes the past.

- Presence: the present moment is experienced as special or ontologically privileged.

- Direction: time appears to flow from a distinguishable past to a distinguishable future.

As purely metaphysical questions considered in isolation from phenomenology, we will answer "no" to all the preceding metaphysical questions. But as we said we still have to explain time as experienced. Furthermore, given the neutral monism defended in chapter 8, we think it is a mistake to bracket experience from the physical or metaphysical. That is, we simply deny that the conscious experience of time (and experience in general) is essentially distinct or different from physical phenomena. This, of course, changes the entire game because it is a false dichotomy to say that time as experienced is either "in the imagination" or reducible to some physical arrow of time. Given neutral monism, these three features of time as experienced are real, and they are not reducible to representations in the mind or any other neural or cognitive process. These three features of time are not explained by fundamental physics but they are real features of the world, not mere projections of the mind/brain. The metaphysical idiom suggests that these features could only be "objectively" real if they could exist in a universe without conscious observers, but we will resist that definition of both objective and real. As it turns out, however, in our view only one of these, Presence, is a universal and intrinsic feature, but not in the sense of objectively picking out present versus past or future events. As this chapter is devoted to explaining the problems, the reader will have to wait until chapter 8 for an elaboration of our account as outlined above. We will now do some conceptual work to qualify the various concerns.

Technically and logically speaking, whether in a metaphysical or phenomenological vein, the question of Direction is distinct from the question of Passage or Presence. That is, one could have a Passage without Direction and vice versa [Price, 2011]. As Prosser states:

> For there to be a Direction it is sufficient there is some kind of asymmetry between earlier and later, as there would be, for example, if the laws of physics were not time-reversible. But there is nothing in the notion of Direction that entails Passage. If this is not already clear, note that we can similarly make sense of an intrinsic Direction to space; we can, for example, imagine a physical process that can take place in an object oriented in one Direction, but not when oriented in the opposite Direction. This would no way suggest that space passes.
>
> [Prosser, 2013, p. 317]

Often in the literature these terms will be used interchangeably or an attempt will be made to ground one in the other. However, if one answers "yes" to the metaphysical Direction question then you have to decide if Direction is an irreducible feature or not. If not, then Direction must be reduced to some other arrow of time such as the thermodynamic arrow. When we are in the metaphysical mode one could try to reduce Direction to the experiential arrow of time, instead of the other way around, but that certainly violates the goal of an "objective" account as defined by Price. Obviously, in the phenomenological mode our job is only to explain experience. We must always keep in mind, however, that whether we are discussing Passage or Direction, and whichever mode we are in, that it is time as experienced that motivates these discussions. We will come back to the metaphysical question of Direction in chapter 8. But typically it is time as experienced that we want to explain and for which we invoke some physical arrow of time. However, if one attempts to ground the experiential arrow of time in some physical arrow of time such as the thermodynamic arrow, then we must be clear to show that this physical arrow has some objective Direction that is thoroughly explained by the physics alone.

For example, as we saw in chapter 3, even granting that the thermodynamic arrow of time has objective Direction—granting that FAPP thermodynamic processes are irreversible—discussions about why the thermodynamic arrow is asymmetric lead most defenders to advance the past hypothesis (PH) which begs questions such as why that initial condition occurred at the Big Bang, etc. Recall that this is because all the dynamical laws allegedly more fundamental than the second law of thermodynamics are all time-symmetric. It is precisely these sorts of questions (e.g., why the Big Bang) that we have argued the dynamical/mechanical worldview is in no position to answer without invoking eternal inflation, the multiverse, and the like, all of which just pushes these kinds of questions into the back-ground.

While there certainly have always been analytic metaphysicians concerned with explaining various features of time, there is now a growing number of people in both foundational physics and philosophy of physics that think time as experienced is a central concern. We believe there are a couple reasons for this. First, the hard problem of consciousness, of which time as experienced is a key subset, is currently a hot topic in philosophy, physics, cognitive science, and neuroscience. Second, an increasing number of people in foundational physics and philosophy of physics worry that physics will be incomplete unless it can explain time as experienced. As Hoefer and Smeenk wrote:

> The idea is that physical science can comprehend and account for all features of the external physical world that we all experience . . . an important goal of philosophy of the physical sciences has been to reconcile the world of common sense and daily experience with the world as described by the sciences. This is important in part because we desire to have an overall consistent set of beliefs about the world. But it also has an epistemological side: if the description of the world given by one or more physical theories appears to radically misdescribe the world in certain special ways, then that theory or theories may be held to be epistemologically self-undermining: if we took the theory to be true, we would have to doubt the correctness of the very experiences (of scientists, in laboratories and observatories) that is the only basis for believing the theory in the first place.

> [Hoefer and Smeenk, 2016, p. 127]

For many, this concern is compounded by the belief that the block universe interpretation of relativity radically increases the difficulty of explaining time as experienced.

> The problem is that both time's flow and time-asymmetry appear to be absent from fundamental physics . . . In the case of time's flow and the related notion of "now" or "the present," not only are these notions absent from physical theories, but relativity theory (special and general) appear to be incompatible with any objective "now" or "present" . . . In so far as our manifest notion of the "now" moving into the future requires the now to be universal, SR renders it perspectival and non-unique . . . neither SR or GR contains anything corresponding to the movement of the "now": both

seem to present spacetime as a four-dimensional block in which past, present and future are as indifferently equally real as the left and right sides of your kitchen.

[Hoefer and Smeenk, 2016, p. 128]

The claim, therefore, that detractors of the block universe make is that a moving Now (Presence plus Passage in our language; what some people call the A-series as opposed to the B-series picture) better accords with our experience. Just as the existence of trees explains why we see them everywhere, the A-series explains why we experience the world in terms of a moving Now, because there really is a moving Now external to in the world, independent of all observers. The argument is that the block universe in which all events are equally real is incompatible with Presence and Passage. Keep in mind, however, that while the block universe is incompatible with dynamical presentism (see below), it isn't necessarily incompatible with either Passage (or so Maudlin claims) or even Presence, given some kind of moving spotlight theory. Nonetheless, this is one of the major reasons why the idea of a block universe is coming under increasing fire from both philosophers and physicists [Dainton, 2016]. As you recall from the first chapter this is one of Smolin's major motivations for rejecting the block universe and citing Passage, Presence, and Direction as physically and metaphysically fundamental in his theory of QG. Presumably the reasoning here is that if all events are equally real, if all events are all just "there" as it were, then our dynamical presentist-like experience of time is inexplicable by fundamental physics. Therefore, according to Smolin, we must somehow make these features of time as experienced a part of our fundamental physics theories. This reasoning will be questioned throughout the rest of this book.

Typically, the tacit assumption in these discussions about the experiential arrow of time is that one can proceed without addressing the hard problem directly. We will argue that this belief is false and that in fact the question of the experiential arrow of time is not only a subset of the hard problem but the very heart of the hard problem and thus can't be ignored. We will also argue that the standard block universe picture actually makes the hard problem harder in a way that ought to be addressed. As we noted in chapter 2, most people involved in the various debates about time, whatever else their disagreements, accept the block universe and ontological reductionism or physicalism about conscious experience. As it is not the professional responsibility of practitioners of foundations of physics to focus on the hard/generation problem of consciousness, we can understand why this might be their default position. But we hope in the future that they will see why neither of these views is the consensus anymore in consciousness studies and why the block universe only exacerbates matters. But basically the state of play in consciousness studies is that neither reductionism nor eliminativism about conscious experience seem very plausible to most people at this juncture (Silberstein, 2017a). We will discuss all this at length in the next section.

Many defenders of the block universe try to shut down the incompatibility of the time-as-experienced objection by arguing that the A-series has no advantage over the B-series in explaining any feature of time as experienced. Perhaps just the reverse is the case, and in fact the A-series and the B-series are at the very least empirically/experientially equivalent, for example, the hypothesis that we are in the Matrix is supposed to be equivalent to the realism hypothesis [Price, 1996; Callender, 2012]. Recall that in the A-theory of time there is a moving Now that is supposed to explain our experience of Passage and Presence, if not Direction. According to this view time really flows and the present really moves from the future to the past, whereas in the B-theory of time there are only a series of events with static temporal relations such as earlier-than or later-than; there is no moving Now or privileged Now. We tend to think of the block universe as the B-series and presentism as an instance of the A-series. However, this equivalency won't do. For example, we can have a moving spotlight theory (see below) in a pseudo-block universe (a kind of presentism) or a static presentism (see below) in which the entire universe is just one moment (a kind of block universe).

As Hoefer [Hoefer and Smeenk, 2016] notes, there is a real sense in which relativity rules out both A-series and B-series given the relativity of simultaneity. Talk about the ordering of events is only well defined with respect to a certain reference frame. With respect to one reference frame, event X may precede event Y while in another reference frame, event Y may precede event X. Hoefer's point is that the the B-series suggests absolute simultaneity which fails to obtain in relativity. However, since the A-series versus B-series distinction is endemic to the literature we will use it, but probably the best way to pose the question is whether or not dynamical presentism, some version of the growing block, or the moving spotlight theory do a better job explaining experience than the block universe. So just to keep it simple we'll count all three as the A-series. However, there is a great deal of debate among philosophers of time as to whether or not the A-series is a clear and coherent notion, and even if it is, whether or not it explains our experience of Passage or Presence. As Dieks states, "[e]ven if we assumed that the A-picture of time makes perfect sense and reflects what the world is like, namely equipped with a moving Now, it would remain obscure how this could play a role in the explanation of our experiences of Passage" [Dieks, 2016, p. 5].

Typically then, defenders of the block universe will make two arguments at this juncture. First, the A-series is not superior with regard to explaining time as experienced—that is it has unique problems—and second that the block universe picture and all the A-series models are experientially equivalent and thus experience does not favor one over the other. There are many ways to argue the case for the first point but the basic idea is, to paraphrase Prosser [2013]: people claim to perceive the Passage of time and the dynamic quality of change directly. However, upon examination the exact nature of these experiences is unclear. By what sense, sensibility, sense organ, or sensory modality do we directly perceive temporal Passage? Is there some irreducible time sense? [Prosser, 2013, p. 323]. Prosser's claim is that "[e]ven if time did pass, no experience could possibly be an experience of time passing" [Prosser, 2013, p. 323]. If there is an experience of time passing, then what makes it the case that it is an experience of the Passage of time, and not of something else such as a succession of visual images? Whatever that something is, it must not be reducible to some other sensory modality for which a B-series account of perception could be given. What can the A-theorist say to explain this given that, unlike visual perception, there is no unique chain of causation from time passing to some brains state or neural mechanism [Prosser, 2013]? In other words, objective Passage is no more justifiable a belief than the ether interpretation of relativity. We agree completely as to both this epistemological and phenomenological conclusion. This is essentially the same point Callender made in chapter 2.

So far we have discussed objective Passage or the A-series in general, but let's now examine dynamical presentism, the growing block, and the moving spotlight theories in more detail. Dynamical presentism is the view in which a Global-Now or hypersurface arises then disappears becoming the past and then another Global-Now arises (what we formerly called the future), and so on. This view is supposed to be the "common-sense" view that best matches experience. We will return to that topic later in the chapter but one basic metaphysical and physical issue is how could a Global Now be said to cause, dynamically lead to, generate, etc., a successive Global Now when the former Global Now no longer event exists anymore. Does it even make sense to ask how long each Global Now exists without invoking some sort of meta-time? That is, do these Global Nows exist in some point-like succession or do they have some duration? How does one establish the relevant transtemporal identity across these Global-Now sheets to even makes claims about the very same object or event undergoing change across them? As Zimmerman writes: "But presentism's utterly empty past creates problems of its own—for there are many cross-temporal relations for which the present seems to lack adequate grounds, and even simple cross-temporal truisms like 'England had three kings named "Charles" prove difficult for presentists to interpret" [Zimmerman, 2011, p. 172].

Price considers what he calls "frozen-block presentism" such as the Barbour view we discussed that reality is just infinitely many three-dimensional snapshots in configuration space with no necessary temporal ordering. So it is only the memories/records of beings on those 3D snapshots that creates the illusion for them of being apart of an entire world in space and time. As Price describes it ". . . what matters most about this is that we seem to have lost the materials for a realist view of Passage, change or temporal transition. All of these notions seem to involve a relation between equals, a passing of the baton between one state of affairs and another. But in this picture, we've lost one party to the transaction. We've lost genuine change, and replaced it, at best, with a kind of fiction about change" [Price, 2011, p. 279]. And why are those memories there? Surely not because they were caused to be so by past events. As Price says, the world is "[n]ot a long series of world-stages, but just a single moment, complete, with its internal representation of a past and future. It is as if we've built just one house in 'Broad street', relying on stories its occupants tell about imaginary neighbours as surrogates for all the rest . . . . There is no such address, and no family who lived there" [Price, 2011, p. 279]. This is obviously even worse an explanation than dynamical presentism both metaphysically and phenomenologically. We will, however, shortly discuss a kind of block presentism we think is plausible in which the entire four-dimensional block universe is all there but from the God's-eye view (at least in principle) (e.g., Dr. Manhattan in Alan Moore's *Watchmen*), the universe constitutes just one Parmenidean moment that is only fractured into multiple moments from the ant's-eye perspective.

The moving spotlight view says that in some sense all events are just there but some sort of meta-process, the ontological equivalent of a moving spotlight illuminates each event in succession thereby conferring nowness or presentness upon it. Of course, if as Weyl suggests, the moving spotlight is just consciousness or minds "crawling" along each individual worldtube then that is simply dualism and/or meta-time thus putting us back to square one. And whatever provides this spotlight must somehow create a Global Now, so it can't be just a bunch of individual minds doing their own thing. If the claim is such a spotlight would exist even in a world with no conscious observers, that it is the spotlight itself that makes people experience the present as special, then what in the (natural) world could this spotlight be? If one is willing to embrace that kind of metaphysical baggage just to save appearances, then why bother feeling compelled to try to save some sort of block universe and the objective present simultaneously? As we saw in chapter 2, it would be a lot easier to just reject the block universe. But perhaps the most damning thing about the moving spotlight theory is that it makes being present and being real orthogonal. This is certainly not a problem for an ordinary blockworlder for whom an event's being present is perspectival/indexical, while all events in the block are equally real because being real has nothing to do with being present (see chapter 2).

But the moving spotlight theory is motivated by more than just phenomenology; it wants to ground a metaphysical notion of the object present. This is just to say that the moving spotlight theory must be motivated by the idea that those events upon which the light shines must be objectively special in some way; that is, they must be more real than those in the shadows, and that is why they feel more real than past events or future events. But the moving spotlight theory is still stuck claiming that all events in the block are *in some sense real*, but what sense of reality is that exactly and why is it an inferior grade of reality? The view that all events are real but only some (at any given time) are Most Real/present seems to be in some kind of tension. When the spotlight is not on me, do I cease to exist or become unconscious? Presumably not. Is the spotlight always on me? Presumably not, but it always feels like the present moment to me. Why is that? Price makes a similar point: "[t]he source of the difficulty is that the moving spotlight view is trying to combine two elements which pull in the opposite Directions. On the one hand, it wants to be *exclusive*, saying that one moment is objectively distinguished, On the other hand it wants to be *inclusive*, saying that all moments get their turn—their Warholian instant of fame, when the spotlight turns on them alone. (Everybody is a star) . . . The inclusive aspect threatens to overwhelm the exclusive aspect, reducing it to an innocuous and uncontroversial perspectivalism" [Price, 2011, pp. 278–9]. And of course, we still have the problem of how and why this metaphysical spotlight causes one to feel that the present is special. Why should these things be correlated or causally related; after all, it isn't obviously true by definition?

This brings us to the growing block or growing–glowing block universe which has similar problems to the moving spotlight theory. The good news is that these views don't require any extra spotlight but the bad news is it still isn't clear that they do any better job explaining time as experienced, which is the goal here. The standard view says that while past and present events are real, future events are not real and grow by accretion from the present in a wave-front of Global Now. So again, the question is, how does the growth of the future explain one's experience of Presence and Passage? How and why does this physical process cause those temporal experiences in us? Dainton raises the same question:

> It seems unlikely that we are aware of the Passage of time itself . . . However, for all that it may initially seem very plausible [the claim that the experience of Passage or Direction can be explained by the physical growth of the future], this claim is difficult to defend in the context of the *weak* Passage-dependence doctrine that we are currently exploring. According to the latter, although temporal Passage contributes (in some as yet unspecified way) to the temporal phenomenal features of our streams of consciousness, it is not itself responsible for the existence of the dynamical phenomenal contents within individual specious presents. Hence the difficulty . . . It is only if the strong dependence thesis obtains [that we would not experience Passage in a universe without Objective Passage] that the contents within individual specious presents are essentially dependent upon temporal Passage.
>
> [Dainton, 2011, p. 408]

The growing–glowing model attempts to answer this objection by saying that events in the present are more real than those in the past and all conscious experience is confined to the present [Dainton, 2011, p. 406]. As Zimmerman explains, "an event is only *really happening* when it is on the cutting edge" [Zimmerman, 2011, p. 170]. He says that events in the past have a "ghostly afterlife" on this view [Zimmerman, 2011, p. 172]. Needless to say, this view escapes few of the objections leveled against the other views. We can still ask, what does it mean to say that past events are not as real as present events? In what sense are past events real? Why do conscious experiences only exist in the present? Is that just a brute fact about reality, a C.D. Broad type bridge-law of sorts that is true independently of any other neural or physical facts about the world? Is it really plausible that past events are real in some sense but all the people in the past are unconscious zombie husks? This view seems like the worst of all three worlds: presentism, moving spotlight and growing block.

From our perspective, while all three of these views may be made compatible with physics given undetermination, unlike the block universe, none of them are *motivated* or *driven* by physics. These three purely metaphysical views are driven by time as experienced, but none of them seems to be in a very good position to explain time as experienced in any naturalistic way. Most importantly, the block universe certainly doesn't seem to be at any disadvantage with respect to these three views when it comes to explaining the experience of time, though we will tackle that question directly in what follows. So, in addition to ultimately explaining time as experienced, we want to next consider these questions:

1. Does the block universe have an inherent disadvantage in explaining the experiential arrow of time?
2. Can the experiential arrow of time be reduced to some physical arrow of time such as the thermodynamic arrow?
3. Can the experiential arrow of time be reduced to some neural mechanism or activity?
4. Is it possible to explain the experiential arrow of time without getting into the hard/generation problem of consciousness more generally?
5. Is the experiential arrow of time strictly "in the head" or (exclusive or) is it a physical, objective, external feature of the world?

So far we have squarely sided with the blockworlder on most issues. But we are going to be in the odd position of blockworlders arguing that the block universe is problematic for conscious experience, but not because we are advocating for any A-series view or because we accept what Dainton called the "strong dependence thesis," but rather because at the end of the day, we believe the standard block universe view can't account for the very existence of experience. The block universe we claim has an extra insurmountable challenge when it comes to resolving the generation problem of consciousness head-on. So we will answer in the negative to all five questions above, whereas the standard blockworlder would likely answer in the affirmative to at least the last three questions. Getting to our argument about the generation problem in the block universe and to our explanation for Passage, Presence and Direction (time as experienced) will take some set-up so please be patient.

We will start with examining the strong dependence thesis. As Price notes, the question of whether or not time as experienced is compatible with the B-series or the block universe is ancient:

> I'll come back to that question, but first to the dispute itself, which is one of philosophy's oldest feuds. One team thinks of time as we seem to experience it, a locus of flow and change, centered on the present moment—"All is flux," as Heraclitus put it, around 500BC. The other team, my clan, are loyal instead to Heraclitus's near contemporary, Parmenides of Elea. We think of time as it is described in history: simply a series or "block" of events, lined up in a particular order, with no distinguished present moment. For us, "now" is like "here"—it marks where we ourselves happen to stand, but has no significance at all, from the universe's point of view.
>
> [Price, 2013]

The block universe view seems equally compatible with the claim that rivers flow, events change, etc. Everything is just one event after another exactly as experienced. It is just that from the God's-eye view all events are real; they are just there. This is the argument from experiential equivalence. As Callender state, "[t]he idea is that there is a one-to-one mapping between experiences in a world of becoming and in a world without. Just by reflecting on experience you can't tell which world you're in" [Callender, 2012, p. 82]. "You experience a relentless, inexorable flow? Fine; but relentless inexorable feelings can be stretched out four-dimensionally too. And the reverse can happen too. Your Zen Buddhism class has you feeling stretched out in time, one with eternity? Fine; but that doesn't prevent you from being shrunk back into a series of presently-feeling—stretched-out-in-time experiences" [Callender, 2012, p. 82]. This is called the "epistemological argument" and it says that experience would be the same regardless of whether the A-theory or B-theory were true, and therefore experience provides no reason to favor one metaphysical theory of time over the other. This presumably goes for Presence, Passage, and Direction, but again, these are distinct experiential features of time. In what follows Prosser elaborates on the "epistemological argument as it focuses on Passage and Direction:

> The epistemological argument can be traced back to Donald William's (1951, 468–9) argument that it is possible to imagine a being whose instantaneous states are exactly the same as those of some normal human being, but occur in the reverse order. Williams suggests that from that being's point of view time would seem to go 'forwards'; yet 'forwards' from their point of view would be 'backwards' from ours. If correct, this would suggest that the nature of experience provides no evidence of an objective Direction of time; and if there is no evidence for a Direction it seems plausible to think that there is no evidence for passage (at least insofar as experiencing time passing entails correctly experiencing its Direction).
>
> [Prosser, 2013, p. 321]

> Price argues that nothing would seem different to us if the A-theory were false and time did not pass. Supposing for the moment that time passes, we would nonetheless be able to imagine a four-dimensional 'block' universe in which everything was exactly the same except that time does not pass. Price argues that events, including mental events, could be mapped one-to-one from the passage world to the block world; in which case our experiences would be exactly the same as they actually are, even if time did not pass. Consequently the nature of experience is compatible with both the A-theory and the B-theory.
>
> [Prosser, 2013, pp. 321–2]

However, Callender worries that many such equivalence claims have been question begging and he acknowledges an argument is needed. Callender attempts to provide that argument as follows:

The reason to think this is that one does not have access to the coming in and going out of existence of events. The popping of events into reality does not make any
sound, emit light or exert itself upon the senses in any way. Mental states do not exist
in a kind of hyper-existence, watching all the other non-mental events come and go
into existence. They are themselves part of what comes and goes into existence. Nor
will the existential "special-ness" of any presently existing events be observable. We
cannot step outside the present and compare present experiences with non-present
ones, if such there be. Suppose that we found out that present events were blue,
whereas past and future ones were green. Since we only (presently) experience present events, we would only experience blue. So the difference between green and
blue wouldn't show up in experience anyway.

[Callender, 2012, p. 83]

Of course, we have already seen this argument now in several different forms.
This argument certainly seems sounds to us. As Callender states, "[e]xperiential
states supervene on the existent" [Callender, 2012, p. 83]. There is no necessary or obvious connection between the experience of becoming and whether
or not new physical events are literally popping into existence from non-
existence. In fact, as many people have noted, even the ordering of events
from a meta-perspective seems irrelevant to experience. One can easily imagine some meta-being taking slices of an individual's worldtube and using some
cosmics cissors to rearrange them randomly. It doesn't seem that this would make
any experiential difference to the individual on any particular slice.

From everything said thus far it should be obvious that we are ready to grant
that the A-series story is neither necessary or sufficient for explaining Passage
and flow, and that there is no necessary incompatibility with time as experienced
and the block universe. But as he acknowledges, what Callender hasn't provided
is an explanation for time as experienced. He states, "[m]y position is that the
metaphysical positions are empirically equivalent and that the differences between
them are explanatorily impotent as regards explaining the temporal data" [Cal-
lender, 2012, p. 83]. That is good as far as it goes, but what does explain the
experiential arrow of time? As Callender says, that is the key question and it still
remains.

There really are an interesting set of problems motivating philosophical study of
time. We do treat time and space very differently, despite their both being modes of
extension treated similarly by our fundamental scientific theories. We imagine time to
have a much richer structure than space. If anything about time calls for explanation,
clearly this fact does . . . Furthermore, inasmuch as I believe explaining the temporal
data is the interesting question in philosophy of time, we should put behind us this
particular metaphysical debate.

[Callender, 2012, pp. 93–4]

How then does Callender suggest we proceed with this debate? As follows:

There is a better way to conceive of the debate. Philosophers of time should model
the debate the way philosophers of mind frame theirs. The natural sciences don't

have sophisticated theories of intentionality and consciousness, for instance. There seems to be an explanatory gap between our experience and the so-far incomplete description of our experience provided by the natural sciences. There is an honest-to-goodness problem over how to explain consciousness, for instance. Philosophers of mind then suggest explanations using naturalistic resources to explain consciousness (which, if picked up, might develop as parts of natural science) or they look elsewhere and supplement the naturalistic resources, either with new 'naturalistic' resources or 'non-naturalistic' resources. They then argue about whether the explanation actually succeeds in accounting for consciousness. Similarly, philosophy of time ought to refine our description of what needs to be explained, carefully examine science and the way it treats time, compare the two, and then try to account for any explanatory gap that arises. The gap may be filled in with scientific or metaphysical resources. However it works out, it's now clear that presentism, possibilism and eternalism need more resources to close the gap. The eternalism debate need not itself be eternal.

[Callender, 2012, pp. 93–4]

We agree wholeheartedly with this advice and will proceed accordingly. From here on, we will argue that the standard moves blockworlders (and others as well) use to explain time as experienced may not be up to the task. And that even if they are, the block universe especially precludes any reductionist or emergentist account of experience generally, thereby leading us to a very different solution to the hard problem of consciousness that is also an explanation of the experiential arrow of time. That is, one can grant everything to Callender and Price about the epistemological argument and still worry not only about how to explain time as experienced in a block universe but also about explaining the very existence of conscious experience in a block universe.

When explaining the experiential arrow of time, blockworlders like to begin by pointing out that given the Minkowski interpretation of relativity, the world isn't a stack of Global Nows and there is no global simultaneity, so any explanation in that regard must be local. They also like to point out that the feeling or experience of Passage, or at least motion, can be generated in cases where presumably none exists. As Dieks writes:

> There is strong empirical evidence that differences in sensory input at different instants of time can result in an awareness of continuous motion. For example, the repetition of a brief sequence of pictures of a road, punctuated by a blank space, creates the perception of continuous motion along the road (Mather 2010). In this case there is no "flow quality" in the input itself: there is just a brief period during which one picture is visible, later a similar period with a second picture, and so on . . . But our response to this sequence is characterized by a feeling of dynamism, of continuous change and flow, at each instant.

[Dieks, 2016, p. 16]

Presumably the argument here is that one can have experience of Passage without anything like objective Passage. That certainly seems true, but is this sufficient to explain the everyday experience of Passage and Presence? Is mere succession with spatial variation sufficient to explain them? Is the feeling of motion in these contrived situations phenomenologically equivalent to everyday experience? Does the fact that, for instance, sometimes we hallucinate trees when none are present, does that explain why we see trees when they are present, does that make my belief that trees exist unjustified? The larger point here is granting that objective Passage is neither necessary nor sufficient to explain the experience of Passage, it does not follow that we don't experience time passing, nor does it follow that experience cannot provide a reason for believing that time passes. Illusions and hallucinations provide lots of examples of non-verfiable experiences that are equivalent to verifiable ones. For that matter, we could all be brains in a vat (even if I can't now refer to real world "brains" and "vats" as I write this, etc.). But given disjunctive theories of perception, epistemological externalism, etc., I might still be justified in believing in objective Passage. And more importantly, we still have not explained the experience of Passage which we must do. Once we have an adequate account of that experience that doesn't advert to objective Passage, then we can fully put that metaphysical debate to rest.

It certainly seems like there are cases where mere succession with spatial variation is present and it doesn't match everyday Passage and Presence. The account on trial here is what philosophers call the "at–at" account and it says a moving object is an object that exists at a sequence of different spatial locations in an entirely static and motionless fashion. But as Dolev [2016] and Dainton [2016] note, the idea that motion or Passage is just the fact that the same bodies are sometimes at one place and sometimes another, and they are at intermediate places at intermediate times, isn't completely satisfying because, to quote Dainton, "motion is a property of objects that we can directly perceive, and moving objects simply look very different from objects that are motionless" [Dainton, 2016, p. 79]. He invites us to think of the experience of watching flame flicker or a turbulent waterfall. As Dainton explains, "[q]uite generally, experienced motion possesses an inherent dynamism that is hard to reconcile with the sequence of immobilities offered by the at–at theory" [Dainton, 2016, p. 80]. Dolev asks if the "at–at" theory is sufficient to explain time as experienced, then why do memories feel different than actual experiences [Dolev, 2016]? It certainly feels like being past, present, or future (Passage and Direction) is, phenomenologically, more than merely being before or after something. On the flip side, is the "at–at" theory necessary for time as experienced? Don't we sometimes experience Passage or Presence even when our visual stimuli or representations are unchanging or non-existent? Some or all of these can be experienced in deep meditation, for example.

In almost every case where we artificially make people experience anything approaching Passage in everyday experience, such as film (imagine an actual reel of film, not digital) or flip-books, *something* is moving. In the case of a film, it is the projector and in the case of the flip-book it is the person's hand moving the pages. If one physically trawls through a reel of film tape from end or turns the pages of a the flip-book slowly, there is no "illusion" of motion created. As Dainton states, "it is a mistake to conclude from this that a continuous stream of consciousness can be formed merely by placing momentary experiences with static contents side-by-side, as it were . . . there is all the difference in the world between watching a movie, and looking at a collection of still images" [Dainton, 2011, p. 389]. So ask yourself this, in the block universe what plays the role of the projector? Obviously, from a God's-eye perspective, nothing is moving in a block universe; there is nothing to play the role of the projector. Brain states are no better off in this regard than any other physical process. As Hoy writes, "But wait. Even if science can achieve an account of brain states that explains how they can be representations of time, how should the temporality of those states be understood: they seem subject to the distinctions of past/present/future and to the passage of time" [Hoy, 2013, p. 22]. Is conscious experience itself the projector? Presumably not, since that is the phenomenon we want to explain after all. Contrary to the Weyl quotation, from a God's-eye perspective conscious experience no more moves than anything else in the block universe, each person's worldtube is allegedly just a continuum of still photographs, as it were, that happen to include conscious experiences and memories as well as physical events. And if conscious experience defies the rules of the block universe then physicalism or reductionism is false.

Many argue that, rather than any arrow of time in physics alone, it must primarily be the brain that ultimately explains Passage and Presence. In other words, they wish to make the brain the projector in the story. Neurobiologist Dean Buonomano expresses this view perfectly with his claim that "the brain is a time machine" that produces the subjective experience of Passage and Presence [Buonomano, 2017]. He calls time as experienced a "mental construct" [Buono-mano, 2017, p. 216]. As he notes, most block worlders and presentists make the same claim about time as experienced, the difference being that presentists think that such mental constructs (i.e., representations or qualia) reflect some actual physical process, whereas block worlders are more inclined to call such constructs "projections" [Buonomano, 2017, p. 216]. As Paul notes, "So, the re-ductionist explanation of our temporal experiences as of passage and change is that the brain manages contrasts between causal impressions of property instances that it receives in quick succession in a way that creates these experiences" [Paul and Healy, 2016, p. 116]. Paul says the brain accommodates and organizes these inputs. As Dieks [2016] and Ismael [2016] are keen to note, this doesn't neces-sarily make time as experienced an illusion or unreal, any more than secondary qualities like "seeing yellow" are unreal [Dieks, 2016] or embodied perspectives from within the block universe instead of the God's-eye perspective are unreal [Ismael, 2016], neither being internal to perceivers nor being perspectival entails being illusory. Like Paul, both authors claim that such experiences are ultim-ately discharged by neuroscience. Dieks [2016] claims, for example, that when the appropriate stimuli are present (appropriately structured specious presents), brains create a feeling of change, motion, and flow. Thus Passage and Presence are secondary qualities that are internal to us on this account. Ismael [2016] makes similar claims invoking representations, memories, and neural processes that she claims are enabled by the thermodynamic arrow of time:

The upshot of all this is that perception and memory working together produce an intricate structure of linked representations of the same moments in time, viewed from different perspectives over the course of a life.

[Ismael, 2016, p. 114]

The phenomenology of flow is a product of the way that the brain processes sensory information . . . The sense of passage arises from the aforementioned poignant awareness of our changing perspective on history. Openness is a feature of the way that the future looks to the decision-making agent. From the perspective of such an agent, the decision process itself resolves a collection of open possibilities into singular fact.

[Ismael, 2016, p. 115]

But if we can identify representational states inside an information-gathering and utilizing device like a robot, which at least have the same functional role as the progression of states that constitute our conscious mental lives, then we can locate something even Einstein will have to recognize falls within the purview of physics, and we will have found some common ground.

[Ismael, 2016, p. 116]

Thermodynamical gradient leads to dynamical asymmetries we experience, same gradient paves way for information-gathering systems and representations in the internal states of those systems leading to experience of passage, flow and openness.

[Ismael, 2016, p. 117]

As Dainton notes, memory is no doubt a part of the story of explaining time as experienced, but it can't be the main story for Passage and Presence because: "from a phenomenological standpoint we seem to experience change or movement with the same immediacy as we experience colour or pain . . . Combining memory-images (in the right sort of way) might conceivably generate a memory of seeing something in motion, but it can never amount to seeing motion first-hand" [Dainton, 2011, pp. 388–9].

Dainton and many others argue that the structure of our temporal experience or representations must be such that while "the present strictly so-called is momentary, the psychological or experiential is not: it extends over a brief interval of time, sufficient to allow motion (and other forms of change or persistence) to be directly apprehended in consciousness" [Dainton, 2016, p. 82]. This interval is typically called the "specious present." The argument for this view is that "motion takes time. If I am directly aware of an object moving from A at t1 to B at t2 then my awareness must extend over this temporal interval" [Dainton, 2016, p. 82]. Dainton defends a version of the specious present he calls the "Extensional model" in which "the successive phases of an Extensional specious present are not confined to a momentary conscious state: they are parts of a temporally extended episode of consciousness" [Dainton, 2016, p. 84]. The point is experience can extend not only experientially but also it can have an objective duration in physical time. Unlike presentism, in which only the momentary durationless present is real, this view is ideal in a block universe because all times and events coexist (not simultaneously) in such a universe: "If everything in the universe, past, present and future, is fully and equally real, then experiences are too. In which case there is no obstacle—posed by time at any rate—to contents separated by a second or so being experienced together" [Dainton, 2016, p. 85]. According to this view, experience of change and motion takes place in a temporally extended specious present and these adjacent and joined specious presents have a union of overlapping shared content along an individual's worldtube, creating the experience of flow and Passage and perhaps Presence as well.

There are certainly neuroscientists as well who accept an extended specious present, but they still assume the mental construct view articulated by Buonomano (i.e., representationalism plus qualia). He is skeptical that even if the block universe model is true, a "slice of the block universe can sustain the phenomena of consciousness" [Buonomano, 2017, p. 176] apart from some of the other past and future slices of that individual brain's worldtube. That is, Buonomano doubts that the experiences of an individual at time t supervenes only on that individual's brain state at time t, but he doesn't deny the supervenience model as such. As he states, "consciousness is something more akin to music than a static image of a movie frame" [Buonomano, 2017, p. 176]. Nonetheless he is convinced that the brain "generates" conscious experience and acknowledges that no one knows how [Buonomano, 2017, p. 175]. As will become clear in the next section of this chapter and especially in chapter 8, we reject the idea that either conscious experience or the block universe are a continuum of objectively countable and frozen slices, as suggested by the analogy using static film frames. Certainly the formalism of relativity and the block universe model suggest such a picture, but given our neutral monism and given the temporally extended, dynamical, or "musical" nature of conscious experience, one cannot disentangle conscious processes from spacetime. Spacetime isn't some substantive arena in which conscious processes/representations occurring inside brains appear to unfold dynamically only because they are structured in the right way.

So we are happy to adopt the extended specious present as a necessary part of the explanation of Passage but we want to note that for us it is part and parcel of our direct realism/neutral monism and not a story about the structure of temporal representations (see chapter 8). In other words, for us the question is this: is the extended specious present sufficient to explain Passage or Presence? And is it sufficient to discharge in full the worry that time is an illusion? We answer in the negative on both counts. As for the illusion question, as Dainton notes himself: "I agree that E-passage [experiential passage] exists in the realm of appearances, and that to the extent that these appearances misrepresent the (non-dynamic) external physical reality they can in this respect be construed as misleading or illusory" [Dainton, 2012, p. 133], but he wants to maintain that experiences themselves really do possess dynamic characteristics [Dainton, 2012, p. 133]; that much is not an illusion contra Paul. We agree on both counts, but we can still ask if Passage and Presence are intrinsic to experiences and that is the end of story, or is there more to be said? We will argue that Passage is a necessary relational feature of a neutral "field of experience" with a subject/object division and that Presence is fundamental and universal; to be clear, our claim is while an event's being present is perspectivial/relational, Presence as in the experience is not (see chapter 8). Furthermore, as we will discuss later, there are experiences of pure Presence without Passage or motion.

Keep in mind as we move on to the next section that the generation/hard problem is precisely the question of why and how certain neural correlates *generate*, in this case, Presence and Passage. Philosophers and scientists like to tell evolutionary stories about the adaptive advantage of having a sense of Passage and Presence: "Perhaps the feeling of the passage of time was even critical to our ability to project ourselves into the far-flung future and engage in mental time travel" [Buonomano, 2017, p. 175]. But even if time as experienced is selected for in evolution, it first had to get "generated" to be selected, and evolution alone doesn't explain how that happens. Second, one way of phrasing the hard problem is why would certain cognitive functions such as mental time travel require a phenomenal and experiential counterpart/correlate in the first place? And if such a counterpart is causally efficacious in cognition then how can neuroscience possibly provide a reductive explanation of that fact? Third, as Buonomano writes, "if we live in an eternalist universe in which the flow of time is unreal, how could perceiving it as flowing have been evolutionarily advantageous?" [Buonomano, 2017, p. 216]. If the future exists then obviously, from a God's-eye view, evolutionary explanations are purely ant's-eye view only, much like the self-reproducing gliders in Conway's Game of Life. But that aside, even from the relatively ant's-eye view, if the future is real, wouldn't it be more advantageous to the organism to "perceive" larger swathes of spacetime such as the near future. Perhaps such a possibility is ruled out by the nature of neural processes or physical processes, but that takes us back to the previous discussion about the spatiotemporal duration of neural and conscious processes. Why couldn't such "processes" have a longer duration that includes what, from the ant's-eye view, we would call the future? We will return to this question in chapter 8.

While we share all the skepticism about the coherence and usefulness of the A-series, just like the Matrix hypothesis (i.e., it is only true that all experience could be in the Matrix if the Matrix can definitely replicate "real" experience), the epistemological argument presupposes that blockworlders have a good explanation for the experiential arrow of time that doesn't refer [or admit] to any features of the A-series. The standard move here is to argue that some combination of successive representations and memories with experiential overlap plus the thermodynamic arrow of time (and perhaps other physical arrows of time) explain time as experienced in a block universe [Dieks, 2016]. We will deal with the thermodynamic arrow in chapter 8. Of course, if one is a physicalist or ontological reductionist, then the representational account must somehow reduce to some neural or physical account.

We will argue that this explanation is possibly insufficient on its face and again, even if it is up to the task, that the standard block universe picture makes the hard problem harder in general, impossible, really. In other words, we will argue that the experiential arrow of time cannot be reduced to either some neural mechanism or to some physical arrow of time in the block universe. As we saw, the defender of the block universe will claim that the experiential arrow of time is if not an illusion, then internal to the perceiver. The experiential arrow of time is neither external nor objective. One might liken it to a "secondary property" [Dieks, 2016] or to being "inherently perspectival" [Ismael, 2016].

But why do such secondary properties or such perspectives exist in the first place? In the next section, we will argue that the standard block universe view is disadvantaged in answering this question. The point is this: if one conceives of conscious experience in terms of qualia and the physical universe as a frozen blockworld of matter as it is typically characterized then there is no hope for resolving the hard/generation problem in terms of ontological reduction, emergence, or panpsychism, and that includes the experiential arrow of time as well.

7.3 The hard problem in a block universe

7.3.1 The hard problem now

Let us begin by reminding ourselves what the hard problem is and why it is hard before we bring in the extra wrinkle of the block universe. We think this is especially important for our readers in the foundations of physics and foundational physics communities who may not be familiar with the literature. The bottom line is that the hard problem has become so hard and the prospects for reductive or eliminativist accounts of conscious experience have grown so dim that radical emergence, panpsychism, and neutral monism seem to be making something of a comeback [Seager, 2016b]. Again, the generation problem (GP) or hard problem is that assuming that matter is fundamental, then how does mere insensate matter *generate* consciousness? For this problem to be as devastating as David Chalmers and others allege (independently of one's judgment about conceivability arguments), one has to assume something like that matter is essentially non-mental. As Barbara Montero writes:

> Instead of construing the mind/body problem as finding a place for mentality in a fundamentally physical world, we should think of it as the problem of finding a place for mentality in a fundamentally non-mental world, a world that at its most fundamental level is entirely non-mental.
>
> [Montero, 2010, p. 210]

Or, as Strawson states:

> That is what I believe: experiential phenomena cannot be emergent from wholly non-experiential phenomena. The intuition that drives people to dualism (and eliminativism, and all other crazy attempts at wholesale mental-to-non-mental reduction) is correct in holding that you can't get experiential phenomena from P phenomena, i.e. shape-size-mass-charge-etc. phenomena, or, more carefully now—for we can

no longer assume that P phenomena as defined really are wholly non-experiential phenomena—from non-experiential features of shape-size-mass-charge-etc.

[Strawson, 2008, p. 20]

Radical emergence (or strong emergence, if you prefer) is the view most closely associated with C.D. Broad which claims, for example, that there are brute psycho–physical bridge laws in the actual world. As Seager notes:

This latter view, that emergence should be understood in terms of a supervenience relation defined via nomological necessity is perfectly respectable and not unfamiliar. In essence, it was the view held by the so called British emergentists (see McLaughlin 1992), notably Alexander (1920), Morgan (1923) and Broad (1925).

[Seager, 2012, p. 147]

Such brute bridge laws are supposed to involve nomological necessity and pertain to the actual world only. The claim is that strong emergence is an answer to the generation problem (GP).

Panpsychism also takes GP at face value and proposes the following resolution:

- Matter is the fundamental building block of reality but has no intrinsic physical properties (fundamental physical properties are relational).
- The physical world is composed of or otherwise determined by basic physical entities and such fundamental physical entities (whatever those may be) have an intrinsic psychical-conscious or subjective aspect (consciousness is ubiquitous).
- Therefore, there is a sense in which all physical composites have a psychical/subjective nature however attenuated (universality of mentality).
- Therefore, the purely physical description of the world is incomplete (consciousness or subjectivity is co-fundamental with matter).

According to panpsychism there can be no consciousness without matter, but it is an intrinsic property of matter. And the latter claim is just an axiomatic fact about the universe, it is not explained by anything else. A panpsychist of a certain sort might object that perhaps only certain physical entities possess consciousness and not all fundamental physical entities. While that is certainly possible, it seems to violate the premise of panpsychism that consciousness is co-fundamental with matter, not radically emergent and not dualism. Strawson makes the same point in what follows:

Micropsychism is not yet panpsychism, for as things stand realistic physicalists can conjecture that only some types of ultimates are intrinsically experiential. But they must allow that panpsychism may be true, and the big step has already been taken with micropsychism, the admission that at least some ultimates must be experiential. 'And were the inmost essence of things laid open to us' I think that the idea that some but not all physical ultimates are experiential would look like the idea that some but not all physical ultimates are spatio-temporal (on the assumption that spacetime is indeed a fundamental feature of reality). I would bet a lot against there being such radical heterogeneity at the very bottom of things. In fact (to disagree with my earlier self) it is hard to see why this view would not count as a form of dualism. So I'm going to assume, for the rest of this paper at least, that micropsychism is panpsychism.

[Strawson, 2008, p. 71]

How did we get to this point? Presumably it is because the prospects of what Searle calls "Biological Naturalism" are dimming even among many of its former greatest proponents. Biological Naturalism states that consciousness or subject-ivity is a biological phenomena generated by the brain in the way that liquidity is a global emergent property of a certain molecular structure, provided one has enough of these molecules interacting in the right way. Conscious experience then, like liquidity, is a weakly emergent property of the whole brain. Searle as-serts that consciousness is completely caused by and realized in the brain but he doesn't deny the fact that humans really are conscious, and that conscious states have an essentially first-person nature. He asserts that, given Biological Natur-alism the scientific study of consciousness requires two steps: first, discover the neural correlate (NCC) of the entire field of conscious experience (not just some particular subset such as visual experience), and second, go from this NCC to a discovery of the actual biological causal mechanisms that give rise to conscious experience. In his paper, "Biological Naturalism," this is what he says about the future prospects of Biological Naturalism among scientists and philosophers:

> *It is worth pointing out that practicing neurobiologists of my acquaintance, such as Francis Crick, Gerald Edelman and Cristof Koch, implicitly or explicitly accept a version of what I have been calling biological naturalism* [emphasis mine]. They look to the operations of the brain to find an explanation of consciousness. It will probably take a long time be-fore Biological Naturalism is generally accepted by the academic profession because we follow a long tradition of teaching our students the mistaken view that this is a philosophical problem of impossible difficulty. But notice that we have to train our students to think there is an impossible mystery as to how neuronal processes could cause conscious states. It is not a view that follows naturally either from reflecting on one's own experiences or from studying brain operations. Once we overcome the mistakes of the tradition, I think the facts will fall naturally into place.

[Searle, 2017, p. 336]

However, at least in certain quarters of the cognitive neuroscience of conscious-ness and consciousness studies more generally, Searle's prediction did not come to pass. Here, for example, is Cristof Koch himself in his 2014 *Scientific American* article entitled, "Is Consciousness Universal," explaining perhaps why Searle's prophecy failed:

> Yet the mental is too radically different for it to arise gradually from the physical. This emergence of subjective feelings from physical stuff appears inconceivable and is at odds with a basic precept of physical thinking, the Ur—conservation law—ex

nihilo nihil fit. So if there is nothing there in the first place, adding a little bit more won't make something. If a small brain won't be able to feel pain, why should a large brain be able to feel the god-awfulness of a throbbing toothache? Why should adding some neurons give rise to this ineffable feeling? The phenomenal hails from a kingdom other than the physical and is subject to different laws. I see no way for the divide between unconscious and conscious states to be bridged by bigger brains or more complex neurons.

[Koch, 2014, p. 2]

What exactly led to Koch's change of heart is debatable but in the preceding passage he seems to have had the epiphany that for conscious experience to be caused by brains and their purely biological properties would require strong/radical emergence, not weak emergence as Searle suggests. And Koch also seems to have sided with fans of panpsychism, such as Galen Strawson, who claim that radical emergence is impossible. Indeed, in this article and others, he defends a version of what he calls panpsychism: "[a] more principled solution is to assume that consciousness is a basic feature of certain types of so-called complex systems (defined in some universal, mathematical manner). And that complex systems have sensation, whereas simple systems have none" [Koch, 2014, p. 2]. He calls his brand of panpsychism more "narrow and nuanced" than the standard kinds of compositional panpsychism, stating:

These century-old arguments bring us to the conceptual framework of the integrated information theory (IIT) of psychiatrist and neuroscientist Giulio Tononi of the University of Wisconsin–Madison. It postulates that conscious experience is a fundamental aspect of reality and is identical to a particular type of information–integrated information.

[Koch, 2014, p. 2]

We should note that it is not at all clear why, exactly, ITT is a form of panpsychism as opposed to a form of radical emergence in which rather than something essentially biological, the emergent base is something essentially informational. He goes on to state that "[a]ny system that possesses some nonzero amount of integrated information experiences something. Let me repeat: any system that has even one bit of integrated information has a very minute conscious experience" [Koch, 2014]. This point is worth exploring for a moment because it will be important to be clear about the essential differences between radical emergence and panpsychism and the motivations and arguments used for each. As Koch's reasoning suggested, the argument on both sides usually runs as follows:

1. Assume either radical emergence (RE) or panpsychism (P) is true.
2. RE is implausible or impossible therefore P (panpsychism).
3. P is implausible or impossible therefore RE (radical emergence).

While it is true that ITT potentially applies to all physical systems and not just brains or biological systems, that doesn't make it panpsychism; that just makes it a twenty-first-century brand of information-theoretic functionalism. Here John Horgan explains why ITT is allegedly a form of panpsychism:

> *Phi* corresponds to the feedback between and interdependence of different parts of a system. In *Consciousness*, Koch equates *phi* to "synergy," the degree to which a system is "more than the sum of its parts." *Phi* can be a property of any entity, biological or non-biological. Even a proton can possess *phi*, because a proton is an emergent phenomenon stemming from the interaction of its quarks. Hence panpsychism.
>
> [Horgan, 2015, p. 2]

However, while both radical emergence and panpsychism agree that conscious experience is fundamental in the sense of being irreducible to physical or biological phenomena (something Searle would grant as well), only panpsychism claims that consciousness or subjectivity is fundamental in the sense of being possessed by whatever the most fundamental physical constituents of the universe are. Note that neither quarks nor protons meet this requirement. But more importantly, panpsychism claims that the only intrinsic properties of matter are mental ones. So whatever the system or entity is in question, if mentality only arises when Phi is instantiated, then mentality isn't an intrinsic property. The fact that ITT entails that certain subsystems of the brain, certain sub atomic particles, etc., may be proto-conscious is not sufficient to make it panpsychism. If what ITT claims is that it is a fundamental law of the universe that whenever, wherever, and by whatever *Phi* is instantiated, that mentality (however limited) arises, then such a view is the very definition of radical emergence. Of course, this is a real problem for people like Koch who wanted to argue for ITT panpsychism on the basis of the impossibility of radical emergence. Nor can one escape this concern by merely appending panpsychism as a distinct metaphysical add-on to ITT because ITT and panpsychism are not logically compatible given the very definitions of panpsychism and radical emergence.

So all this begs the questions: how did the cutting edge of the cognitive neuroscience of consciousness come to find itself in such dire (or at least weird) straights? Why have some of the thought leaders in the field so quickly abandoned Biological Naturalism? After all, if one looks at the rhetoric of Koch, Crick, and other later twentieth-century arch reductionists [Crick and Koch, 1990; Crick, 1994; Koch, 2004] the message was clear: "neuroscience will now resolve the ancient mind/body problem and put thousands of years of ancient metaphysical speculation to rest." It certainly doesn't look as if Biological Naturalism has prevailed over ancient philosophical speculation. On the contrary. So what happened? The answer, it turns out, is in the difficulty of making the very first step in Searle's method, namely, trying to find the NCC:

> However, there is still no consensus on whether any of these signs can be treated as reliable "signatures" of consciousness. In particular, there can be consciousness without frontal cortex involvement . . . gamma activity without consciousness . . . such

as during anaesthesia . . . and consciousness without a frontal P300, for example, during dreaming sleep . . . Moreover, it is likely that many of the signatures proposed as possible NCC may actually be correlates of neural activity that is needed leading up to a conscious percept . . . or for giving a report following a conscious percept . . . rather than for having an experience. A major challenge is to keep constant cognitive functions such as selective attention, memory, decision making and task monitoring, in order to isolate the 'naked' substrate of consciousness at the neuronal level . . . Finally, NCC obtained in healthy adults may or may not apply to brain damaged patients, to infants, to animals very different from us, not to mention machines.

[Tononi and Koch, 2015, p. 3]

What we learn from Tononi and Koch here is not just the difficulty of finding the NCC (even if it exists) but the acknowledgement that it might be completely different in different individual humans, the same individual at different times, and in different animal species. In other words, even if the NCC model is true, conscious experience, like so many other processes in complex biological systems, might be multiply realizable in many different ways, radically plastic, etc. So from this perspective it is perfectly reasonable to search for some universal factor such as *Phi* that might be shared by diverse conscious systems. But of course, information integration (e.g., complexity) of which there are many measures and varieties might be multiply instantiated in many ways as well and might also inherit many of the other problems of the NCC program mentioned by Tononi and Koch. Therefore, consciousness might once again also be multiply instantiated in many ways.

More importantly, the point is that such information theoretic schemes while perhaps perfectly reasonable to pursue are not instances of Biological Naturalism. Rather, they are instances of an updated kind of functionalism—abstracting away from biological details/mechanisms to look for "higher-level" structures and processes realized by these biological systems. As may be recalled, functionalism waned in part precisely because one can't functionalize conscious experience. So again, at best, ITT and the like are really kinds of radical emergence that are stuck with one of the most counterintuitive implications of panpsychism, namely, that some particles and goodness know what else might be proto-aware. It is also important to add that even if one had the NCC or even THE theory of ITT (which is really just a more abstract correlate of consciousness, call it ICC), that alone would not rule out any metaphysical position on the mind/body problem. Such correlations are compatible with everything from substance dualism to neutral monism. To start to believe that science had made non-physicalist/non-naturalist alternatives much less probable would require, at the very least, making good on Searle's second step which is to reveal to us convincingly the *causal biological mechanisms that generate* conscious experience. Of course underdetermination could never be fully defeated here, but we are a long way from that being the upper limit of our knowledge.

Here then is an argument to summarize our situation when it comes to explaining conscious experience:

1. Assume ontological and methodological naturalism are true and therefore substance dualism, subjective idealism, and the like are not options.
2. Conscious experience can't be simply reduced to biological or physical processes.
3. Biological Naturalism is a not an option as a distinct alternative because it entails radical emergence.
4. So for all practical purposes, we are left with radical emergence or panpsychism.

In what follows it will be argued that both these alternatives are fatally flawed. In section 7.3.2 the "generation problem," as Seager and others call it, will be defined because it is that problem that drives both radical emergence and panpsychism. And then radical emergence and panpsychism will be defined in terms of their respective reactions to this problem. In section 7.3.2 it will be argued that both radical emergence and panpsychism (in general and in their current incarnations) are untenable. Ultimately it will be argued that, especially in a block universe, only neutral monism truly defeats the generation problem by deflating it, and neutral monism also explains the experiential arrow of time. The overall thrust is that we are driven to radical emergence or panpsychism because of the standard physicalist/materialist assumptions shared by both: that matter is fundamental, the nature of reality is therefore compositional, and therefore that material and mental features are essentially different and distinct. Radical emergence and panpsychism are therefore just "riders" or "patches" for physicalism, whereas neutral monism is the cure.

7.3.2   The generation problem: Radical emergence versus panpsychism

If we take GP seriously then consciousness must be fundamental in some sense, no matter what your assessment of conceivability arguments. Radical/strong emergence attempts to answer GP in terms of some brute fundamental psycho–physical bridge law of the sort described earlier. However, if we take GP seriously, it would seem to rule out the very (physical or nomological) possibility of such a bridge law. How can the GP problem be true and strong emergence (psycho–physical bridge laws) also be true? That is, given GP, it can't be true that such bridge laws possess only nomological necessity. Such a law is beyond the bounds of naturalistic explanation and is therefore not a natural law. So given GP, radical emergence must be hold that such psycho–physical bridge laws are stand-alone, one of a kind, brute/fundamental, and metaphysically necessary; presumably such a law isn't ruled out by conceivability arguments as long as they do not entail identity relations. The point is, one can't simultaneously hold that the physical is fundamental and essentially non-mental and that there is nonetheless some fundamental law in our world alone that necessitates that if matter, and matter alone, is in the proper configuration then conscious minds "pop" out. Such a law can not be conceived as mere physical/nomological/natural necessity but must somehow be modally necessary. Such a metaphysical law must somehow transcend the purview of science or the merely empirical. Indeed, such a law at least borders on the supernatural.

A stand-alone, one of a kind brute/fundamental metaphysically necessary law is a deus *ex* machina, miraculous affair. In other words, if matter is fundamental and essentially non-mental, then radical emergence must be some sort of occasionalism that replaces God with a miraculous law of nature. For those who want unity, such psycho–physical bridge laws are deeply disunifying, no matter how you construe the nature of their necesssity. Again, such laws are a very strange thing to have the status of fundamental facts, given ontological reductionism or physicalism for all other facts, that is, given the right physical, functional, informational structure, etc., and miraculously, conscious experience appears! As many people have pointed out, a fundamental feature of the universe should be efficacious but strong emergence is consistent with the causal closure of the physical—consistent with epiphenomenalism. Worse, radical emergence is only motivated by GP to begin with, which in turn is only motivated by physicalism or ontological reductionism, which means that consciousness (which itself can't be analyzed functionally) must be either epiphenomenal (a fact which Chalmers [1996] accepts and embraces) or it undercuts physicalism and ontological reductionism.

Proponents of strong emergence will surely say that the foregoing argument is question begging. As O'Connor states:

> It is sometimes suggested that there being metaphysically emergent capacities would be 'spooky,' not amenable to empirical investigation. But this is simply not the case. While they are basic features of reality, emergent capacities may nevertheless be fruitfully studied and eventually explained in detail in nonreductive fashion, by spelling out the basic inventory of emergent properties, detailing the precise conditions under which organized physical systems give rise to them, and isolating the precise behavioral impact their presence has on the system.
>
> [O'Connor, 2014, pp. 30–31]

Here then is the dilemma that faces strong emergence: either such laws involve only nomological necessity or they involve modal necessity without identity. If the former, then what kind of law are such laws? What makes them laws and how do they explain the phenomena in question? If laws explain by necessitating, and necessitating requires showing why some phenomena must obtain given certain conditions, that is, by showing that things could not have been otherwise, then such brute bridge laws don't explain anything. Perhaps one has a perfect correlation in such a case but that doesn't constitute an explanation, as we all know. Even proponents of the neural correlate of consciousness approach such as Searle's Biological Naturalism [Searle, 2017] agree that finding such correlates would only be a first step. One can not stop there and say a new fundamental law has been discovered, one must then use that correlation to discover the causal mechanism that generates conscious experience. If one asks the question "but why does conscious experience always arise in the actual world when certain material or functional conditions obtain?," no answer will be forthcoming; it is just a brute fact/brute necessity about the actual world. That is not a "non-reductive explanation," that is no explanation at all. This is especially troubling if such a psycho–physical bridge law is the only law in the "basic inventory of emergent properties," as Chalmers alleges. Even if one is willing to reject causal closure of the physical (CoP) as O'Connor is, there is still the question of how the "presence" of conscious mental states "affects" brains states?

On the other horn of the dilemma one can opt for strong emergence as a modal necessity without identity (assuming that even makes sense). But again, such a stand-alone, one-of-a-kind brute/fundamental metaphysically necessary law is a deus ex machina—"and then a miracle occurred" kind of affair. That is, if matter is fundamental and essentially non-mental then radical emergence must be some sort of occasionalism that replaces God with a miraculous law of nature for every possible world. This may be explanatory for some, but it isn't a natural or scientific explanation, nor is it a law of nature. It is a brute law of meta-nature, surprising to not only the Mathematical Archangel but perhaps to God herself.

We might stand accused of just begging the question about what constitutes a natural law, about what can be natural. It might also be argued that GP implies not the impossibility of such bridge laws but that in order to explain conscious experience there must be something that is not "essentially non-mental," and this is precisely what bridge laws are. Perhaps positing the naturalness of laws is not quite the right way to make the point. Perhaps the point is that bridge principles are ad hoc and brutish in a way that is worse than regular laws because they don't map any (non-trivial) modal relations, unlike regular laws of nature. Such bridge laws are just stipulative and added after the fact to allow for an account of the world in a God-of-the-gaps-type fashion. They couldn't be used to generate a hypothesis. They aren't laws in the understood sense of the practice of science. They are constructs to link up parts of the world described by science, but that linking isn't itself the same practice as the science.

Why is reductionism or physicalism true and satisfactory for everything over the course of billions of years but then suddenly fails and is trumped in the end by psycho–physical bridge laws? Such a jarring violation of reductionism or unification makes a mockery of those very ideas. People worry about the status of the collapse postulate in quantum mechanics and this disrupts unity more than that. From the perspective of physicalism, ontological reductionism, etc., a stand-alone, one-of-a-kind-brute/fundamental bridge law is something of a deus ex machina, miraculous kind of affair.

However, there is a more recent causal or dynamical, dualistic account of strong emergence that rejects CoP and physicalism as it pertains to mental properties up front. In other words, it rejects the following supervenience-based, synchronic, non-reductive physicalist conception of strong emergence that we were just criticizing:

> Earlier emergentists did not give very clear accounts of the relationship between the necessary physical conditions and the emergents, apart from the general, lawful character of emergence. Given the requisite structural conditions, the new layer invariably appears. Recent commentators have suggested that we think of this in terms of synchronic supervenience, specifically "strong" supervenience. So, for example, McLaughlin (1997) defines emergent properties as follows: "If P is a property of w, then P is emergent if and only if (1) P supervenes with nomological necessity, but not with logical necessity, on properties the parts of w have taken separately or in other combinations; and (2) some of the supervenience principles linking properties of the parts of w with w's having P are fundamental laws" (p. 39). (A law L is a fundamental law if and only if it is not metaphysically necessitated by any other laws, even together with initial conditions.) And though he does not say it explicitly here, it's clear that he thinks of this supervenience synchronically: given the 'basal' conditions at time t, there will be the emergent property at t. Van Cleve (1990) and Kim (1999, 2006a, 2006b) also think of the relation as a metaphysically contingent but nomologically necessary form of (synchronic) strong supervenience.
>
> [O'Connor and Wong, 2015]

On this causal account of metaphysically emergent properties, "it will be natural to suppose that they are caused to be by the object's fundamental parts, which have latent dispositions awaiting only the right configurational context for manifestation" [O'Connor and Wong, 2015]. In turn, these newly arisen emergent properties such as mental properties can causally effect biological and physical processes. This conception of strong emergence is certainly a rejection of any kind of CoP, rejecting realization, and embracing downward causation. Therefore, it doesn't conflict with physicalism, reductive or otherwise. It is very important to understand what is meant by the word "caused" here:

> We do not use the term in this neutral manner. Our usage corresponds to the first of these: a power to produce or to generate, where this is assumed to be a real relation irreducible to more basic features of the world. Our favored technical term for this is "causal oomph"! So understood, causation is not amenable to analysis in non-causal terms, but instead involves the exercise of ontologically primitive causal powers or capacities of particulars. Powers are either identical to, or figure into the identity conditions of, certain of the object's properties, which are immanent to those things as non-mereological parts.
>
> [O'Connor and Wong, 2015]

The type of causation specified here requires that under certain conditions, when a physical or biological system reaches, say, a certain degree of complexity of some sort, it has the irreducible causal power to produce or generate a new causally efficacious emergent property such as mental properties. Such properties are, by hypothesis in this case, essentially different and completely novel with respect to their emergent base. This causal power is not reducible to any other feature of the universe, contrary to what are called "Humean" accounts of causation which reject any irreducible causal glue in the universe in favor of "constant conjunction" and the like. Humean accounts of causation reduce causal talk to certain reliable regularities or patterns in the universe, though these accounts vary greatly otherwise. We agree with the causal strong emergentist that Humean causation could not possibly explain mental phenomena popping out of brains. We agree that rejecting Humean causation is a necessary condition for causal strong emergence but it hardly seems sufficient. So obviously anyone inclined toward Humeanism about causation cannot possibly accept causal strong emergence, but being a realist about causation doesn't obviously make causal strong emergence of minds from brains naturalistically acceptable or explanatory. If this irreducible kind of causation is the norm in the world or at least common, then why are mental properties the only properties where this sort of causation seems even remotely plausible to most people? For example, the analog of this sort of explanation for the emergence of life from a chemical base doesn't seem at all natural or explanatory; it wouldn't count as a viable scientific hypothesis. So again, if mental properties are the only case of this sort of causation, then we are back to the objections raised earlier.

Does this causal account of strong emergence escape the concerns leveled against the former conception? As we said, this causal account of strong emergence has an advantage in the sense that it rejects the basic tenants of physicalism, at least as regards mental phenomena, but it is largely still in the same position in that mental properties are the only plausible phenomena for which this sort of strong emergence might even be considered. This is just to say that such causal strong emergence is equally damning for ontological and explanatory unity. After all, causal strong emergence cannot really alter the conception of matter as intrinsically non-mental without veering into pan(proto)psychist territory, and it still maintains that while causally efficacious mental properties are irreducible, matter is in some sense fundamental. While the causal account of strong emergence has no concerns about CoP and while this is property and not substance dualism, it still has to explain how non-dual essentially mental phenomena and essentially physical phenomena can causally interact. In short, moving from talk about laws to talk about causation isn't a cure-all for strong emergence. And while we agree wholeheartedly with the rejection of CoP, etc., on the part of the strong causal emergentist, on their account, mental properties are the only real exception to ontological reductionism or physicalism. This schism is the source of many of their ontological and explanatory problems. Let us bring this point home now.

With regard to emergent property dualism versus substance dualism, why is it more believable and more probable that causally potent qualia-baring immaterial souls/selves/subjects POP out of brain processes under the right conditions than the claims of substance dualism? Interestingly, in what follows, O'Connor and Wong in considering the argument from realism about consciousness to theism, argue that the hypothesis of theism increases the probability that the admittedly potentially implausible claim of strong emergence of mental properties is true.

> We have suggested that the phenomenal realist may reasonably suppose the existence of basic, general laws connecting neural-state types and families of phenomenal-state types (corresponding more or less directly to distinct sensory modalities). Such laws will encode in part facts about specific emergent dispositions of fundamental physical particulars. Here is where we see the potential for design-style reasoning. It seems plausible that there are a variety of ways things might have been with respect to the fundamental constituents of the world. We do not have in mind the Humean claim that the very particulars there are might have interacted in fundamentally different ways. We mean, rather, that there might have been ever so many different sorts of entities having different sorts of basic dispositions from the ones that are manifested in our world. And in particular, it seems a priori rather unlikely that fundamental physical entities should have emergent dispositions toward phenomenal qualities. (that this is a plausible claim is suggested by the fact that many brash but otherwise reasonable philosophers judge the emergentist view to be an utterly implausible hypothesis about our own world, and some are tempted to declare it outright impossible.) Yet, given theism, it seems more to be expected, since we may reasonably suppose the conditional probability of there being agents capable of the kind of experiential life that we enjoy on the hypothesis of theism to be at least not very low, since it is reasonable to think that one of the goods a purposive world designer would wish to see in its creation are creatures of just that sort.
>
> [O'Connor and Wong, 2015]

Here, O'Connor and Wong seem to us to be leveling/acknowledging much the same criticisms against strong emergence as we, and in response claim that the assumption of theism makes it a much more probable and believable doctrine. We don't deny their inference as such, but we just think it is telling that defenders of what is supposed to be a naturalistic account of conscious experience feel compelled to back it up with theism. For example, one cannot imagine Searle making the same claim about his Biological Naturalism. This is in keeping with precisely the line of attack we have been pursuing here in our discussion about strong emergence. To relate all this back to the last paragraph, causal strong emergence seems like the worst possible combination of materialism and dualism. We are stuck with brute, downwardly causal powers that make minds appear from insensate matter and we are still stuck with dualism. Why not just give up the ghost and embrace the package of theism and substance dualism? After all, doesn't the hypothesis of theism make substance dualism more probable (and vice versa) than causal strong property emergence? The answer, of course, is because substance dualism violates both ontological and methodological naturalism. The problem, of course, is so does causal strong emergence, or if not, it fails to be explanatory either in terms of unity, laws, or even causation as those schemas are typically conceived in the rest of science.

Strong causal emergence would still have us believe that physicalism or ontological reductionism is true for everything except consciousness, that everything else in the universe is a nomological, logical, or metaphysical consequence of the fundamental physical facts, whatever they may be. To be fair, science is in no position to rule out the very possibility of radical emergence and the doctrine is certainly not incoherent. However, again, this is a weird law or causal process by the lights of science given the stipulated nature of the rest of the universe. Such a law is beyond the bounds of naturalistic explanation. As Strawson writes, in defense of panpsychism versus radical emergence:

> Assuming, then, that there is a plurality of physical ultimates, some of them at least must be intrinsically experiential, intrinsically experience-involving. Otherwise we're back at brutality, magic passage across the experiential/non-experiential divide, something that, ex hypothesi, not even God can understand, something for which there is no reason at all as a matter of ultimate metaphysical fact, something that is, therefore, objectively a matter of pure chance every time it occurs, although it is at the same time perfectly lawlike.
>
> [Strawson, 2009, p. 52]

Strawson is right that such psycho–physical bridge laws or causal processes are for all practical purposes, supernatural, or worse that causal strong emergence just sneaks in pan(proto)psychism by claiming that matter has a disposition to manifest mental properties.

Does panpsychism fare any better then radical emergence? No. The best arguments for panpsychism are generally taken to be what Seager [2009] calls the "argument from analogy," the "genetic argument," and the "argument from the dispositional nature of fundamental physical properties." The first argument claims there is some feature of fundamental matter such as quantum entanglement that is mind-like and therefore maybe quantum entities are at least proto-conscious. The second argument is really just the claim that radical emergence is impossible. The third argument claims that fundamental physical properties are not intrinsic and that they must nevertheless possess intrinsic properties and the only intrinsic properties are mental, for example, qualia. None of these arguments are decisive and they have all been heavily attacked (see Silberstein, 2010; 2014).

The standard arguments against panpsychism are as follows:

1. Combination problems.
2. No sign and not-mental problems.
3. Unconscious mentality problem—pan(proto).
4. Causal completeness problem.

As for the first problem, panpsychism may not have a generation problem but it does have several combination problems [Chalmers, 2016], namely, how do all those simple minds combine to make conscious agents such as ourselves? If one thinks fundamental physical entities possess free-floating qualia, then how do all those very tiny discretized quales come together to make one of us? If one thinks subjective experience requires an experiencer, then how do those very tiny conscious beings combine to make one human conscious agent with a unified experiential field? The second problem says that, contrary to the argument from analogy, there is absolutely no evidence that fundamental physical entities have mental properties or minds and therefore panpsychism is simply unjustified. The third problem says that pan(proto)psychism only makes the first problem worse because now we can't even conceive of what proto-mentality might be or if it is even coherent. So it threatens to turn the combination problem back into the generation problem. The fourth problem is that causal closure of the physical—CoP (or microphysical closure) would seem to render mental properties epiphenomenal. But CoP aside, the point is that we never have to bring mental properties to bear to explain the behavior of purely physical, chemical, or biological systems. This brings us back to the second problem, of course.

To many of us, any one of these problems is enough to reject panpsychism. But of course ever-hopeful defenders have their responses. Here we will only consider two recent responses to the combination problem, show them wanting, and then a very general argument against any form of panpsychism will be given. The first possible solution comes from Seager [2016b] and it combines the following two claims:

1. Deferential Monadic Panpsychism (DMP) solves the generation problem.
2. "Infusion" (such as QM entanglement or some mental analog) could solve the combination problem.

DMP is just panpsychism with relaxed criteria for what fundamental physical entities could have mental properties, so not just fundamental particles but whatever physical entities turn out to be fundamental; for example, the wave function of quantum systems. Infusion posits that there is something like Humphreys' fusion account of quantum entanglement for fundamental mental properties or conscious subjects. In the fusion process individual entities and their properties are "fused" together such that they are no longer individuals and a new system/entity is created [Humphreys, 1997].

The first issue here is the implausibility of "fusion" for quantum entanglement extends to "infusion" of tiny minds. As many of us writing on quantum entanglement and emergence have noted, fusion is much too strong a claim even for the quantum let alone anything classical [Wong, 2006; Silberstein, 2009]. For example, one can still manipulate individual particles (whatever they are) or quantum systems even when they are entangled. The second issue is that quantum mechanics is open to interpretation, and entanglement and nonlocality get different ontological descriptions and explanations in different interpretations.

According to at least one interpretation, called retrocausal accounts of quantum mechanics, both separability and locality are preserved [Stuckey et al., 2015], so there is no fusion there. In the Bohmian interpretation, the wave function always obeys the Schrödinger equation and never undergoes a collapse. It is, rather, the particle positions (point particles with definite position are fundamental on this view) and not the wave function that create measurement outcomes. For example, in the twin-slit set up, the particle follows one of these trajectories, and therefore passes through one slit or the other only. The wave function, however, passes through both slits, and the two components interfere. The resulting peaks and troughs in the wave function push the particle around, according to the Bohmian dynamical law [Lewis, 2016]. So according to Seager's theory, what part of all this possesses consciousness? Is it the individual particles? Is it the wave function (or pilot wave, if you prefer)? If the latter then perhaps vast swathes of the universe are conscious because the wave function somehow informationally connects many, many particles even at spacelike separation (faster than light). And of course, in general, entanglement can hold between multiple particles to varying degrees at great spatial distances. So where is the consciousness located here? How "big" is it, if such a question even makes sense. Most importantly of all, there appears to be nothing like fusion in any of this. Take the Everett or Many-Worlds interpretation: there are no collapses and no hidden variables; thus the system is described completely in terms of the evolution of the wave function as dictated by the Schrödinger equation. The most basic physical entity in this interpretation is the unitary wave function of the entire universe evolving in configuration space. It is true that said wave function has a branching structure of "many worlds" but they are all just aspects of the universal wave function. What does Seager's theory say here? Is the entire universe now conscious? And again, there appears to be nothing like fusion here either.

Third, we can also reasonably ask, given all this quantum weirdness why does consciousness appear to be correlated with and limited to individual brains? Why don't we see evidence of consciousness EVERYWHERE? What, according to this view, explains the apparent intimate connection between brains, embodied agents, and consciousness? is hard to see how panpsychism or infusion explains any of that or makes it even remotely plausible. If ontological reductionism is true then quantum mechanics had better underwrite neuroscience in some sense. Unlike, say, Biological Naturalism, panpsychism seems be scientifically moribund, a purely metaphysical theory designed to do one thing: solve the generation problem. Even though Seager bases infusion on quantum entanglement, it doesn't really seem to help with this problem.

The fourth worry is the following dilemma: either conscious experience according to this view is supposed to be a quantum mechanical property subject to all the same issues as other quantum properties or it is supposed to be more classical-like, immune from quantum weirdness. If the former, then Seager's theory now has a version of the measurement problem for conscious experience, namely, how do we get from the quantum-mind to the classical mind. How does the classical mind of our everyday experience which seems not at all like a quantum system in terms of its unity of experience, commutative nature, continuousness (as opposed to discreteness), etc., get generated by the essentially non-classical quantum mind? For example, the classical world of experience seems not to have superposition of mental states. Note that appealing to infusion won't help here, it is rather the problem, because the classical world of experience seems void of quantum weirdness such as fusion and infusion allegedly are. The point is that on this horn of the dilemma consciousness now inherits all the foundational problems that were unique to matter in quantum mechanics. This doesn't seem like a step forward. On the other horn of the dilemma there are also problems. If as some theories contend that consciousness is immune from quantum weirdness, perhaps it even collapses the wave function for example, and consciousness is the only intrinsic property of quantum systems then why is there quantum behavior at all? If every quantum system in the universe has a tiny subject or conscious observer associated with it and observing it then why doesn't it always behave classically? As an aside, if mental properties are intrinsic as panpsychism insists, then it's hard to see how something like entanglement, infusion or fusion could happen to them anyway, the right analogy would be something like position in Bohmian mechanics; in short, consciousness would perhaps be more akin to a hidden variable.

Fifthly, while we don't have a consensus on what quantum particles and fields are, there is a consensus that they are not fundamental and therefore we can say per Seager's DMP that consciousness is not directly associated with quantum systems from non-relativistic or relativistic quantum theory. So appealing to essentially quantum processes may not help with the combination problem. And while we don't have a consensus on what is fundamental the contenders from dynamical theories of quantum gravity as we know are things such as: m-branes, super strings, causets, etc. So if m-branes (imagine 3D sheets with matter on them moving around in higher dimensional spaces) are fundamental and therefore conscious according to DMP, the question now becomes how that will help solve the combination problem and explain human consciousness? Good luck.

Now to go from small length and time scales to very large ones, recall that according to panpsychism every fundamental physical entity has subjectivity associated with. But what if quantum systems per se are not fundamental but the entire universe is fundamental as say "Priority Monism" claims [Schaffer, 2010]. This is precisely what Goff's version of panpsychism, which he calls Cosmopsychism, claims [Goff, 2013, Goff, 2017]. It holds that the universe is the one and only fundamental entity and it is a subject of experience. Leaving aside the question of what would or should or could motivate someone to believe this, the view is still stuck with lots of problems. Instead of combination problems it now has the decomposition problem, why are there individual minds. The only scientific motivation for Priority Monism that is widely acknowledged is the aforementioned Many-Worlds interpretation of quantum mechanics but there are a couple problems here with regard to the decomposition problem. First, it isn't clear that the Many-Worlds interpretation, even if true, entails Priority Monism

[Lewis, 2016] and second said interpretation has notorious problems explaining the existence of conscious individual minds, such as the "mindless husks" problem [Felline and Bacciagaluppi, 2013]. It also seems to make the No sign and not-mental problem and the causal completeness problem even worse. In addition, what does Cosmopsychism even mean? Does it mean that the universe is the body of a supermind that is to the universe what the mind of an individual human is to their body? If so, it is hard to see this view as consistent with naturalism or scientific investigation. Cosmopsychism seems more like subjective idealism, theism, or at best, panentheism.

There is a more general argument against any form of panpsychism that to our knowledge no one has thus far mentioned: the argument is that panpsychism is not a logically consistent position. According to panpsychism, the only intrinsic properties of fundamental physical entities are psychical. But here is the problem: most theorists hold that essential properties must be intrinsic properties because those are the only non-relational and non-contingent properties an individual possesses. According to the SEP entry on "Essential vs. Accidental Properties":

> David Denby (2014) defends a similar version of the modal characterization, appealing to the more familiar distinction between intrinsic properties and extrinsic properties: P is an essential property of an object o just in case (1) it is necessary that o has P if o exists, and (2) P is an intrinsic property. Roughly, an intrinsic property is a property that an object possesses in isolation, while an extrinsic property is a property that an object possesses only in relation to other objects.
>
> [Robertson and Atkins, 2016]

So if the intrinsic and thus essential properties of matter are mental as panpsychism claims, then it just isn't true, as panpsychism asserts, that matter is the fundamental building-block of reality and the physical world is composed of or otherwise determined by basic physical entities. One cannot consistently assert that matter is fundamental or co-fundamental and that its essential properties are mental. In short, panpsychism is not a stable position; it reduces to idealism of some sort.

Let us summarize what we have learned thus far. It is hard to see how radical emergence could be true and the GP is real. If matter is essentially non-mental and yet there is some basic physical law that says under the right configurations of matter that consciousness arises from it, then such a law must be either impossible or beyond naturalistic explanation. That is, if matter is fundamental and essentially non-mental then radical emergence must be some sort of occasionalism that replaces God with a miraculous law of nature. On the panpsychism side, in the service of explaining conscious experience on the length and time scales of embodied creatures on Earth with at least some sort of sensory apparatus and some sort of central nervous system, we have seen people appeal to the very small (quantum systems) and the very large (the universe itself), but everything we have ever experienced tells us that only the middle porridge is, "Ahhh, just right" when it comes to the processes associated with conscious experience. Panpsychism is also fraught with several other well-known problems, none of which are easily discharged, even to this very day.

As noted, neither radical emergence nor panpsychism seems like very stable positions, they are at best patches for physicalism or ontological reductionism, the views that drive them. The best argument for either position seems to be the claim that they are the lesser of evils with respect to the other, but there are alternatives. One thing we know for sure: on pain of contradiction, panpsychism cannot invoke radical emergence to get out of its various combination problems and radical emergence cannot invoke panpsychism to resolve the GP. So what assumptions led to us to this absurd situation where we think it is either one or the other? The assumptions are as follows:

1. The generation problem is real (matter is essentially non-mental and it is fundamental).
2. All fundamental entities must have intrinsic properties.
3. Fundamental physical entities don't have intrinsic physical properties.
4. Consciousness is an intrinsic property (qualia) and by elimination must be the intrinsic aspect of fund physical entities.

Where does this leave us? We agree with the third assumption but the rest we reject. Given the problems with both radical emergence and panpsychism it's high time to question at least some of their shared assumptions. This brings us to the generation problem in a block universe.

7.3.3   The generation problem in a block universe

Without even discussing the block universe we have seen that the hard/generation problem is so daunting that the equally problematic "solutions" of causal strong emergence and panpsychism have come to the forefront. Both are problematic precisely because they accept the GP on its own terms. We have seen that iron-ically, materialism, etc., can't escape one or the other of these views and they are both non-starters. We can see why at this point the reader might say, "assum-ing all your arguments are good, how can the block universe possibly make the GP any worse?" But things can get worse as a matter of principle. If one takes dynamical explanation as fundamental (whether it be dynamical or causal) and one takes matter (as essentially non-mental) as fundamental, then the question is: how do neural processes cause or dynamically *bring about* or *generate* previously non-existent conscious processes? Philosophers have long debated the ontological status of dynamical laws and the causation relation. But basically, the question is this: are either of these reducible to some other feature of the universe (i.e., Humeanism) or is there really some fundamental nomological or causal "glue" in the universe that explains why new events come into being? It should be clear that in a block universe, whatever else one wants to claim about either dynamical laws or causal relations, they certainly can't literally be bringing new events into existence that never existed before, at least not from a God's-eye perspective. As we will see, it is especially hard (if not impossible) to explain how matter so described could *generate* essentially distinct conscious experience (conceived as qualia) given only a Humean conception of causation. In fact, in the previous section, we already saw that problem without even bringing in the block universe. After all, on this view causation talk isn't about some pseudo-magical, irreducible phenomena-producing glue in the universe that makes new orders-of-being POP into existence. And this is especially true in a block universe where, from a God's-eye point of view, even all the Humean talk about stable patterns and counterfactuals is all just talk from the ant's-eye perspective. Recall what Price and Wharton said about causation talk in a block universe (chapter 4), it is all perspectivial, there is no objective and global arrow or Direction of causation in a block universe. From a God's-eye perspective talk about laws and causes is ultimately time- and determination-symmetric [Price, 2011, p. 279]. All events and conscious experiences are just there, at once in a block universe. We agree. But here the question is why are there conscious perspectives in the block universe to begin with? Obviously, you can't invoke conscious perspectives to explain why there are conscious perspectives. It should also be clear that all this goes double for RBW because the adynamical global constraint (AGC) is fundamental and the key dynamical laws are derived from it, given spatiotemporal contextuality.

Let us now return to time as experienced: a subset of the generation problem. Recall that the standard answer that proponents of the block universe give to this question is to claim that cognitive neuroscience will one day explain the phenomenology of time perception by appealing to purely cognitive and ultimately neural mechanisms. So, the claim is that while physics does not have the resources to explain the experience of Passage and Presence fully, cognitive neuroscience will someday possess this ability. The reason that physics presumably does not fully have the resources is not just the fact that the A-series fails to obtain (i.e., objective Passage, Presence, and Direction) in a block universe interpretation of relativity, but as Dainton points out:

> The arrow of phenomenal time has considerable autonomy in relation to the others [e.g., physical arrows such as entropy] that we have encountered. If you were to fly a spaceship into the time-reversed half of a Gold universe and visit an inhabited Earth-like planet, you would perceive all manner of bizarre things going on, but you would continue to feel your experience flowing in the usual manner, *forwards*, even though the world around you seems to be running in reverse.
>
> [Dainton, 2001, p. 106]

Whether one agrees with Dainton or not in this particular case, the point is that it isn't at all obvious that there is any necessary connection between a physical arrow of time, such as the thermodynamic arrow, and the experiential arrow of time (we will return to this question later in the chapter). Of course, one can raise the same concern for neural correlates as well. "However, it has been argued that the existence of temporal integration in the brain does not necessarily entail a uniquely distinguished present. . . . In other words, evidence for the discreteness of temporal processing does not necessarily entail the existence of discreteness in the subjective experience of time" [Wittmann, 2016, p. 513]. At best, certain physical arrows of time are necessary but not sufficient for the experiential arrow of time.

We have already seen some of the claims about brain-based explanations of the experiential arrow of time but here is another:

> They [detensers] argue that our experience of time passing, our experience of the presentness of now derive from our subjective point of view and not from time objectively passing or from events having the property of being present. Change on these views is just a matter of an object having different properties at different times. If change requires the coming into being and passing out of being of events this is entirely due to a dynamism to be found within experience. Detensers usually take experience to supervene on brain processes. They argue that the dynamism we find in experience is a product of the active interpretation of our brains and is projected onto the world Whatever continuity and change we find in experience is not passively imposed on the brain by the world but is actively constructed.
>
> [Kiverstein and Arstila, 2013, p. 454]

Now there are certainly strictly empirical concerns one can raise here, such as:

> As noted above, there is as yet no consensus on brain areas involved in time, and there are reasons to doubt that timing is a central dedicated function of any particular brain area. In any case, at the neural level there is a mystery where time is concerned. In at least some situations, time passes in settings where nothing (else) is changing, and yet we detect its passage. How does that work?.
>
> [Arstila and Lloyd, 2014, p. 662]

But we have a number of fundamental questions raised here. First, even granting that the generation problem has been solved, *why* does the brain in a block universe project Passage, Presence and Direction in particular onto the block universe? People love to give evolutionary answers to these sorts of questions, but first, that confuses cause and effect—the brain has to generate these experiences *before* they can be selected for. Second, this begs the question why the ant's-eye view is more adaptive than, say, the the view of the heptapods. Third, from the God's-eye view looking at the entire block, invoking natural selection is like being told about the rules in Conway's Game of Life that generate processes like self-reproducing gliders and still trying to maintain that such patterns are *fundamentally* explanatory, when talk about processes like gliders are all strictly ant's-eye. In fact, in RBW it is even worse because the dynamical laws of physics themselves are not ultimate explainers either, much less natural selection and reproduction.

More importantly, why is there conscious experience *of any sort* in a block universe in which matter is fundamental? What we want to know is if time as experienced supervenes on brain states and brain states are every bit as "frozen" as any other physical states in the block universe, then *how* do brains "actively interpret," "actively construct," or project Passage and Presence onto the world? How do the worldtubes of brains interpret anything and for whom is this interpretation constructed? The brain itself? This is just another version of the generation problem of course. How do the worldtubes of brains create the temporality we find in the world if they themselves have no more temporality than any other physical phenomena? After all, the only thing we can appeal to here is the before, during, and after of certain brain states and their correlations with certain experiences. How is this explanatory? And if it is: we could have made the same claim for lots of other physical processes other than neural ones.

One immediate problem is that such biological mechanistic explanations involve causal processes unfolding over time—as we saw earlier, proponents talk about neural processes generating conscious experiences—but in a block universe neural processes can no more generate or produce anything than physical processes can. After all, those neural states and their 4D worldtubes live in the block universe as well. The point is that any timelessness we want to attribute to physical processes is going to be inherited by biological ones also. So, if there is a problem here for physics, then there is a problem here for biology, especially given that most proponents of this view believe that biological processes are determined by physical processes.

Buonomano comes very close to appreciating the problem in what follows:

> Deciphering whether the flow of time is a fiction created by the mind or something that eludes the current laws of physics is a uniquely complex problem that lies at the interface of physics and neuroscience. And if this mystery were not sufficiently challenging as it is, there is a further wrinkle to consider: the laws of physics and the human brain are not independent of one another. It is not simply that the inner workings of the human brain must obey the laws of physics, but that our interpretation of the laws of physics is filtered by the architecture of the human brain. If we must question whether we can trust our brain's account of something as self-evident as the flow of time, must we not also question the brain's impartiality in interpreting the current laws of physics? . . . . Do we gravitate towards certain interpretations of the current laws of physics because of the way the brain represents and thinks about time?
>
> [Buonomano, 2017, p. 178]

Even having thought this far, however, Buonomano still assumes that the brain "constructs" and "generates" time as experienced. He is still wedded to a qualia plus representationalism picture in which *the brain* is interpreting and projecting time as experienced. Of course, the reigning paradigm in cognitive science and cognitive neuroscience is computationalism/representationalism supported by underlying neural mechanisms. The current fashionable incarnation of representationalism or "internalism" are representation-heavy models based on prediction-driven processing that often involve Bayesian inference [Kirchhoff, 2015]. In these models "the material basis of conscious experience is inferentially secluded and neurochemically brain bound" [Kirchhoff, 2015, p. 68]. However, as we will make clear in chapter 8, there is another growing paradigm at work in cognitive science that rejects representationalism with a capital "R" and internalism in favor of direct realism and extended cognition wherein it is the unified brain–body–world complex that is the unit of cognition.

One standard argument for representationalism is the argument from perceptual illusion. The argument in a nutshell is that since people often misperceive in various ways, the brain must be internally representing the world. Buonomano makes this argument about time as experienced on the basis of various temporal illusions [Buonomano, 2017, p. 217]. The debate about perceptual illusions is very old and has many convolutions. For now, the important point is that the argument from illusion to internalism has many well-known weaknesses and the defenders of direct realism have written a great deal on how to account for perceptual illusions [Kirchhoff, 2015], including temporal illusions [Kiverstein and Arstila, 2013].

Furthermore, from the God's-eye view, conscious experiences/states are no less fundamental than physical or biological states, while there may be in principle certain correlations we can work out from the ant's-eye view, such as the alleged neural correlates of conscious experience (NCC), from the God's-eye perspective it is all just "there." As we saw, the hard problem of conscious experience was hard enough when the future didn't exist, matter was objectively in motion and capable, at least in principle, of being causal in the generative or productive sense that the radical emergence view was advocating for, but it seems much harder in a block universe. This is a serious extra challenge to ontological reductionism or strong emergence about conscious experience.

The point is this: in a block universe one is stuck not only with a very potent brand of Humeanism about causation from the ant's-eye perspective but also with the fact that it is all just there. Not only are there no irreducible causal powers or causal glue in the universe, from a God's-eye perspective, every event is just there, including conscious experiences. From a God's-eye perspective, nothing generates or produces anything. Referring to our earlier discussion, how can the brain be "where the magic happens," how can it be the projector of conscious experience, when it too is nothing but a material 4D worldtube? So much for strong causal emergence.

But as Dainton notes, panpsychism fares no better at explaining the experience of time in block universe:

> And if it should turn out that phenomenal properties—in their familiar, unreduced form—are as much a part of the material world as mass or charge [as panpsychism asserts], the problem of reconciling their dynamic character with the eternal character of the Block universe will be all the more pressing.

[Dainton, 2011, p. 387]

The problem here is caused by assuming that matter, as conceived in the dynamical/mechanical worldview, is fundamental, and treating conscious observers as if they were somehow distinct from the physical universe, an extra element that needs explaining. This view is consistent with Hermann Weyl's talk about consciousness "crawling along the worldline" in a block universe like a moving spotlight "lighting up" events so as to mark those as present events for the observer [Wilczek, 2016, p. 37]. Proponents of this view do not tell us where consciousness comes from, how it can perform this trick, or how it relates to the worldtubes of neural processes. And as Dainton notes:

> Since (ordinary) time makes up one of the Block universe's four dimensions, to make sense of a consciousness (or point of apprehension) moving along the block—and so moving through time—we need to posit an additional temporal dimension, a meta-time . . . we are being required to embrace an unusually radical form of psycho–physical dualism . . . A further difficulty is perhaps less obvious. What sort of time is the proposed meta-time? If it too has the character of a Block universe then from the point of view of explain the temporal appearances absolutely nothing has been gained.
>
> [Dainton, 2011, pp. 389–90]

According to the ontological reductionist or strong emergentist, to explain our experience of time and change we are going to have to, at least in part, appeal to conscious brain processes whether it is a blockworld or not. Natural science such as cognitive neuroscience wants to explain the very existence of phenomenal conscious experience (such as the experience of a special "moving now") by appealing to dynamical brain processes, such as the mechanism of "temporal binding" that makes coherent experience possible, the brain's various clocks, the "processing time" of all sensory modalities, the saccadic suppression mechanism, etc. All these mechanisms and all the models of brain mechanisms in general are essentially dynamical in nature. In neuroscience, in every case such as cognition, perception, memory, etc., the attempted explanation of these functions appeals to the dynamics of cell assemblies, neuronal firing rates, etc. Explanation of specific conscious states and specific cognitive functions in cognitive neuroscience is always in terms of "underlying causal mechanisms" in the brain or the "neural correlate of consciousness or cognition," both of which are inherently diachronic and dynamic conceptions of explanation. Thus, the fundamental working assumption of cognitive neuroscience and much of philosophy of mind is that matter in general and brain processes in particular are more fundamental than consciousness, both ontologically and by way of explanation.

Again, from the God's-eye view in the blockworld there is no absolute/objective motion or change, no objective dynamical/causal processes actually exist. The block universe "is," "was," and "always will be" as it "is." In a blockworld in which all events are equally real, any explanations proffered for any event (including those pertaining to conscious brain processes) that appeal to causal mechanisms/processes, dynamical laws, or more generally "becoming," "change," "generation," etc., must be error theories or merely compatibil-ist/perspectival accounts of such processes, all of which implies that brain "processes" do not literally cause (as in bring about or give rise to something that did not exist before) conscious "processes," and conscious processes are every bit as fundamental as brain processes. There is no absolute or objective or frame-independent sense in which the better part of the universe's history unfolded without phenomenal consciousness and then conscious processes sprang into be-ing and then became more sophisticated over time as the result of dynamically evolving brain processes. So given all this, why persist in claiming that matter is ontologically and explanatorily fundamental?

Thus we find ourselves in the following dilemma: either attempt to explain phenomenal consciousness by appealing to dynamical brain processes as we do in cognitive neuroscience, or explain the projection of a dynamical world that appears to have Passage, Presence, and Direction by appealing to the bracketed phenomenology of conscious experience. There are only two possibilities: either conscious experience is being explained or doing the explaining. If the former, then the ontological reductionist or strong emergentist must answer the gener-ation problem. If the latter, then clearly that might be useful phenomenology but it doesn't answer the generation problem.

There is one well-known physicist who, tacitly at least, acknowledges the in-escapable brute nature of conscious states and their correlations with brain states in a static world but nonetheless still tries to tell a story whereby brain states some-how determine, explain, or are identical to conscious states. We have in mind Julian Barbour who, in his book, *The End of Time*, advocates what is arguably an even more radically static conception of the universe than our own RBW. As we know, Barbour's interpretation of the Wheeler–Dewitt equation is that the universe is an N-dimensional configuration space wherein each point is a static three-space with one of infinitely many possible static configurations of matter-energy embedded in it. There is no temporal axis in Barbour's world. Barbour calls each of these points in configuration space a "Now." Each "Now" will, of course, have its conscious observers such that "any human experience is deter-mined by that human's neurological state at a particular Now. A person will have different experiences at different Nows. Some of these will include representa-tions of others, integrated in such a way as to be experienced as having happened earlier. Others will be integrated in such a way as to be experienced as perceived motion" [Healey, 2002, p. 295]. This is the view Price was critiquing earlier as the "frozen-block." In spite of his radically timeless universe, Barbour says things that make him sound like a crude kind of mind/brain identity theorist when explaining, for example, the experience or "illusion" of motion:

> Could all motion be a similar deception? Suppose we could freeze the atoms in our brains at some instant. We might be watching gymnastics. What would brain spe-cialists find in the frozen pattern of the atoms? They will surely find that the pattern encodes the positions of the gymnasts at that instant. But it may also encode the

positions of gymnasts at preceding instants. . . . The brain in any instant always contains, as it were, several stills of a movie. They correspond to different positions of objects we think we see moving. The idea is that it is this collection of 'stills', all present in any one instant, that stands in psychophysical parallel with the motion we actually see. The brain 'plays the movie for us,' rather as an orchestra plays the notes on the score. . . . If we could preserve one of these brain patterns in aspic, it would be perpetually conscious of seeing the gymnasts in motion.

[Barbour, 1999, pp. 28–29]

This is, of course, exactly the kind of "at–at" view we have been critiquing throughout this chapter. And all the same concerns apply but even more as in his scheme the only possible explanation for such neural correlates of consciousness and their conscious correlates is brute synchronic relations or brute psycho–physical bridge laws. But we want to know *why* these correlations obtain, and merely invoking supervenience of some sort (of any sort) or brute bridge laws explains nothing. Barbour must realize of course that given his claim that "Nows" are fundamental elements of reality, then brain states are no more fundamental than or explanatory than conscious states. Brain states do not exist prior in time to conscious states, nor do they "give rise to them" or cause them. Barbour's insinuation that brain states explain by merely "coding for" or being "isomorphic to" the conscious state they correlate with would be considered crude neuroscience even in a dynamical world; it would be at best a neural correlate and, as we noted earlier, that is at best only a start to finding a causal mechanism. But the claim is even less well-motivated in his timeless world in which there is at best a brute correlation between a particular conscious state and a particular brain state; there is really nothing more to say about such correlations in such a world. In his more cautious moments Barbour [1999] appreciates that at best his view can support a kind of naturalized psycho–physical parallelism:

Nothing in the material world gives us any clue as to how parts of it (our brains) become conscious. However, there is increasing evidence that certain mental states and activities are correlated with certain physical states in different specific regions of the brain. This makes it natural to assume, as was done long ago, that there is psychophysical parallelism: conscious states somehow reflect physical states in the brain. Put in its crudest form, a brain scientist who knew the state of our brain would know our conscious state at that instant. The brain state allows us to reconstruct the conscious state, just as musical notes on paper can be transformed by an orchestra into music we can hear.

[Barbour, 1999, p. 26]

We have already seen that this way of thinking is fading fast even in the cognitive neuroscience of consciousness. However, even in the preceding passage embracing psycho–physical parallelism, Barbour cannot resist making the additional claim that particular brain states code for particular conscious states such that in principle a super future neuroscience could read off the latter from the former. But again, in Barbour's timeless world the correlation or parallelism between particular brain states and particular conscious states is just a static brute fact; neither kind of state "explains" the other in any way. If neuroscience really could read off conscious states from brain states in Barbour's timeless world, that would just mean that we discovered that, in a purely a posteriori fashion, it just so happens that certain conscious states are always correlated with certain brain states, say across all "Nows."

We are absolutely puzzled by the widespread belief that there is any simple mapping (whether causal, nomological, or modal) between some brain state or neural pattern and some particular conscious experience. What in the world (ours or any other) would lead someone to believe that if a certain brain state is induced in any individual anywhere, at any time, she would automatically be experiencing as in Barbour's example, a gymnast frozen or in motion? Regardless of whether one spins this in terms of mere brute nomological necessity (strong emergence) or metaphysical necessity (physicalism), why in the world would such correlations exist? This just looks like psycho–physical parallelism or Occasionalism all over again, but without God to provide the mechanism of correlation. The point of all this, and what Barbour does not fully appreciate, is that at least from a God's-eye view, whether it be his timeless universe or the blockworld, the best one can say about the relationship between a particular conscious state and a particular brain is that they are correlated or, if you prefer, "parallel" to one another.

Maudlin, in his debate with Price over the Objective Direction and Passage of time, makes an argument similar but importantly different to the one we made here. Price essentially gives the epistemological argument for the conclusion that we are not justified in believing there is an Objective Direction or Objective Passage and that therefore it must not make any real difference [Price, 2013]. Maudlin, ever the defender of Objective Direction and Objective Passage, retorts that there might not even be any conscious experience in a world without Objective Direction or Objective Passage:

> Tim Maudlin (2002) objects that Price's argument begs the question by assuming that the 'block' world would contain equivalent experiences, or indeed any experiences at all; for the physical processes of the 'static' block world would be quite unlike those of the 'dynamic' passage world. Assuming that mental states are, or supervene on, physical states there is therefore no justification for assuming that the mental states would be the same in both possible worlds. Given that Price claims only that experience does not favor the A-theory over the B-theory however, Maudlin's objection fails. For although the A-theorist may hold that only a world in which times passes contains experiences, the B-theorist will disagree and hold that a block world contains experiences. . . . Consequently each theory, on its own terms, expects the world that it describes to contain experiences; and the presence of experience therefore cannot settle the question of which theory is correct without begging the question by presupposing the truth of the theory for which experience is supposed to provide evidence. Not, at least, without further arguments concerning the significance of passage for the physical–mental relation.

[Prosser, 2013, p. 322]

Here is what Price claims: "there's a parallel hypothesis—generated by an obvious mapping between underlying brain states as described in the flow picture and the corresponding brain states as described in the flowless picture—offering an explanation of the same phenomena *without* invoking flow. . . . The phenomena [Passage and Presence] do not support the existence of flow, because—at least for a physicalist—any flow-invoking explanation of the phenomena is easily matched by a flowless explanation" [Price, 2013, p. 301].

It should be clear that for us, how we adjudicate this debate will depend on whether it is about: 1) the need for Objective Passage or Direction to explain the experience of Passage or Direction, or, 2) does the hypothesis of a block universe given the fundamentality of matter make it impossible to answer the generation problem in a satisfactory way? We are already on record as siding with Price in the first debate. We would say, based on everything said herein, that we are on the side of Maudlin in the second debate. However, we can't quite say that, since both Price and Maudlin accept the block universe and physicalism. So really the only argument here is if Objective Passage or Objective Direction *within a block universe* are necessary for explaining the very existence of experience. Maudlin says "yes" and Price says "no," based on physicalism. Since physicalism defined in a certain way does entail what Price claims and since Maudlin appears to be a physicalist of some sort, we would say Price appears to win this debate as well.

What we want to impart to both Price and Maudlin's thinking, however, is that physicalism is an empty metaphysical dodge in this debate and once they let go of the empty metaphysics and think about the question from a more this-worldly perspective, they will hopefully see that the second debate is the really fundamental one. First to physicalism. There are many, many different varieties in the literature but, however it is defined, it is supposed to underwrite the *metaphysical* claim that everything is physical (an identity relation that is true in all possible worlds). Of course, what primarily motivates this metaphysical claim is the *empirically* based belief that science is largely a chain of successful inter-theoretic reductions or part/whole compositional reductions. It should be obvious from chapters 4 to 6 that we reject this claim, but forget about RBW. Ontological and explanatory disunity and pluralism are perhaps the consensus now in philosophy of science [Silberstein, 2002, 2012; Horst, 2007]. The point is that this view of the unfolding of science is highly debatable at every step of the way including, of course, in foundations of physics and foundational physics. Nonetheless, even though there is a great deal of skepticism about theoretical and ontological reductionism now, many still persist in being physicalists. And what is telling about at least a minimal conception of physicalism, it neither entails ontological reductionism nor is entailed by it. As Seager notes:

> In the abstract, physicalism thus demands that there be a dependence relation of the non-fundamental upon the fundamental. In order to sustain the claim of monism, this relation has to be pretty strong in at least two ways: logical and ontological. As to the first, the dependence relation must be of maximal logical strength: physicalism requires that it be absolutely impossible for two worlds to be identical with respect

to the properties, laws and arrangement of the physical fundamentals and yet differ with respect to anything else. This relation is that of logical supervenience.

Logical supervenience is, in principle, consistent with non-monism if there is a maximally strong necessitation from the fundamental physical domain to some putative non-physical domain. For example, traditional epiphenomenalism can be made consistent with logical supervenience if the modal relation between the physical base and the supervenient mental state is "bumped up" from the standard relation of causation to one of maximally strong necessitation.

[Seager, 2016a, p. 170]

Of course, if this is all there were to physicalism, then, as many have noted, it would be compatible with all sorts of mind/body relations including substance dualism. Clearly then there are many metaphysical accounts of the relationship between mental states and physical states that would make so-called physicalism true. Obviously, physicalism thus defined will be true in a block universe, a universe with or without objective Passage, or almost any other sort of universe. So if this is all Price has in mind then he is correct. We called this version of physicalism an empty metaphysical dodge because it doesn't explain or tell us *why* inviolate psycho–physical correlations exist.

Maudlin does argue, however, that in a world in which the objective Direction of time were reversed or absent, people would not have the same or any conscious experiences:

The physical processes going on in the Doppelganger's brain are quite unlike the processes going on in a norma brain. Nerve impulses do not travel along dendrites to the cell body, which then fires a pulse out along the axon. Rather, impulses travel along dendrites to the cell body, which then fires a pulse out along the axon. . . . There is no reason to belabour the point: in every detail, the physical processes going on in the Doppelganger are completely unlike any physical processes we have ever encountered or studied in a laboratory, quite unlike any biological processes we have ever met. We have *no reason whatsoever* to suppose that any mental state at all would be associated with the physical processes in the Doppelganger.

[Maudlin, 2002, pp. 272–3]

Maudlin thinks a world in which there is no Objective Passage or Direction would be even worse in this regard:

in a world *with no objective flow of time at all*, i.e. . . . to a world in which there is no time at all, perhaps a purely spatial four-dimensional world. . . . Nothing happens in this world. True, there is a mapping from bits of this world to bits of our own, but (unless one already has begged the central question) the state of this world is so unlike the physical state of anything in our universe, that to suppose there are mental states at all is completely unfounded.

[Maudlin, 2002, pp. 273–4]

So Maudlin would say that such worlds are not physically identical to our own and therefore Price's appeal to physicalism will fail. But as Price notes, this isn't so much an argument for Objective Direction or Passage as it just assumes that the world has these features and therefore this absence of consciousness weirdness would ensue:

> Why is this argument unconvincing? Essentially, because we regard spatial translation as a fundamental physical symmetry, *and therefore expect that it holds in biology and psychology, too* . . . the processes in question are *exactly alike*, by the similarity standards embodied in the fundamental symmetries. It *could be* that these symmetries break down for life, or consciousness, But that would be a huge surprise, surely. And similarly for T or CPT symmetry . . . . we have every reason [to believe in doppelganger experiences] rounded on an excellent general principle: physical symmetries carry over to the levels that supervene on physics.
>
> [Price, 2011, p. 299]

Of course, if Maudlin is right and there really is Objective Direction and Passage that somehow underwrites all the key causal and nomological relations even in a block universe in which all events are "just there," then these really wouldn't be physically equivalent worlds and perhaps various symmetries would fail, etc. But we still need an independent reason to believe there is Objective Direction and Passage and what reason could there be in a block universe other than experience. So if you can explain experience (Passage, Presence and Direction), then you are done. But do either Price or Maudlin have any such explanation for conscious experience in a block universe other than the invocation of physicalism? The only discernable difference between the two is that Maudlin believes the physical supervenience base must contain Objective Direction and Passage, and Price does not.

To be fair, we can see that perhaps there is some case to be made for the fact that both sides beg the question here. But we would say that focusing on Objective Direction and Passage is wrong if according to Maudlin these features are compatible with a block universe. It is very clear from what Maudlin says that he doesn't believe there would be conscious experiences in the Barbour world and it seems pretty clear that Price would say there could be assuming minimal physicalism. Yet, Maudlin isn't bothered by the generation problem in a block universe, but only if that block universe possesses Objective Direction and Passage (recall his own formal account of this from chapter 2).

For us the question here is what magic are Objective Passage and Direction adding in a block universe? Clearly whatever that magic is, it isn't generating new events that never existed before. So let's examine Maudlin's view a little more deeply. He writes: "Insofar as belief in the reality of the past and the future constitutes a belief in a 'block universe', I believe in a block universe" [Maudlin, 2006, p. 109]. But he quickly goes on to state that "I am one of those unusual defenders of the block universe who does not deny that there is any objective flow of time" [Maudlin, 2006, p. 109]. Maudlin provides the following three reasons for believing in what he calls the objective passage of time:

1. The world seems (everywhere and every-when) to be a world in which things change (i.e., where one thing comes out of or is produced from another).

2. Change appears to affect everything, not just some select group of objects (unlike, for instance, pain).

3. No good arguments suggest that change is not real and fundamental [Maudlin, 2006, p. 127].

Maudlin is very clear that he takes the notion of *production* seriously:

> But the passage of time connotes more than just an intrinsic asymmetry: not just any asymmetry would produce passing. Space, for example, could contain some sort of intrinsic asymmetry, but that alone would not justify the claim that there is a 'passage of space' or that space passes. The passage of time underwrites claims about one state 'coming out of' or 'being produced from' another, while a generic spatial (or indeed a generic temporal) asymmetry would not underwrite such locutions.
>
> [Maudlin, 2006, pp. 109–10]

But again, from a God's-eye view, literal *production* or *generation* seems to be exactly what is ruled out by the block universe. It is hard to imagine holding any other view of causation in a block universe other than a Humean/perspectival one. This brings us to Maudlin's views about dynamical laws. "Counterfactuals, and Explanations," the first chapter of *The Metaphysics Within Physics*, begins almost immediately with what he considers to be one of the greatest oversights within the philosophical community; for Maudlin, a proper analysis of natural laws must take into account that these laws describe how something changes. Maudlin writes that "[w]hat is most obvious about these laws is that they describe how the physical state of a system or particle evolves through time" (11). Given the fact that Maudlin takes all laws to represent this kind of evolution, it is no surprise when he says, "the principle of temporal change is the motor of the enterprise" (12). It is these dynamical laws in particular that Maudlin claims are *ontologically primitive* [Maudlin, 2007]. Maudlin's understanding of the "primitiveness" of dynamical laws is explained as "the idea that a law of nature is not logically derived from, and cannot be defined in terms of, other notions" [Maudlin, 2007, p. 15].

In short, even in a block universe, Maudlin thinks dynamical laws are basic and irreducible. Obviously, this is exactly what RBW denies. We admit that Maudlin's position is coherent and logically consistent but again it strikes us as being at some odds with a blockworld. Presumably what Maudlin would say is that one can imagine many different blockworlds but not all will have the dynamical and temporal structures that allow us to explain patterns in the blockworld dynamically, and that is the sense of production or generation he has in mind. We agree that this is possible and reasonable, but that claim is equally consistent with either a reductive or non-reductive account of dynamical laws. As Price says, "[t]he contents of time—that is, the arrangement of physical stuff—might be temporally asymmetric, without time itself have any asymmetry" [Price, 2011, p. 292]. We would make the same point about dynamical laws. As long as the relevant patterns in the "stuff" are there, that is all one needs to ground dynamical and causal explanations for relations between physical events. So we still need an argument from Maudlin about why we should prefer primitive dynamical laws *even in a block universe*. In RBW we take blockworld thinking very seriously and underwrite dynamical laws with an adynamical global constraint and thereby discharge these worries. We also discharge all concerns about how to explain the fundamental dynamical laws and the initial conditions, while Maudlin is still stuck with all these problems.

Of course, Maudlin wants to say that it is the primitive dynamical laws that explain why the "stuff" has the appropriate patterns. But how? Another way to put our criticism of Maudlin's account of explanation is that it raises as many questions as it answers. For instance, though we can practically explain the state of events at time t_n by appealing only to the laws of nature and the "initial conditions" present at t_{n-1}, there is still the question of how exactly it is that the laws and initial conditions explain the outcome; that is, what is it about laws that allows them to explain future events, and how exactly do these laws "govern" the evolution of states? This seems to be precisely the question that Maudlin seeks to avoid by positing the laws of nature as primitives, but it is intuitively unsatisfying to give an account of explanation that does not provide any sort of answer to the question raised here. Surely the "governing" metaphor doesn't really hold up in a block universe. In short, we still need an argument from Maudlin as to why the physical supervenience base needs Objective Passage and Direction, and primitive dynamical laws to resolve the generation problem and what exactly he thinks that might mean in a block universe.

Let us return to physicalism however, because as Seager states, physicalists typically have something stronger in mind than logical or metaphysical supervenience:

> The most frequent metaphor philosophers use to express what they mean by ontological dependence is a theological one: once God created the physical world, set the physical laws and the arrangement of the fundamental physical entities in the world, there was nothing left to do about the non-fundamental things [including conscious experience]. They would follow of necessity as a metaphysical "free lunch" (another common expression used to denote ontological dependence). If some physically non-fundamental entity was a radical emergent, then God's job would not be done with the creation of the physical. God would have to add in the "laws of emergence" in order to ensure the generation of such emergent features.
>
> [Seager, 2016a, p. 170]

So the generation problem in a block universe is to explain why one gets conscious experience for free given all the physical facts and given that, from a God's-eye perspective, both physical and mental features are all just there. This is what a robust version of physicalism entails, but that isn't an explanation itself. And it is hard to see how either Price or Maudlin offer any answer to this question, and hard to see in a block universe how adding Objective Passage and Direction makes the problem disappear. In a way we are back to Price's point. Maudlin would claim that one can imagine two physically (at least in terms of classical and quantum physics) and mentally experientially/empirically identical block universes, one of which nonetheless has irreducible dynamical laws *and* has Objective Passage and Direction. Maudlin would say that only the latter can underwrite claims about causal production, dynamical determination, solve the generation problem, etc. But again, given that these block universes are otherwise identical (like the ether interpretation of relativity, there is no experiment one could perform that would indicate which universe one is in) on the ontological side it is hard to see how such features could or would make any real difference, and on the epistemological side it is hard to see how we could ever be justified in believing or knowing that these features make a difference. To reiterate, the deeper point here is that we think what Price and Maudlin, both blockworlders and both physicalists of some sort, ought to really be focused on when it comes to conscious experience is the generation problem in that setting.

Now one can easily imagine Maudlin saying, "wait a minute, you yourselves have just spent pages arguing that the block universe needs something extra to solve the generation problem so how can you criticize me for agreeing and trying to provide the something extra?" But again, our point is that Maudlin is right that merely invoking physicalism (as Price does) as an answer as to why there is conscious experience is insufficient, but wrong that adding Objective Passage/Direction and irreducible dynamical laws can solve the problem *in a block universe*. We would say that once one sees the deeper problem one ought to either become suspicious of the block universe hypothesis or seek another way to think about the generation problem.

We can easily imagine Price saying that our argument proves too much, that it isn't unique to conscious experience, that it equally damning for claims about strictly physical processes, that it casts as much doubt about purely physical causal or dynamical claims. As we said, we agree with Price that Humeanism of some sort and perspectivialism about physical causation is perfectly fine in a block universe, though we make it very clear that such explanations are not fundamental in the sense of being strictly ant's-eye relative, but the generation problem is unique because the claim here is that *somehow* fundamental dynamical and causal physical processes *generate* essentially different orders of being in the form of conscious experience. Again, perspectivalism about causal claims is fine for everything except explaining such perspectives themselves. We saw that Humeanism on causation can't make good on this even in a dynamical world and the block universe just makes it worse.

So again, just as we were led to think differently about the physical world, and embrace the fundamentality of adynamical global contraints, we should take this opportunity to start re thinking consciousness and its relationship to the physical world. This is what we do in chapter 8.

8

Relational Blockworld: Experience, Time, and Space Reintegrated

The whole duality of mind and matter . . . is a mistake; there is only one kind of stuff out of which the world is made, and this stuff is called mental in one arrangement, physical in the other.

Bertrand Russell [Russell, 1913, p. 15]

8.1 Introduction

As we saw in chapter 7 the block universe model makes the generation problem insurmountable. For some this might be a good reason to reject the block universe model or become an eliminativist about consciousness, etc. For us it is a good reason to question the dynamical paradigm once again, in this case the idea that insensate matter is fundamental and must somehow *generate* conscious experience. We think it is high time to rethink the nature of both conscious experience and matter, and their relationship to one another. After all, if one is a realist about conscious experience and believes that matter is fundamental, at the end of the day the hard problem of consciousness is really this: how do neural processes *cause* or *generate* conscious experience? Fortunately, RBW gives us a way to deflate this problem. We will argue that given how RBW reconceives matter, rejiggers the adynamical block universe and fundamental explanation, RBW naturally admits a kind of neutral monism that simultaneously deflates the generation problem and explains time as experienced. And then we will then have refuted the claim [Bardon, 2012] that the block universe is incompatible with time as experienced. Obviously our account here is highly speculative, but then at this point in history so is every proposed resolution to the hard/generation problem.

In the next two sections of this chapter we will set the stage by showing in broad strokes how RBW enables a kind of neutral monism that deflates the generation problem and explains time as experienced. These sections will focus on the Passage and Presence of time in particular and show in detail how the neutral monism of RBW explains Passage without making it either a mental projection

Beyond the Dynamical Universe. Michael Silberstein, W.M. Stuckey and Timothy McDevitt,
Oxford University Press (2018). © Michael Silberstein, W.M. Stuckey and Timothy McDevitt.
DOI 10.1093/oso/9780198807087.001.0001

or some fundamental arrow of time in foundational physics. The fourth section is about the Direction of time and argues that while there is no Objective Direction in time, explaining the Passage is sufficient for explaining Direction. The fifth section argues that embodied, embedded, and extended cognitive science and phenomenology support the neutral monism of RBW and its explanation for the experience of time. The sixth section focuses on freedom, spontaneity, and creativity in the relational block universe. Many people object to the block universe picture because they think it negates free will in human action, and negates creativity and spontaneity in the universe. We argue that the reality of the future entails none of those things, and that RBW has all the freedom, creativity, and spontaneity anyone could reasonably desire. Finally, in section seven we characterize Presence in detail, and its relation to time as experienced.

8.2   Setting the stage: Neutral monism, the generation problem, and time as experienced

How one thinks about the relationship between experience and the physical world of course depends in part upon how one conceives of the physical and how one conceives of experience. For example, if the physical is a material substance that is inherently/essentially non-mental/non-experiential in nature and the physical is fundamental, then the result is the hotly debated hard problem of consciousness [Chalmers, 1996], which, as we saw in chapter 7, is exacerbated by the block universe model. Philosophers often debate the nature and very existence of conscious experience, but an equally important question is: what is matter beyond its mathematical characterization in foundational physics? For example, when we talk about the properties of an electron, are those intrinsic properties that explain why the electron behaves as it does, or rather, is talk of the electron's properties just a shorthand way of codifying its dispositions to behave in certain ways? This is an ancient debate about the very nature or essence of matter. We favor the latter argument.

Let us recall that in RBW the fundamental physical entities are spatiotemporally contextual 4D spacetimesource elements that "occupy" (not *in* space and time, they *are* spacetime-matter) multiple interacting spatial and temporal scales, having neither intrinsic properties nor individual autonomy (what some philosophers call "primitive thisness"). On this view, "macroscopic" or "classical" phenomena do not reduce to, are not "realized" by, nor "determined" by some microscopic entities with intrinsic properties and individual autonomy such as superstrings. Also recall that the fundamental explanatory principles in RBW are global constraints and not dynamical laws. Therefore, fundamentally speaking, from a God's-eye perspective, macroscopic phenomena, including conscious experience, don't "emerge" from the dynamical, causal, or temporal evolution of some fundamental microscopic or pregeometric entities. In short, the so-called physical elements of RBW are nothing like matter as characterized by physicalism/materialism. There is no suggestion either that spacetimesource elements are essentially non-experiential or that they are fundamental to conscious experience. As we made clear in the last chapter, we are not hinting at panpsychism, but rather neutral monism.

As many writers have recently noted [Dolev and Roubach, 2016], continental philosophy (e.g., Bergson, Husserl, Heidegger, Levinas, and Deleuze) tends to take time as experienced (and experience in general) as fundamental in some way, and analytic philosophers are more likely to try to reduce or explain away time as experienced (and experience in general) in terms of some physical, cognitive, or neurobiological phenomena. Certainly there have been other philosophical traditions that have focused on time as experienced such as pragmatism (e.g., William James), Hinduism, and Buddhism, but until recently those traditions held little sway over analytic philosophy. Consequently, until recently thére wasn't much cross-talk between continental philosophy and analytic philosophy on the experiential arrow of time, nor much cross-talk on this subject between physics, philosophy, and cognitive science. However, this trend is changing of late [Bardon, 2012; Arstila and Lloyd, 2014; Unger and Smolin, 2015; Dolev and Roubach, 2016].

More generally, we think it is fair to say that there has been something of a rapprochement between continental and analytic philosophy, at least with regard to the phenomenological tradition and its offshoots. Certainly, phenomenology is a key player in the foundations of cognitive science these days, especially as it bears on consciousness studies. We would say the same is true about cross-fertilization between various Eastern and Western traditions; relations have definitely improved [Thompson, 2015; Windt, 2015]. Since continental philosophy and various Eastern traditions have been more focused on the phenomenology of temporal experience, it isn't surprising they have much to offer here, as we will see. The point is that while in the Western philosophical tradition the kind of neutral monism on offer, while most closely associated with William James and Bertrand Russell, also has convergent counterparts in continental philosophy, Hinduism/Buddhism, and most recently in cognitive science. We will borrow from all these traditions to paint a picture of neutral monism more generally and what it tells us about time as experienced in particular.

We will argue that, at least regards Passage and Direction, time as experienced is in fact neither "in the head" as a mere quale or representation, nor is it purely external or objective in the sense of existing independently of conscious subjects. We will argue that the physical world as conceived as RBW allows for (doesn't entail) a different way to think about the relationship between the physical world and conscious experience, and it allows for a direct realist account of the experiential arrow of time. RBW allows for a kind of neutral monism in which ultimately the hard and fast dichotomies between the physical/mental, objective/subjective, outer/inner, external/internal, object/subject, self/world and the like, become if not spurious, then perspectival or merely pragmatic [Silberstein and Chemero, 2015; Silberstein, 2016, 2017]. As Thompson says, "such an understanding would replace our present dualistic concepts of consciousness and physical being, which

exclude each other from the start, with a nondualistic framework in which physical being and experiential being imply each other or derive from something that
is neutral between them" [Thompson, 2015, p. 105]. The type of neutral monism defended here is very close to that articulated by William James and Bertrand
Russell at various points in their careers. Take the following, for example:

> Pure experience, in itself, "is no more inner than outer . . . It becomes inner by be-
> longing to an inner, it becomes outer by belonging to an outer, world."
>
> [James, 1988, p. 217]

> Subjectivity and objectivity are affairs not of what an experience is aboriginally made
> of, but of its classification.
>
> [James, 1905b, p. 1208]

> Things and thoughts are not at all fundamentally heterogeneous; they are made of
> one and the same stuff, stuff which cannot be defined as such but only experienced;
> and which one can call, if one, wishes, the stuff of experience in general.
>
> [James, 1905a, p. 110]

> "Subjects" knowing, "things" known, are "roles" played. Not "ontological" facts.
>
> [James, 1905a, p. 110]

> Consciousness, as it is ordinarily understood, does not exist, any more than does
> matter.
>
> [James, 1905a, p. 109]

> Just so, I maintain, does a given undivided portion of experience, taken in one context
> of associates, play the part of the knower, or a state of mind, or "consciousness";
> while in a different context the same undivided bit of experience plays the part of
> a thing known, of an objective "content." In a word, in one group it figures as a
> thought, in another group as a thing.
>
> [James, 1904, p. 251]

We will have much more to say by way of characterizing neutral monism but we
hope these Passages start to convey the flavor of the idea, namely that matter and
conscious experience are non-dual and thus not essentially distinct.

What does all this have to do with time as experienced? At least as regards Passage and Direction, neutral monism allows for an adynamical global constraint-
type explanation for the experiential arrow of time that is also at the heart of the
resolution to the hard/generation problem of consciousness. Much as Kant and
other continental thinkers in his wake anticipated in variously different ways, the
Passage of time is the result of the aforementioned "split" between the subject
(the "minimal self") and object (the minimal external world) in the neutral "field
of experience," as James would say. This relational phenomenological structure is

not in anything nor located anywhere; rather, it is why there are things in time and space as experienced. Given neutral monism, self/world are two sides of the same coin; therefore, the dynamical character of experience/thought and the world are two sides of the same coin as well. It will become clear that while this may be a transcendental argument of sorts, this is not transcendental idealism, subjective idealism, etc. The conscious mind is not some utterly distinct entity that imposes or projects time and change onto a static, material universe as if the mind were just some virtual reality machine that we are stuck behind and the physical world some unknowable frozen noumena. Nor is mind or self some immaterial ghost shining the light of presentness at it moves along its frozen worldtube as Hermann Weyl suggested in his 1949 book. No; in our view the very existence of the co-dependent, co-extant conscious self/world-in-space-and-time unity is yet another adynamical global constraint. You simply cannot have one without the other.

Neutral monism also solves the problem of Presence or nowness, but in a very different way. As Seager says:

> On the other hand, neutral monism is a radical doctrine from the usual physicalist stand-point. It knocks the physical, as scientifically understood, from its perch of ontological preeminence. It suggests that any effort to reduce everything to the physical is fundamentally misguided. Neutral monism has to accept a notion of "*presence in experience*" (what James called "pure experience"). This presence is not labeled as "consciousness" by the neutral monists, since they regard consciousness, and its subject, as a very sophisticated feature of the constructed mental realm. Nonetheless, presence is, I believe, what funds the hard problem of consciousness. Presence is what constitutes the "what it is like" of conscious experience. This is quite explicit in the neutral monist's alignment of the neutral with the qualities of experience, and especially perceptual experience (the paradigm case for explaining the "what it is like" aspect of consciousness). Speaking for myself, I do not think that presence can be denied. The neutral monist claim that it forms the bedrock of reality is surprisingly powerful and fertile, and may yet help us understand reality and our place within it.
>
> [Seager, 2016b, p. 326]

Presence in this view is not reducible to anything else on the subject or object side of the divide. Presence is neither a quale "in the head" nor a special external marker. Presence isn't moving over events in spacetime like a spotlight. Rather, Presence or Nowness is the neutral in which all else is grounded. Nowness is universal and unchanging. Adopting the God's-eye perspective here, one might say that all of spacetime is just one moment. The universe is only carved into various foliations or series of moments from the perspective of each individual subject and their worldtubes—all the ant's-eye views. In the words of James:

> "Natural experience" may be just a "timemask," shattering or refracting the one infinite Thought which is the sole reality into those millions of finite time streams of consciousness known to us as our private selves.
>
> [James, 1898, p. 1110]

Keep in mind that for an event to be present for an individual as opposed to past or future is a separate issue from the specialness of Presence as experienced which pervades all moments/events. Bertrand Russell here expresses much the same idea, in his "On the Experience of Time":

> [P]ast, present and future arise from time-relations of subject and object . . . In a world in which there was no experience there would be no past, present or future
> . . . .

> [Russell, 1915, p. 212]

But in such a world there would still be Presence. The point is that while the presentness of an event in spacetime is relative to a subject and their worldtube, Presence and Nowness are not; they are universal.

In short, neutral monism is a rejection of physicalism and reductionism, and it is a rejection of any dualism as well, even the more subtle forms of property dualism such as panpsychism and strong/radical emergence. Experiential features of the world and physical features of the world are non-dual and they are grounded in something neutral, namely Presence. This means that neural monism is a rejection of any explanation of time as experienced that is predicated on any of those other views.

As hinted at by the preceding quotation from James, once we have articulated all the preceding claims we will end this chapter with even wilder and more unfounded speculation about what all this might mean for the relationship between conscious experience and spacetime from an adynamical global constraint perspective. If you are a dyed-in-the-wool physicalist or reductionist, none of this will be very palatable. But, if you think there is some merit to our master argument and you can at least allow yourself to imagine other possibilities without undo bias, just as you have presumably done thus far with dynamism about the physical world, then other relatively new and admittedly weird possibilities are open to us.

Keep in mind as you read this chapter that RBW embodies a kind of ontological relationalism/contextualism for both spacetime and entities (matter), that is, we have spacetimesource as a background independent non-duality. And keep in mind too that given neutral monism, there is ultimately a non-duality between conscious experience and the physical world as well: much more on this below. We issue this reminder to guard against the mindset of trying to project or insert conscious experience such as qualia into a noumenally given physically substantial block universe. As we have ourselves invoked on occasion, it is easy to imagine a God's-eye view from some perspective outside the block universe where we look "down" upon the frozen block. Here is a Passage from Alan Moore's new novel, *Jerusalem*, that tries to convey this:

> If Einstein is right, then space and time are all one thing and it's, I dunno, it's a big glass football, an American one like a Rugby ball, with the big bang at one end and the big crunch or whatever at the other. And the moments in between, the moments making up our lives, they're there forever. Nothing's moving. Nothing's changing,

like a reel of film with all the frames fixed in their place and motionless till the projector beam of our awareness plays across them, and then Charlie Chaplin doffs his
bowler hat and gets the girl.

[Moore, 2016, p. 61]

Notice the tacit invocation of the moving spotlight theory that suggests somehow,
in some unspecified sense, the events are all there but lack Passage or Presence until conscious experience sweeps over them; that is the notion of God's-eye we want
to warn the reader against. In the novel there are ghosts who inhabit a supernatural realm or dimension that they call "Mansoul" and the entire bejeweled block
universe in all its continuum glory is frozen within this realm like a fly in amber.
The ghosts are free to roam anywhere in or outside the block universe and view it
from numerous spatiotemporal perspectives as if one were viewing a paused 4D
movie from various angles; but again, notice that these entities are still viewing
the block universe from some perspective or the other. But, precisely what relativity does is tell us how to relate various spatiotemporal perspectives (reference
frames that happen to have conscious observers on them) in a universe with no
preferred foliation, not even the cosmic time frame is preferred, however useful it
may be for cosmology. We could imagine it the way Alan Moore does, but what
if there were no conscious observers with any ant's-eye views in the block universe or any outside of it like the ghosts, no experiential-perspectives as it were,
what would remain? We will return to this question again because we think that
once neutral monism is fully appreciated, what remains is not an external world
in time and space in the sense that we experience these; indeed, not a world at all,
but only James' field of experience denuded of subjectivity. Moore would have us
think of the block universe from the outside as a kind of 4D museum that is just
there exactly as it is independently of experiencers. But given neutral monism,
subject/external world are co-dependent and co-creating.

On a very related note, we agree entirely with the later Dainton who says that

[e]ven if we do live in a universe of the block variety, a universe where there is no
moving metaphysically privileged present, a universe in which all events are fully and
equally real, it is wrong to conclude from this that our universe is entirely static or
"frozen" . . . This characterization is misleading because our experience is itself a
part of the universe, and it is difficult to conceive of any experience existing which
does not exhibit phenomenal Passage in some form or other.

[Dainton, 2016, p. 92]

Of course, it is trivially true on any account of anything that experience is a subset of the universe. As Dainton acknowledges, what matters for fully dispelling
the claim that time is an illusion is exactly how experience is characterized and
exactly how it relates to the physical world. He argues that experience is part
of the universe, experience is dynamical, and therefore so is universe. We agree
wholeheartedly. However, as he notes, "the precise extent to which the dynamism
and Passage that exists in experience infects or pervades the universe as a whole

depends on the precise relationship between experience, the properties we find in our experience, and the rest of reality. Different views on this issue have very different consequences" [Dainton, 2016, p. 92]. For example, if all experience has secondary qualities, then dynamism is confined to our consciousness and is not extended to the physical world external to our consciousness. Dainton considers other possibilities as well such as Russellian monism, panpsychism, and substance dualism.

He also mentions anti-representationalist views about conscious experience—some form of direct realism. He does not mention neutral monism explicitly, which is what we will be defending here. [For a broader defense of direct realism, extended accounts of cognition and conscious experience, and their relationship to neutral monism see Silberstein and Chemero [2012], Silberstein and Chemero [2015].] What will become clear is that given neutral monism, Passage and Presence are as much a part of the "external" world as anything else. The point can be put this way: neutral monism is a complete rejection of the phenomenal/noumenal distinction in any guise, and a complete rejection of the primary/secondary property distinction. It should be obvious that if there are no primary properties then there are no secondary ones either. We think if Moore had been looking at things this way, he might have described the block universe quite differently.

It is also worth asking, given the block universe of RBW and the Extended specious present account, why does the duration of the Extended specious present only encompass the subset of events in spacetime that it typically does? As Dainton writes, "According to the B-theorist, we find ourselves in an analogous position in time. What stretches only a short distance is not light through space but consciousness over time: the temporal span of direct awareness is very brief. And as in the analogous spatial case, the fact that we are not aware of experiences occurring at other times does not mean that these experiences are not there" [Dainton, 2001, p. 30]. Our question is why is the temporal or spatiotemporal span of direct awareness so attenuated? Sometimes people ask the related question, why in a block universe don't people "remember" or experience the future? As we noted in chapter 1, science fiction certainly asks us to imagine that there are beings, such as the heptapods, the Tralfamadorians, and Dr. Manhattan in Alan Moore's *Watchmen* whose "span" is much greater. These beings experience the entire block universe or at least their own entire worltubes as one moment. That is, they do not hold the ant's-eye view, but rather some God's-eye view of the world that is somehow still *within* the block universe. These are beings who have taken off the "timemask," as James would have it. By the time we get to the end of this chapter, we hope it will be clear why this isn't such a crazy question to ask.

Ismael would certainly say that while the block universe theory or Minkowski spacetime is a theoretical God's-eye or non-perspectival view, in reality there is only the ant's-eye view "of the everyday experience of time as a view of history through the eyes of the embedded, embodied participant" [Ismael, 2016, p. 107]. No doubt this is because she is a physicalist and reductionist of sorts and will appeal to the limits of certain physical signals such as light, the limits of certain record-keeping mechanisms in the brain, etc.

> The claim is that in the view of History through the eyes of the embedded, embodied participant, events are ordered by their practical and epistemic relations to the viewer at different points in her life so that when they are strung together in a temporal sequence, they produce a changing image of a world with a fixed past and open future, in the process of coming into Being. Passage, flow, and openness arise as artifacts of changes in perspective, relative to the fixed backdrop of History.
>
> [Ismael, 2016, p. 115]

We agree with Ismael that many important truths about the universe are inherently perspectival and that doesn't make them illusory or unreal. But of course that also applies to time as experienced in the Matrix or in dreams. The question is, can the phenomenological structure of experience that she adverts to and so many continental writers have alluded to be explained by neuroscience or physics? Here the hard problem looms again: why do information gathering and storage devices of a certain type have *experiences* such as Passage and Presence at all, as opposed to having mere representations, or "access-consciousness" only? About Presence she states, "Now is not an object but a fixed point in a series of frame-dependent representations of time that has different values for different elements in the series" [Ismael, 2007, ch. 10]. But how does this explain the experience of Presence? Different values for whom? A subject of experience? We agree with her in principle when she writes

> To think that accepting the Block Universe as an accurate representation of time as it appears sub specie aeternitatis means rejecting Passage, or flow, or openness, as illusory is like thinking that accepting a map as a non-perspectival representation of space means that you are under an illusion that anything is nearby. As we develop an increasingly absolute conception of the world, more and more of the structure at the forefront of our experience of the world is revealed to be perspectival.
>
> [Ismael, 2007, p. 119]

But what we want to know is what perspectives are possible (can actually be had in principle and in practice) in the block universe? Given a certain sort of physicalism or biological reductionism about conscious experience, the constraints will be relatively severe. But what if, as we are suggesting, internalism, physicalism, and reductionism are false? We will explore these questions as we progress.

All of this, including the neutral monism we have just introduced, is a consequence of transcending the dynamical/mechanical worldview in which neither the Cartesian conception of matter nor mind obtains, not even the ghost of these conceptions remains. As anyone can easily confirm these days on YouTube, coming from the physicalist, ontological reductionist, or eliminativist perspective, it is common to hear that "time is an illusion" and "self is an illusion," according to some even that "conscious experience itself is an illusion." The idea presumably is that these things are an illusion because, in virtue of being reduced to or even eliminated by some truly fundamental neurological or physical phenomena, they are not things-in-themselves. It should be obvious why all these are "illusions" within these various dogmas, because they can't be easily explained away or easily reduced in those explanatory schemas. It should be clear that this kind of ontological reductionism/eliminativism simply fails to obtain in RBW.

In terms of fundamental principles, RBW is certainly a unified and unifying theory from quantum gravity (QG) on up, but there is no part/whole reductionism because there are no fundamental physical entities with individual autonomy and intrinsic properties residing at small spatial and temporal length scales or in some funky pre-geometric space. In RBW there are no things-in-themselves; spatiotemporal relationality and contextuality are fundamental. So the idea of equating "realness" with being a thing-in-itself is dubious; either that or everything is an illusion, take your pick. Though nothing reduces to it, there is one fundamental in neutral monism, and thus RBW, that defies relationality and contextuality: Presence or Nowness. So in RBW if we were going to go nuclear and drag out the "I" word, what would count as an illusion? The idea that the universe is fundamentally compositional, atomism in the broadest possible sense, is an illusion. The idea that only the present moment from the ant's-eye perspective is real, that is, dynamical presentism, is an illusion. The idea that matter and conscious experience are ultimately essentially distinct in nature is an illusion. Subjectivity and the minimal self, time as experienced, these are not illusions: they are contextually/relationally emergent, but not illusory.

We fully acknowledge that while RBW certainly saves many of the aforementioned features of common sense and everyday experience from the chopping block of science, in other ways it is probably as much a blow to "the manifest image" as atomism and its many iterations. But while RBW is a blow to certain reductionist dreams, it is nonetheless driven by the same goal which is unification. Those still clinging to the reductionist dream of materialism and the idea of a physical Theory of Everything will not like the picture of fundamental reality we paint here, but we think RBW is the most unifying game in town. In constructing RBW we tried to own up to what can and can't be explained by fundamental physics. And, we hope to have made the case that, irony of ironies, our best physics (not to mention experience) is telling us that every major tenet of the dynamical/mechanical worldview is false. So from our perspective, RBW is science self-correcting, which is of course the very spirit of scientific inquiry.

8.3 Relational Blockworld and neutral monism: The generation problem and time as experienced explained

Both strong emergence and panpsychism agree to attempt to answer the generation problem directly. To deflate the generation problem and get around the master argument for radical emergence and panpsychism we need another alternative. As long as matter is conceived of as essentially non-mental and experience is conceived as qualia we are stuck with these problematic views. So the alternative must reconceive matter and mind. This is exactly what, properly understood, neutral monism does. Given that the key defenders of neutral monism such as William James and Bertrand Russell also defended panpsychism at various points in time and given that "the avowed neutrality of neutral monism tends to slides towards some kind of panpsychism or idealism" [Seager, 2007, p. 28], people can certainly be forgiven for thinking that neutral monism and panpsychism may not be completely distinct doctrines, but it is very important to see that they are distinct. As James saw very clearly, we need to deflate the generation problem, not answer it directly. We need to deny that everyday conscious experience is an entity, that it is qualia-like, it is not intrinsic.

With regard to time as experienced in particular, we certainly agree that there is no Objective Passage, Direction, or presentness (something that marks a particular event as present as opposed to past or future) in the block universe. But as we said, one still has to explain not only time as experienced but experience more generally. We have seen that even without invoking the block universe, the leading contenders for resolving the generation problem, strong emergence, and panpsychism are deeply flawed. We have also seen that the block universe model further exacerbates the generation problem. Ironically, both of these troubled views are in effect driven by a kind of matter fundamentalism. Ironic given that both these views entail that physicalism and ontological reductionism fail for conscious experience, and are therefore incomplete. We will argue that neutral monism completely deflates the generation problem, doubly so in RBW where *generation* is the wrong metaphor even for "physical" relations and events.

As James notes in *A Pluralistic Universe*, it is "intellectualism" or rationalism that got us into this mess in the first place and that is what we must reject:

> Intellectualism in the vicious sense began when Socrates and Plato taught that what a thing really is, is told by its definition. Ever since Socrates we have been taught that reality consists of essences, not of appearances, and that the essences of things are known whenever we know their definitions. So we first identify the thing with a concept and then we identify the concept with a definition, and only then, inasmuch as the thing is whatever the definition expresses, are we sure of apprehending the real essence of it or the full truth about it . . . Intellectualism does not stop till sensible reality lies entirely disintegrated at the feet of "reason."
>
> [James, 1912, p. 218]

Bergson makes a similar point in what follows: "Metaphysics would not have been sacrificed to physics, if philosophy had been content to leave matter half way between the place to which Descartes had driven it, and that to which Berkeley drew it back to. They should leave it, in fact, where it is seen by common sense" [Bergson, 2002, p. 82].

When it comes to the mind/body problem, James' point is that said problem is an artifact of "rationalism." Neutral monism is just the undoing of this odious essentialism. According to this view, conscious experience (subjectivity) is not an add-on, it is as much a part of the fabric of the universe as spacetimesource elements. The generation problem is a cognitive illusion generated by the inference or projection that experience is inherently or essentially "internal" and mental, and the "external" world is inherently non-mental. The claim that the world is carved at the joints—physical/mental; outer/inner; object/subject, etc.—is not a datum, but rather an inductive projection. This assumption is an ancient mistake that creates the whole mind/body problem.

Therefore we will be advocating for a kind of neutral monism account of the relationship between mind and matter as two non-dual modes of a non-dual blockworld. It is clear that even in a standard blockworld (as opposed to RBW) there is an obvious sense in which the blockworld is a fundamental singular entity rather than something that can be decomposed or partitioned into discrete and autonomous parts in any absolute sense. Again, from an ant's-eye view not only is it open to multiple foliations, from a God's-eye perspective, even without a preferred foliation, it is all "just there." This is certainly true in RBW. We say it is time to abandon matter fundamentalism and adopt the neutral monism of William James and Bertrand Russell.

Let us start with what we actually know: everyday conscious experience appears to be intimately related to embodied organisms with certain complex internal structures maneuvering an environment. How does all this connect with the experience of time and change? How does all this connect to neutral monism? It turns out that in neutral monism the explanation for why there is experience (subjectivity) at all and why that experience manifests itself spatiotemporally (objectivity) is the same explanation because these turn out to be the same question. The existence of experience (subjectivity) just is the "bifurcation" in what James (unfortunately) called "the field of experience" between the so-called subject and the so-called object. One part of that field gets treated as the experiencer and the other the experienced, but it is really just one thing. So again, James was on the right track as well when he said

> [t]he individualized self, which I believe to be the only thing properly called self, is a part of the content of the world experienced. The world experienced (otherwise called the 'field of consciousness') comes at all times with our body as its centre, centre of vision, centre of action, centre of interest. Where the body is is 'here,' where the body acts is 'now'; what the body touches is 'this'; all other things are 'there,' and 'then' and 'that.'
>
> [James, 1912, p. 170]

What James is suggesting is the following:

- There is no conscious experience without a subject.
- Where there are perceptions there is a perceiver and vice versa.
- There is no subject/self without an object/world and vice versa.
- So experience is inherently relational!

More generally neutral monism holds that:

1. Mental and material features are real but, in some specified sense, reducible to or constructable from a neutral basis in a non-eliminative sense of reduction.
2. The neutral basis is generally not conceived as substance.
3. Mental and material features are not separable or merely correlated: they are non-dual, indeed, they are not essentially different and distinct aspects.

Quoting James himself, here is how Stubenberg characterizes his neutral monism [Stubenberg, 2009]:

> Prior to any further categorization, pure experience is, according to James, neutral—neither mental nor material:
>
> The instant field of the present is at all times what I call the "pure" experience. It is only virtually or potentially either object or subject as yet. For the time being, it is plain, unqualified actuality, or existence, a simple that.
>
> Mind and matter, knower and known, thought and thing, representation and represented are then interpreted as resulting from different groupings of pure experience.
>
> [James 1904b, 23]

What follows is a Passage from James in which he rejects the conception of consciousness as something inside the mind and representing the external world. This is an oft-quoted statement of James' brand of neutral monism or radical empiricism:

> As 'subjective' we say that the experience represents; as 'objective' it is represented. What represents and what is represented is here numerically the same; but we must remember that no dualism of being represented and representing resides in the experience per se. In its pure state, or when isolated, there is no self-splitting of it into consciousness and what the consciousness is 'of.' Its subjectivity and objectivity are functional attributes solely, realized only when the experience is 'taken,' i.e., talked-of, twice, considered along with its two differing contexts respectively, by a new retrospective experience, of which that whole past complication now forms the fresh content. The instant field of the present is at all times what I call the 'pure' experience. It is only virtually or potentially either object or subject as yet. For the time being, it is plain, unqualified actuality, or existence, a simple *that*.
>
> [James, 1904, p. 256]

From the perspective of neutral monism, the hard problem gets it wrong from the start. The apparent division between the mental and the physical as separate domains is a cognitive illusion of sorts. One has to keep in mind that the division of phenomenal mental experience and external material world is not a datum of experience that cannot be questioned; it is, rather, an inductive leap, an interpretation of our experience—that is James' point in the Passage. Neutral monism is a rejection of the absolute binary distinctions that carve nature up at the joints: mind/matter, inner/outer, self/world, subjective/objective, etc. These allegedly categorical divisions have infected traditions as diverse as metaphysical realism, physicalism, and even phenomenology with its bracketing of experience. Seager [2013] states that both mind and matter are abstractions, derived from the more basic neutral "stuff" of experience. The so-called subjective and objective are only designated subjective and objective in relation to one another, and only "as described." Of course, invoking James has its drawbacks because sometimes James talks of the "stuff" of neutral monism as "pure experience," making it sound like phenomenalism or idealism. However, nothing should hinge on James' misleading names. Again, Seager writes:

> The neutral monists all choose to use terms that are decidedly non-neutral in tone. Mach: sensations; James: pure experience; Russell: sense-data or sensibilia, but they had other, ostensibly more neutral, terms as well such as "elements" (Mach) or "events" or "momentary particular" (Russell). A charitable interpretation depends on looking for a non-mentalistic understanding of their canonical terms.
>
> [Seager, 2013, p. 6]

It helps to look at a possible way of understanding neutral monism that is inspired by [Merleau-Ponty, 1962; Schopenhauer, 1969; Heidegger, 1996; Kant, 1998; Husserl, 2001], and a variety of Asian traditions such as Advaita Vedanta [Gupta, 1998]. This is not the place for historical details, and important differences between these thinkers will be glossed over. Most of them do not necessarily self-identify as neutral monists, but following Zahavi [2005], who is writing about Husserl, Merleau-Ponty, and others, we can say that for all these traditions of thought, the minimal subject and the external world (the object) self-consistently coexist; you cannot have one without the other. As Zahavi writes, "a minimal form of selfhood is a built-in feature of experiential life," and "rather than speaking simply of phenomenal what-it-is-like, it is more accurate to speak of what-it-is-like-for-me-ness" [Zahavi, 2014, p. 88]. As we would put it, there is a self-consistency relation such that the subject and the external world are both necessary and sufficient for the other. For a visual analogy, think of the faces–vase illusion (Figure 8.1). There are no faces without a vase and vice versa. Following Kant's unity of apperception and Schopenhauer's will and representation, we can go a step further: it is only when the subject/object division exists that one gets a world in space and time. So subject/object and world in space/time are both necessary and sufficient for one another. One can of course find similar ideas in the works of Husserl and Merleau-Ponty in their respective accounts of temporal experience. Zahavi, approvingly channeling Dainton, expresses this idea beautifully in what follows:

Figure 8.1 *The faces–vase illusion.*

Moreover, experiential processes are intrinsically conscious and hence self-revealing. Given that phenomenal unity is an experienced relationship between conscious states, we do not have to look at anything above, beyond, or external to experience itself in order to understand the unity we find within experience; rather, experiences are self-unifying both at a time and over time.

As Dainton puts it, consciousness does not consist of a stream running beneath a spot of light, nor of a spot of light running along a stream. Consciousness is the stream itself, and the light extends through its entire length.

[Zahavi, 2014, p. 67]

Admittedly we are not scholars of continental philosophers who write on the metaphysics of time or the experience/phenomenology of time. For a detailed comparative analysis of such views see David Couzens Hoy [2009]. In what follows we are synthesizing the views of various continental thinkers, or more bluntly, cherry picking certain features without regard for attribution or historical context. Of course none of the aforementioned thinkers shared all these claims, but we embrace them all as follows: 1) a rejection of any *essential* dichotomy or dualism between the subjective and objective. That distinction is a projection or reification onto, in James' words, "the field of experience." There is no subject or subjectivity without an environment, that is, a world conceived as external, and vice-versa. As Couzens Hoy puts it, ". . . the very distinction between subject and object is derived from the more primordial structure of Being-in-the-world" [2009, p 68], 2) there is no subjectivity without temporality and vice-versa. They are each the interdependent and co-determining source of the other, though some emphasize one over the other as prior for various purposes, we do not, 3) thus, *ultimately,* there is no *essential* dualism between time as experienced and the time of the "external" universe. While time as experienced can deviate from clock time in certain cases, clock time is grounded in time as experienced in the normal or ordinary case (providing the baseline); of course relative to human constructed clock time, processes at different scales happen more or less quickly "objectively" speaking, 4) thus, time cannot merely be a succession of numerically distinct and Objectively divided "Nows" or events like a film strip or flip book, divorced from existential significance that is rooted in the past and projecting into or from the future—subjects project a division of events onto the world in a perspectival fashion. Events as such only exist for subjects and only happen to subjects, 5) therefore, even though there is no objective Direction in time, for the subject the past and future are always experienced differently, they have a different character because of their relation to one another from the subject's perspective in the present, and 6) thus, while some thinkers emphasize the significance of the past, the present or the future respectively (or our reactions to one or more of these) for explaining time as experienced, we want to emphasize their inherent interdependence—much like the answers to a crossword puzzle, they co-determine and interpenetrate each other, and you can't identify one without the other. From all this it follows that, "We should not look to physics, then for an account of our temporality and our experience of Being-in-the world. Where there is a world there is consciousness already at work, which is not to say that first there is consciousness and only then is there a world" [Hoy, 2009, p 74]. Couzens point here is that none of this entails subjective or transcendental idealism, but rather, what we are calling neutral monism.

What we do here is marry this way of thinking with neutral monism. Recall that in neutral monism not only are self/world two sides of the same coin as expressed earlier, but they are co-extant, non-dual aspects of a neutral and fundamental Presence (note this isn't the same thing as presentness). It is worth noting that Presence is neutral in three important respects. First, there are no qualia, no fundamental self, no material substances, and more generally no categorical distinction between the mental (the subjective) and the physical (the objective). Therefore, there is also no dualism between the qualitative and the intentional. Second, the subject/object division is a self-consistency relation; there is only one reality (the field of pure 'experience' as James calls it). Third, there is no ontic priority of Presence over the external world in the sense of reducibility; they are co-fundamental in that sense. Given that Presence is fundamental, it cannot be defined in terms of other concepts, of either a material or mental nature. Presence cannot be ignored if one is going to take experience at face value. Moreover, Presence is not the product of neural processes, physical processes, or action (as the sensorimotor view sometimes suggests).

Many of the aforementioned Western thinkers, however, make a mistake that neutral monism corrects. Namely, they place the explanation for this grand self-consistency relation between subject/object/world in space and time, in the head of individual experiencers. For Schopenhauer it is his "representations," for Kant it is his "categories," and for Husserl it is his "inner representations" or presentations of temporal experience. Husserl very famously decouples subjective time from objective time with his "bracketing of experience" [Yinon, 2016]. Kant famously argues that the unity of experience in time and space requires a unity of self and vice versa; otherwise, there is no manifold of successive representations. As Dainton writes:

> Duration and succession are inherent in our experience—so much so, indeed, that it is tempting to think that this sort of dynamic character is an essential attribute of conscious states . . . it also seems plausible to suppose that our impression that time itself passes or flows is heavily bound up with the dynamic, flowing, changing character of our ordinary everyday experience—the sort of experience we have during all our waking hours.
>
> [Dainton, 2011, p. 385]

For Kant, time is an *a priori* condition for experience, no subjectivity, no time or space. Kant here is providing a transcendental analysis in mentalistic terms. This means that the dynamical character of thought/experience and the world are two sides of the same coin. Kant's transcendental arguments from *The Critique of Pure Reason* are supposed to show that we must conceive of the world in a certain way, structure it internally according to certain categories such as time, space, and causation. Those arguments are fraught with many interpretative perils and controversies, but the basic idea is that experience is possible for me only if some experiences are conceptualized as being of enduring objects, enduring through time and space. Likewise, to experience a world of enduring objects there must be some sense of an enduring self. You can't have one without the other [Dicker, 2013, chapter 24].

Kant and the others are right that we do not experience things in time and space but rather we experience them temporally and spatially. But they are wrong to say that this is an imposition of individual minds and their categories upon some unknowable noumenal world (the thing-in-itself). The mistake, in one form or another, which we find in both the analytic and continental tradition, is representationalism. Once we take neutral monism on board we can immediately see that there is no need for (and no sense for) representationalism to explain the experience of space or time [Silberstein, 2014; Silberstein and Chemero, 2015]. The point here is that subject and object co exist as a subject in a world in space and time, so the agent is not trapped behind "a veil of perception" but is directly part of the world, and the external world is not some external container onto which the subject projects a virtual reality. Given neutral monism, (transcendental) phenomenology cannot be and should not be divorced from natural science, and experience cannot be separated/bracketed from the natural world.[1] See also Dainton [2017] where he argues, correctly in our view, that Bergson was not a presentist and that what bothered Bergson about Minkowski spacetime wasn't

---

[1] We are no experts on Bergson, but we believe he held a very similar view. See for example Bergson [1896, 1913].

primarily the reality of the future but the denuding and cinema-like spatialization of temporal experience.

One can also find excellent expressions of this idea in the Advaita Vedanta tradition as illustrated by the following Passage:

> The goal of Advaita Vedanta is to show the ultimate non-reality of all distinctions; reality is not constituted of parts (p. 1). When pure consciousness individuates itself into subject and object, there results knowledge—the distinction between "knower and known" . . . In talking about Brahman [pure consciousness], it is not a subject or an object, but neither and both; the distinction is not real. Because reality is nondual, the known and the knower come to be recognized as one: brahman and atman, the objective and subjective poles of experience, are nondifferent (p. 31). Atman is pure distinctionless, self-shining consciousness, which is non-different from brahman. It is that state of being in which all subject–object distinction is obliterated. It does not have a beginning or an end; it is eternal and timeless. Time only arises within it.(p. 34)
>
> [Gupta, 1998]

It is easy for some Westerners to be put off by the Advaita Vedanta terminology, but we urge the reader to set aside that reaction and think of such claims from a purely phenomenological perspective. What Advaita Vedanta calls "pure consciousness," following neutral monism, we would call Presence. As Thompson notes, some version of neutral monism can also be found in some Buddhist traditions:

> Take a moment of visual awareness such as seeing the blue sky on a crisp fall day. The ego consciousness makes the visual awareness feel as if it's 'my' awareness and makes the blue sky seem the separate and independent object of 'my' awareness. In this way, the ego consciousness projects a subject–object structure onto awareness. According to the Yogacara philosophers, however, the blue sky isn't really a separate and independent object that's cognized by a separate and independent subject. Rather, there's one 'impression' or 'manifestation' that has two sides or aspects— the outer-seeming aspect of the blue sky and the inner-seeming aspect of the visual awareness. What the ego consciousness does is to reify these two interdependent aspects into a separate subject and a separate object, but this is a cognitive distortion that falsifies the authentic character of the impression or manifestation as a phenomenal event.
>
> [Thompson, 2015, p. 61]

Recall our previous discussions of Smolin's temporal fundamentalism. The neutral monism characterized here shows where Smolin goes wrong in building time/change into fundamental physics is in his thinking that there is a hard and fast division between subjective and objective time. Now ask yourself this: what special mathematical or physical marker could possibly explain the phenomenology of the specialness of the present moment without just sneaking experience itself in at bottom? Smolin assumes that the experience of time can be explained if there is some corresponding aspect of the physical world that might causally account for or correspond to such experiences, contra the block universe view, he inserts a moving, universal/global Now into fundamental physics. But as we discussed earlier, there is no obvious necessary connection between any such physical arrow of time and Passage or Presence.

Recall the Merleau-Ponty quotation from chapter 1:

> French philosopher Maurice Merleau-Ponty argued that time itself does not really flow and that its apparent flow is a product of our "surreptitiously putting into the river a witness of its course." *That is, the tendency to believe time flows is a result of forgetting to put ourselves and our connections to the world into the picture* [our emphasis]. Merleau-Ponty was speaking of our subjective experience of time, and until recently no one ever guessed that objective time might itself be explained as a result of those connections. Time may exist only by breaking the world into subsystems and looking at what ties them together. In this picture, physical time emerges by virtue of our thinking ourselves as separate from everything else.
>
> [Callender, 2010a, p. 65]

Callender goes on to note that this idea is taking shape as a concrete research project in quantum gravity (QG):

> The universe may be timeless, but if you imagine breaking it into pieces, some of the pieces can serve as clocks for the others. Time emerges from timelessness. We perceive time because we are, by our very natures, one of those pieces . . . Historically, physicists began with the highly structured time of experience, the time of a fixed past, present and future. They gradually dismantled this structure, and little, if any, of it remains. Researches must now reverse this train of thought and reconstruct the time of experience from the time of nonfundamental physics, which itself may need to be reconstructed from a network of correlations among pieces of a fundamental static world.
>
> [Callender, 2010a, p. 65]

Recall in chapter 1 how Asher Peres avoided the "problem of time" in the quantization of GR "provided that one degree of freedom is kept classical, so that it can be used as a clock" [Peres, 1997, p. 3]. In neutral monism one part of the neutral "field of experience" is designated as the subject and the rest is designated as the object or external world, thereby creating the Ur-"clock."

In this section we focused on the Passage of time. In the next section we will focus on the Direction of time.

8.4   The Direction of time

As we noted earlier the question of the Objective Direction of time is logically distinct from both Objective Passage and Presence. People often talk as if Objective Passage would automatically give you Objective Direction and vice versa, but it isn't necessarily so. The problem here is supposed to be that the fundamental dynamical laws are time-reversal invariant, but experience is temporally asymmetric. People often invoke the second law of thermodynamics, the entropic arrow of time, to underwrite Objective Direction, but as Maudlin pointed out [2011] (and we explain later), if the direction of time is given by increasing entropy then the second law of thermodynamics becomes an analytic truth. Thus, we have no dynamical explanation for why the entropic arrow obtains. As we saw, this typically leads people to invoke the Past Hypothesis (PH) which for many of us just begs the question as to why the PH is true, that is, why those initial conditions at the Big Bang.

We saw in chapter 3 that the trend these days is to invoke the PH (the initial conditions) to explain more and more things, such as the fact that there is a cosmic time frame in our universe. Recall that there are solutions of GR with no global foliation possible, and certainly nothing like a cosmic time frame. For those who associate the Objective Direction of time with the cosmic time frame and further believe that the initial conditions at the Big Bang are contingent, there is a worry that Direction can't really be explained by (is not built into) GR because it is contingent upon initial conditions. We addressed those concerns in chapter 3 and we will take much the same approach here.

Our view is that there is no (nor need be any) Objective Direction or Passage in a block universe of any metaphysical or epistemological significance. Our claim is that once the experience of Passage is explained there is nothing more to say about Direction. When it comes to the idea of a global and fundamental notion of Direction, we agree with Price: "At best, there's a fact of the matter relative to a particular temporal perspective, or choice of coordinate frame" [Price, 2011, p. 283]. He says Direction is like being up versus being down and the right question to ask is what extra feature does or could the world have that could prove there is an objective "earlier/later-than" relation? We agree that there is no physically or empirically deeply interesting answer to that question.

We claim that in a block universe from a God's-eye point of view there is no intrinsic Objective Direction or Passage *in themselves*; there are just spacetime and the distribution of matter–energy, those plus perspective is all there is to direction-talk. Therefore, we need not find a physical or metaphysical basis for this familiar aspect of our phenomenology. We simply deny that one direction is more Objectively deserving of the label "forward in time." Nor, as we have made clear, is there any Objective Passage. For those who hope that Objective Direction (even if it existed) will yield Objective Passage, as many have noted, the *rate of* Passage depends on the relative state of motion in relativity theory, as illustrated by the twin paradox.

Let us begin by deconstructing the concern about Objective Direction and Passage in RBW much as we did in chapter 3 regarding the origins of the Big Bang. In order to be vexed by the fact that the dynamical laws of classical physics (including GR, of course) and quantum physics are time-symmetric, one has to believe that the dynamical laws really are the most fundamental explanations in physics and that dynamical laws don't merely describe patterns but somehow "govern" or provide an irreducible dynamical/causal explanation for the changing universe we experience. Maudlin, of course, even though he is a blockworlder, is someone who does accept all of this. But obviously, RBW simply rejects these claims about dynamical laws. In RBW there is no worry about time-symmetric dynamical laws because for us those laws are neither fundamental (the AGC is) nor do they "govern." We could just stop there. Given the AGC and given spatiotemporal contextuality neither the dynamical laws nor the initial conditions are in any way mysterious; at least not empirically mysterious.

As we said in chapter 3 we address the mystery of the Big Bang in much the same way. The worry is if the initial conditions at the Big Bang are going to do so much explanatory work, then how do we explain those initial conditions? But as we saw in chapter 3, much like a particular word or clue in any crossword puzzle, the Big Bang is no more mysterious or important than any other point in spacetime. The Big Bang, like every other point in spacetime, is explained in an adynamical global constraint fashion. In RBW we close the explanatory loop, it isn't turtles all the way down, and the only terminus is the AGC itself which is what closes the loop. Another way to make this point is as follows. In the mechanical or dynamical worldview, the dynamical laws and initial conditions are completely physically and metaphysically orthogonal. Furthermore, as we saw in chapter 1, in the mechanical worldview regarding initial conditions, there is nothing to stop one from forever asking why those laws? Why those initial conditions? But in RBW, when it comes to the universe as a whole, the dynamical laws/initial conditions distinction isn't a very telling one because the dynamical laws are just descriptions of regularities from the ant's-eye view that are ultimately explained by the AGC. In RBW we could just as well have picked any point in spacetime as the "initial condition" and applied the AGC from there. In RBW you really can't vary so-called initial conditions and vary so-called dynamical laws completely independently of one another.

So dynamical laws are clearly not fundamental and there is no Objective Direction of time in RBW. Just as Smolin predicted, when it comes to the universe as a whole you have to give up this mechanical or dynamical way of thinking. Put another way, in RBW the dynamical laws and initial conditions, taken as a package, really aren't contingent given the AGC. Of course, one could still persist in asking questions, such as why this block universe as opposed to some other, or why this AGC as opposed to some other? But in asking such questions one is either in the realm of metaphysics/theology or one just can't transcend the mechanical universe way of thinking.

But let's put RBW aside for a moment. On the face of it, could invoking the entropic arrow of time really provide Objective Direction? The answer as Earman [1974], Maudlin [2007], and Price [2011] all note for various reasons is "no." Let us start with the most obvious problem. The idea is that the second law of thermodynamics—the entropy of any isolated system always increases over time—will underwrite the Objective Direction of time, most generally on cosmological scales where one is dealing with the universe as the isolated system. In other words, the direction of increasing entropy determines the forward arrow of time. But, as pointed out by Maudlin [2011], if one claims the forward direction of time is the direction of increasing entropy, then the second law of thermodynamics is just an analytic truth. The second law of thermodynamics tacitly contains a notion of "forward time" in order to make sense of "increasing entropy over time." So, the second law cannot be used to underwrite the arrow of time. In other words, one needs to assume a Direction of time to make this claim about entropy in the first place. As Hemmo and Shenker state, "the Second Law and the Past Hypothesis in statistical mechanics cannot yield arrows of time since they assume an arrow of time" [Meir and Orly, 2016, p. 156]. Correct, and we concur.

We believe there are a growing number of people who appreciate all this, but for those such as Maudlin who still insist there is an Objective Direction of time then they must embrace this argument which we paraphrase and embellish from Hemmo and Shenker [Meir and Orly, 2016]:

1. Time as experienced suggests there is Objective Direction.
2. Given physicalism, ontological reductionism, or Biological Naturalism, the explanation of time as experienced or even Objective Direction if it exists must reduce to *some* physical or biological arrow of time.
3. But none of the known physical or biological arrows of time can underwrite Objective Direction; rather these arrows all presuppose Objective Direction.
4. Therefore, both the Objective Direction of physical and mental processes must be underwritten by some fundamental physical arrow of time (the arrow of Objective Direction).

It should be obvious what RBW says about this argument. First, it denies our experience of Direction is good reason to believe in Objective Direction, in part because we deny the antecedent condition of premise 2. We claim that explaining the experience of Passage is sufficient. One might ask why, given the experience of Passage, is there is no Objective Direction? Why is GR time-symmetric? Why is the AGC fundamental? Why do we experience the universe as expanding as opposed to contracting? Why does cosmology make this claim? Why do we describe the current state of affairs as an expansion rather than a contraction? As Price says, the answer to those questions is perspectival and that includes our physical perspective as living in the "expanding phase" of the universe per the cosmological solution, and our perception and memories which are explained by the experience of Passage which in RBW is beyond physics.

We want to be clear that our rejection of the antecedent condition of premise 2 is based on the arguments in this book and is grounded in neutral monism. We have argued that the antecedent condition of premise 2 is false because neither physicalism, ontological reductionism, nor Biological Naturalism can answer the generation problem. And given neutral monism, the assumption built into premise 2 that the experiential and physical arrow/Direction of time are essentially distinct such that we must tell a story about how the latter explains/reduces the former is also false. Just as with Passage, this is a game changer because it implies that the experience of Direction is as real, objective, and external as anything else in the universe. Direction is not "just in the head." Yet, without subjectivity and without the subject/object division, there would be neither Passage nor Direction.

But again, just to be clear, we are not saying that consciousness itself grounds Objective Direction because the arrow of conscious experience is always unidirectional in time. We agree with Price, Boltzmann, and others that nothing prevents the existence of conscious observers who have the opposite orientation to us in time [Price, 2011, p. 308]. Again, in RBW there is no Objective Direction in time as defined by people like Maudlin.

Of course, as we noted earlier, Maudlin just accepts the preceding argument as sound and attempts to provide the fundamental arrow of Objective Direction. He wants to admit both Objective Direction and Passage and he believes the added topological machinery outlined in chapter 2 achieves those goals. Note that all Maudlin's topological tinkering is underdetermined by the geometry; there is nothing about the geometry of spacetime, no experience, experiment, or intervention we could perform that would tell if we were living in the Maudlin universe or the ordinary Minkowski universe.

And Maudlin's account has the same problem as all such attempts to make correspondence between some physical feature of the universe and some feature of time as experienced, in this case Direction. How and why does that particular physical feature, whatever it is, bring about that particular experience and why should there even be such a law-like correspondence? It is hard enough to answer this question for some physical/geometric feature of the universe such as being on the "cutting edge" of a growing block universe, but Maudlin's correlate is purely topological, making the question even tougher. Maudlin will reply that if the extra topological machinery were not present, then the physical world would contain no Objective Direction and therefore experience would contain no such Direction, but there is no reason to think that is true.

8.5 Neutral monism, time as experienced, extended cognitive science, and phenomenology

The neutral monism of RBW, which allows us to reject representationalism, reject the neural correlate conception of conscious experience, embrace direct realism and an embodied/extended account of conscious experience, already has a dynamical analog in cognitive science as well. We will discuss this at some length lest the reader think that the conception of experience and mind defended here is somehow contrary to some established cognitive science or neuroscience. Philosophers and cognitive scientists working in various traditions such as ecological psychology, radically embodied cognition, and enactivism have tried to express similar ideas [Silberstein and Chemero, 2016]. These traditions differ in many respects but they also share many features. For example, "[e]nactivism, after all, gives explanatory pride of place to dynamic interactions between organisms and features of their environments over the contentful representations of such environmental features" [Hutto and Myin, 2013, p. xi]. This could be said for all the relevant traditions. These are traditions that focus on embodied, embedded, and extended accounts of cognition, but some explicitly discuss extended accounts of consciousness. The argument in a nutshell is that, contrary to computationalism (CTM), representationalism (RTM), biological naturalism, etc., cognition is environmentally extended in physical and social space and in time. Contrary to CTM and RTM there is no cognition without conscious experience; they go hand-in-hand. Therefore if cognition is extended, then so is conscious experience [Kiverstein, 2016; Silberstein and Chemero, 2015].

Silberstein and Chemero argue that in order to resolve or deflate the hard problem, the hypothesis of extended consciousness needs to be understood as an expression of neutral monism and can be naturally understood that way [Silberstein and Chemero, 2016]. They agree with critics that the enactive, sensorimotor, and ecological accounts do not by themselves deflate the hard problem of consciousness. While there are important differences between these accounts, they all "front-load" phenomenology [Gallagher and Zahavi, 2008]; that is, they all focus on the perception and action of organisms-in-their-environments, they all utilize dynamical models as a crucial explanatory tool, and they are skeptical of the explanatory usefulness of mental representations. Silberstein and Chemero argued that proponents of these accounts should explicitly adopt neutral monism. The conception of cognition and experience held by these views lends itself naturally to neutral monism, and neutral monism properly conceived really does deflate the hard problem once and for all, especially in RBW. The key is in rejecting the idea that matter and mind are different from the start. This is what extended mind properly conceived enables one to do in a naturalistically acceptable way.

Käuffer and Chemero [2014] note that in the 1930s Gibson, the father of ecological psychology, considered himself to be both a Jamesian radical empiricist and, like most empirical psychologists of his era, a behaviorist. Both of these identifications might account for Gibson's realism about the contents of experiences. As a radical empiricist, Gibson did not believe in a distinction between the world as experienced and the world-in-itself. As a behaviorist, Gibson was quite skeptical about internal mental causes, and thought that psychological research needed to be done in terms of publicly observable stimuli and responses, combined with learning theory. Importantly, although Gibson remained a radical empiricist to the end of his days, he became dissatisfied with behaviorism as his career progressed. Indeed, by the time his last book was published [Gibson, 1979], he argued that stimuli are not the causes of behavior.

Gibson's psychology involves both an epistemological claim and an ontological one. The epistemological claim is that we perceive the world directly, without adding information in mental representations or projecting meaning onto it. The ontological claim is that meaning, in the form of affordances, is a feature of the world that we perceive. We perceive opportunities for acting, and we perceive them by acting in the world. In both these claims, Gibson's views are similar to Merleau-Ponty's existential phenomenology of perception, minus the existential ontology. Gibson is a realist about the objects of experience, including affordances, and he aims to give a naturalistic account of perception. The world contains enough information to support our experiences and our actions. Gibson [1979] says that "the rules that govern behavior are not like laws enforced by an authority or decisions made by a commander: behavior is regular without being regulated. The question is how this can be." His solution to this problem was to invoke the environment surrounding the animal. He claimed that the information in the surrounding environment was sufficient to control behavior (without, that is, mentally added information, computation, or inference, and without a mentally represented plan).

Recent work by Noë [2004; 2009], for example, applies ideas from Merleau-Ponty and Gibson to a host of issues. Although it is often also called "enactivism," the strong sensorimotor approach associated primarily with Noë, O'Regan, and Hurley [O'Regan and Noë, 2001; Hurley, 2002; Hurley and Noë, 2003; 2004; 2009; O'Regan, 2011] is rather different from enactivism as promoted by Thompson and others. Although this approach is influenced by Merleau-Ponty, it is more closely related to the Gibsonian approach. To avoid confusion, we will call it the "sensorimotor approach." The key to the sensorimotor approach, as we have seen, is that perceiving, seeing, experiencing, and the like are things that we do, not things that happen inside us. The enactivism of Varela *et al.* [1991] in contrast focuses on organisms, as lived bodies, that enact or "bring forth" worlds; that is, a biological, living organism is also a phenomenological, lived body. On the relationship between life and mind, Thompson says, "[m]y point is rather that to make headway on the problem of consciousness we need to go beyond the dualistic concepts of consciousness and life in standard formulations of the hard problem" [Thompson, 2007, p. 224]. The idea of bringing forth a world is called "sense making":

> Organisms regulate their interactions with the world in such a way that they transform the world into a place of salience, meaning, and value—into an environment (Umwelt) in the proper biological sense of the term. This transformation of the world into an environment happens through the organism's sense-making activity. Sense-making is the interactional and relational side of autonomy.
>
> [Thompson and Stapleton, 2009, p. 3]

This sense-making, a form of direct realism, is the activity through which organisms learn about, think about, and experience the world. Indeed, it is the activity through which they have a world.

Heidegger, Merleau-Ponty, and Gibson argued explicitly against rationalist assumptions, and against the assumption that intelligence and understanding belong to the province of the mind, to which the character of the body and world is irrelevant. Beginning in the 1960s, Hubert Dreyfus used these arguments against the assumptions made in cognitive science and, especially, artificial intelligence. Those of us who largely reject CTM and RTM read the state of research in artificial intelligence as a vindication of phenomenology and ecological psychology, and of Dreyfus' arguments. Dreyfus argues that the rationalist tradition ignores the role that the body plays in experience and intelligent behavior. Ignoring the body is an obvious consequence of taking the brain to be a computer and the mind to be computer programs, making it natural to think of the body as peripheral: our sensory surfaces are analogous to keyboards; our muscles are analogous to monitors and printers. Enactivists and sensorimotor theorists have all been influenced by Dreyfus.

One interesting thing to note from the history is that both behaviorism and neutral monism were and are included within enactivism, the sensorimotor account, and ecological psychology. Perhaps this is why one can hear those notes in these traditions even when people do not intend them to be there or did not place them there consciously. Perhaps these notes sometimes get jumbled up. The important thing now is to explicitly throw out the behaviorism and more fully develop the connection between neutral monism and extended experience, that is, extended neutral monism. This is because extended neutral monism provides an unambiguous account of what it means for conscious experience to be extended that does not shade off into behaviorism, eliminativism, obscurantism, panpsychism, type-B materialism, wide supervenience-ism, wide realizationism, wide computationalism, and the like.

Given this account of extended human cognition and action, Silberstein and Chemero [2011b; 2012] argued that we can now eliminate qualia (understood as intrinsic trope of experience that exist independently of cognitive processes). We can understand conscious experience and cognition as inseparable and complementary aspects of coupled brain-body-environment systems. Experience is cognition and cognition is experiential. Our cognitive, conscious, and behavioral capacities co-explain and co-determine each other dynamically. The systems that cognitive scientists have identified as extended cognitive systems are in fact extended phenomenal–cognitive systems. Taking this position has advantages. For our purposes the most important advantage is that one need not reify conscious experience in the form of qualia nor deflate it to the point of reduction or elimination. The phenomenological world of experience is neither in the "head" nor in the "external world": it is fundamentally relational. It should be very clear according to this approach that brains are necessary but not sufficient for conscious experience, that the brain is not some sort of virtual reality machine that generates a matrix world internal to the brain. The very idea of a neural correlate of consciousness as a sufficient condition for some conscious state is a misnomer. Thompson offers an enactivist version of what we call neutral monism:

On one side lies a unique sensorimotor perspective that constitutes the subject of perception and the agent of action. On the other side lies the environment as the meaningful locale of perception and action. In this way, sensorimotor I-making and sensorimotor sense-making arise together and are inseparable; they're dependently co-arisen.

[Thompson, 2015, p. 334]

What has all this got to do with time as experienced? As Gallagher and Zahavi point out:

Thus, the enactive character goes all the way down, into the very structure of time-consciousness, and one doesn't get this enactive character without an integration of all three components. What we are suggesting here is that experience has an enactive character, not only on the act or action level, but in its most basic self-constituting, self-organizing level, in its very temporal microstructure. Consciousness is not simply a passive reception of the present; it enacts the present, constituting its meaning in the shadow of what has just been experienced and in the light of what it anticipates. . . . This means that the temporal structure of consciousness should be considered as in-the-world, and in very pragmatic terms.

[Gallagher and Zahavi, 2014, pp. 95–6]

In other words, we are talking about direct realism; that is, it is time to take the brackets off phenomenology and let it be unbound. As we stated earlier, the experience of time is neither in the head as such (the subject) nor the external world as such (the object), the experience is fundamentally relational. It is the self-consistency relation between subject and object that allows for the experience of time. As we stated earlier, this relation or structure is not in anything nor located anywhere; rather, it is why there are things in time and space as experienced. Given neutral monism, self/world are two sides of the same coin; therefore, the dynamical character of experience–thought and the world are two sides of the same coin.

Once again, the conscious mind is not some utterly distinct entity that imposes or projects time and change onto a static universe as if the mind were just some virtual reality machine that we are stuck behind. No. In our view the very existence of the co-dependent conscious self/world-in-space-and-time unity is yet another adynamical global constraint, you simply cannot have one without the other. So the fundamental explanation for subjectivity and the "subjective experience" of time isn't a dynamical one. In fact, the world is really not spacetime plus matter with experience as a mysterious add-on, but actually the world is spacetimesourcesubjectivity. Indeed, in RBW there are no mysteries of existence that demand or require a dynamical explanation at bottom, and that includes conscious experience.

There is related work in embodied cognition and phenomenology specifically relevant to the experiential arrow of time. The suggestions made about the explanation for the experience of time are not merely phenomenology however;

they also have grounding in the recent cognitive science of time perception. For example, recall the quotations in chapter 1 from cognitive neuroscientist Marc Wittmann confirming the embodied and extended nature of time perception [Wittmann, 2016]. Here is another:

> On the phenomenal level, consciousness—and the self-consciousness deriving from it—is distinguished by spatial and temporal presence. Consciousness is tied to corporeality and temporality: I experience myself as existing with a body over time . . . I experience myself now as an embodied entity, but I also experience myself as persisting in time, beyond the moment: for the perceiver, perception is . . . bound up with the necessary certainty of having to be a spatial something in time and a temporal something in space.
>
> [Wittmann, 2016, pp. 103–4]

The phenomenology here is striking which, as we saw, has been noted by thinkers in both the Buddhist and Hindu traditions for centuries [Thompson, 2015]: self/world-in-space-and-time co-arise together upon waking from deep sleep, and they disappear together when deep sleep is entered. The same holds true for deep meditation:

> Ultimately, the goal of very experienced meditators during meditation practice is the joint modulation of the perception of self, space and time culminating in the feeling of "selflessness," "spacelessness," and "timelessness" (Berkovich-Ohana, Dor-Ziderman, Glicksohn, & Goldstein, 2013; Wittmann, 2015). This peak experience attainable during meditation is describable (in retrospect) as joint modulation of the notion of self and time. That is, during this kind of ASC the feelings of the self and of time are intensely modulated, in the extreme leading to the feeling that time stands still—a reported universal experience in mystical states where time is not experienced at all and the self becomes one with the world (Achtner, 2009; Ott, 2013). The notions of self and time disappear.
>
> [Wittmann, 2015, p. 5]

One thing to note about Presence in particular, it does remain even in deep meditative states; it is all that remains [Thompson, 2015]. But what about deep sleep? Of course, the difference between the deep sleep state and deep meditation is presumably that in the former there is no experience whatsoever and in the latter exactly what remains is pure Presence. So if Presence is fundamental then why does it disappear in deep sleep? Interestingly, maybe it doesn't:

> According to these traditions [Yoga, Vedanta, and Tibetan Buddhism], deep and dreamless sleep is a mode of consciousness, not a condition where consciousness is absent.
>
> [Thompson, 2015, p. 233]

> According to the Tibetan Buddhist sleep yoga teachings . . . Besides ordinary deep sleep there's lucid deep sleep. Ordinary deep sleep is called the 'sleep of ignorance';

awareness is void or blank and in total darkness. Lucid deep sleep is called 'clear light sleep'. It occurs when the body is sleeping but the practitioner is neither lost in darkness nor in dreams, but abides in pure awareness . . . The Tibetan Buddhist and Yoga and Vedanta descriptions of deep sleep coincide in other ways . . . Being able to witness lucidly deep sleep means that the concealment of ignorance is removed and pure awareness [Presence] shines forth.

[Thompson, 2015, pp. 265–6]

Of course we understand all of this violates the axiomatic assumption of Western consciousness studies that precisely what we are explaining is the contrast between the absence of consciousness in the off-state (deep sleep) and the presence of consciousness in the on-state (waking experience). But this way of thinking, while perfectly reasonable, hasn't yielded much success regarding the hard problem even though presumably we now know all kinds of neural correlates associated with the on-state versus the off-state. As Windt notes in her book, *Dreaming* [Windt, 2015], the consensus about dream experiences now is that they can happen in both REM states and non-REM states of several sorts. The relationship between both waking and dreaming experiential states and neural states is most probably many-many, which is not to say we shouldn't look for patterns and generalities linking such states. The point is that there is not going to be some simplistic demonstration that the identity theory is true, the brain for all we know right now could be some rheostat-like mechanism that modulates or reflects conscious experience rather than generating it.

But all this aside, positing the idea of lucid deep sleep also helps us to appreciate the nature of Presence and its relationship to the subject/object cut in a world of time and space:

there remains a kind of awareness in dreamless sleep . . . In deep sleep, however, this awareness doesn't witness any object separate from itself–no waking world of perceptible things and no dream world of images . . . dreamless sleep is described as lacking the obvious or gross subject/object duality that's present in the waking and dreaming states. In the waking state, the subject appears as the body and the object appears as what we perceive. In dreams, the subject appears as the dream ego or self-within-the-dream and the object as the dream world. In deep sleep, consciousness doesn't differentiate this way between subject and object, knower and known.

[Thompson, 2015, pp. 5–6]

Obviously this is all reminiscent of the kind of neutral monism on offer here. If true, what it shows is that the dream state reflects neutral monism as much as the waking state, raising questions about how to think about waking consciousness versus the dream state. Many in consciousness studies have argued that the dream state is a good model for conscious experience, that waking consciousness like the dream state, is really just a kind of brain generated virtual reality [Metzinger, 2003; Revonsuo, 2016]. Thus, if we can explain how the brain generates these virtual worlds and selves in the dream state then we can likewise explain waking consciousness.

As quoted in Windt [2015, chapters 11 and 12], thinkers like Revonsuo [2006] and Metzinger [2003] also agree that the experience of Phenomenal selfhood requires the experience of a self/world boundary even in dreams. They agree experience is always situated and immersive such that the "field of experience" in the dream is carved into experiencer (the dream self) and experienced (the "environment," the part of the field of experience in the dream that is not counted as part of the dream self). To be conscious or aware is to have a sense of presence (to be here and now) in an "external" environment: an environment that is meaningful in the Heideggarian sense of being a world-for-me. Of course neither Revonsuo, Metzinger, nor anybody else who claims that brains generate conscious experience have anything remotely like a solution to the hard/generation problem. What is interesting from our perspective is that such thinkers are absolutely right about the phenomenology and the fact that it extends to dreams, but they have the explanatory order backward; dream states are a reflection of the neutral monist nature of the waking state, not the other way around. And therefore none of these experiences, waking or dreaming, are *generated* by brains.

Thompson notes that while lucid deep sleep may not yet be affirmed by Western sleep scientists, lucid dreaming has been studied more thoroughly and might provide an inroad to lucid deep sleep:

> Another way to enter lucid deep sleep is the practice of seeing through the dream state in a lucid dream . . . After recognizing the dream state and transforming the dream, you can try to see through the dream by dissolving it completely. You release imagery and thoughts and rest in the awareness of being aware. To be aware this way is to experience lucidly the state of deep sleep.
>
> [Thompson, 2015, p. 267]

But one might worry that whether a dream is lucid (i.e., the dreamer knows they are dreaming while in the dream state) or not, the very existence of dreams refutes embodied and extended accounts of conscious experience. Revonsuo [2016] takes this position: if all this stuff about embodied, embedded, and extended experience is true then why do dream experiences exist at all in a state where the body is not engaging in action and the vehicles of perception are largely shut down, such that there is less connection to the external world. Don't dreams suggest that the virtual reality model of experience must be true?

There are two things to note here. First, the body and senses are not as disconnected from the external world in the dream state as you might think:

> A majority of dreams are weakly phenomenally–functionally embodied states, as well as states in which cognitive activity is only weakly expressed and incompletely differentiated from dream cognition. This view has important consequences. For instance, it casts doubts on the widely accepted claim that dreaming arises completely independently of external and bodily inputs.
>
> [Thompson, 2015, p. 515]

Second, as we said earlier, the experience of being a distinct self immersed in an external world exists even in the dream state. The dreamer tends to identify with the dream self or dream avatar. In deep sleep there is neither self nor world. Windt makes the same point: even in dreams phenomenal selfhood (the dream you) requires the experience of a self-world boundary and phenomenal selves are always spatiotemporally situated even in dreams. As she states, "[s]patiotemporal self-location is a highly invariant property of phenomenal experience: though the phenomenal here and now as such are constantly shifting, they consistently form the center of the phenomenal world" [Windt, 2015, p. 521]. Though she doesn't use this language, her observation about the phenomenology of dreams is a perfect expression of the kind of neutral monism advocated for here:

> The experience of being in a world only makes sense if we assume that someone, a phenomenal self, is experienced as being located in this world. Phenomenal self-hood requires the experience of a self-world boundary; phenomenal selves are always situated selves, located at some point in the experienced world and distinguished from other persons or object in it. They are always firmly located in the present and bear a particular relation to remembered past and anticipated future selves . . . Spatiotemporal situatedness thus is a plausible candidate not only for identifying the phenomenal core of dreaming but also for understanding conscious experience and phenomenal selfhood as such.
>
> [Windt, 2015, p. 522]

Thompson makes a similar point:

> The feeling of being a distinct self immersed in the world comes back in the dream state. We experience the dream from the perspective of the self within it, or the dream ego. Although the entire dream world exists only as a content of our awareness, we identify our self with only a portion of it—the dream ego that centers our experience of the dream and presents itself as the locus of our awareness.
>
> [Thompson, 2015, p. xxxi]

We agree completely with Windt and Thompson but stress that for us this isn't some "brain-fiction" generated by neural processes. According to neutral monism, this is the very nature of reality, which must be reflected in the dream if there is to be a subject having experiences. Thus, while the nature of waking-state experiences and dream-state experiences are surely different in many ways, given neutral monism they are not as different in kind as many believe. After all, in neither case is the brain "projecting" or *generating* the experience of subject-in-an-environment. And in both cases the explanation is the same in that the non-dual "field of experience" is apportioned into the experiencer and the experienced. We are all aware that lucid dreams can be very like waking-state consciousness and that waking-state consciousness can be very dream-like. But obviously, however, there are important differences, for example, one shouldn't try flying without a plane in the waking state. It is interesting to note, however, that in the dream state the point of view of the dreaming self or dream avatar can shift to other dream characters quite quickly over time. Yet from the perspective of the lucid dreamer or the waking state, the dreamer (the God's-eye view in this case) can see that she is all of the dream characters and everything else in the dream as well.

8.6   Relational Blockworld: Time, freedom, and spontaneity

One of the things that bothers people about a block universe is not only the loss of free will in the libertarian sense of freedom (i.e., that no matter the antecedent conditions or in the case of RBW future boundary conditions, one really *could*, in every sense of the word, have acted differently than one did), it is the loss of "spontaneity" or "creativity" for the universe as a whole. Of course, how one ought to define these terms is an interesting question, but many people clearly feel that the reality of the future is sufficient to rule out these properties in a block universe. We disagree for reasons we will try to make clear in this section. In the famous debate between Bergson and Einstein this complaint about the block universe was allegedly Bergson's main concern [Canal, 2015]. But in fact Bergson raised several concerns and, as it turns out, they are logically distinct. Here is a short list of his concerns about Minkowski spacetime from Canal, 2015:

1. The reality or foreclosure of the future means the universe has no spontaneity or creativity, which it appears to have.
2. The block universe entails dynamical determinism which again robs the universe of spontaneity or creativity, and negates human free will.
3. Time as experienced is just an illusion or projection, merely "psychological" in Einstein's terminology. It isn't a real feature of the world despite appearances.

The point is that while the future does exist in RBW, it doesn't entail that the universe has no spontaneity or creativity (unless that is supposed to be an analytic truth). RBW certainly doesn't entail dynamical determinism as we will show. And finally, given the brand of neutral monism embedded in RBW, it in no way entails that time as experienced is an illusion or mental projection.

Though it may not be immediately obvious yet, what is shaping up here is that the block universe of RBW given the kind of neutral monism we espouse is a beautiful and consistent blend of phenomenology, existentialism, and the natural sciences. We have seen Gallagher, Zahavi, and others provide an account of the connection between subjectivity, temporal experience, and action that is historically a theme in various corners of continental philosophy such as existentialism and phenomenology. That is, to be is to be in time. As Heidegger [1925, p. 169] wrote, "[h]uman life does not happen in time but rather is time itself." Time is interwoven with consciousness. Bergson, who very famously argued with Einstein about the nature of time, was deeply bothered by the implication of relativity theory that time was just all in the mind. His claim, much like Heidegger's, was that "[w]e are time." "Bergson . . . believed this distinction could never be absolute; that we could never establish a fixed boundary between matter and mind" [Canal, 2015, p. 340].

Here is the point: neutral monism does what Bergson wanted, "to turn subjective time into something objective." Neutral monism tells us that Einstein's "time of the universe" and Bergson's "lived time" are one. It is very interesting to note according to Canal how Einstein's own views changed toward the end of his life with regard to time:

> Contrast this confession with Einstein's words the evening he confronted Bergson, when he firmly insisted on a clear separation between subjective and objective factors, ascribing to psychology the study of the subjective realm, to physics the study of objective events, and to philosophy simply nothing—at least when it came to the study of time. Yet later in life he admitted that he did not think that the division between the subjective and objective could be established once and for all, or even that between physics and metaphysics . . . By then, he was similarly skeptical about the difference between objectivity and subjectivity.
>
> [Canal, 2015, p. 346]

Somehow, people decided that in a block universe, the existentially essential, fundamentality, externality, and reality of Passage and Presence must be an illusion. At least from the ant's-eye perspective, neutral monism allows us to accept everything about the centrality of time that certain figures in continental philosophy have emphasized and to also accept a block universe of a sort. Nothing Heidegger says about time as experienced is incompatible with the block universe of RBW. For example, that in the living present, past and future do not appear as independent moments but rather as dimensions of the present, forming a meaningful unity only because the present is colored by the "having beeness" of the past and the "coming toward oneself" of the future, all sandwiched between life and death. Time as experienced isn't just succession but must be meaningful in some way [Dreyfus, 2015]. Events must make a difference to some subject, such as the anticipation of an agent about the future. The experience of time is narrative, active, and perspectival just like memory and perception more generally. The neutral monism of RBW accepts all of this.

As Heidegger would say, this is the datum of experience prior to being distorted by any philosophical or scientific theory [Dreyfus, 2015]. One rarely finds it stated in this way, but the idea is that, for humans in the ant's-eye perspective, to be is to feel like one is in a story of sorts, in part of one's own making of course. We have the sense we are in a very particular story in fact. This includes all the narrative, historical, and social notions of self and identity that piggy back on the minimal self we discussed earlier. The experience of struggling with possibilities and ultimately actualizing some and foregoing others is equally potent in a block universe. To see modern analytic philosophers and cognitive scientists adopting these insights from ancient Asian traditions and continental philosophy is certainly heartwarming, but for all the reasons given, we are very skeptical that these insights are going to be underwritten by neuroscience or physics. We don't doubt the value in finding physical and neural correlates that perhaps we can manipulate, along with studying and developing techniques for manipulating consciousness directly such as meditation and dream yoga. But hopefully, RBW and neutral monism provide a kind of unity nonetheless.

Just as Heidegger states, being-in-the-world is the nature of our experience and reality, not some separate Cartesian ego. Given the non-duality of space, time, source, and subjectivity according to our view, and given the background independent nature of our view, it should be clear that spacetime is not some autonomous cosmic theater in which separate physical events unfold that in turn are perceived/represented by disconnected conscious agents acting on the stage. Rather, the phenomenological world of experience, the "living present," is both inherently relational and fundamental. As Heidegger claims, for Dasein to be in the world is to be with the world, to be in relation to other entities, to find itself in a complex of possible enterprises. "Dasein is thrown into a world with which it is essentially involved, past possibilities are taken up and projected futurally, and so on" [Cooper, 1993, p. 32]. In the block universe of RBW it is still true that we are not objects; time isn't a substance or an endless string of "Nows." Existence is process because given neutral monism, subjectivity, which is the very fabric of reality, is non-dual and process-like.

These given features of existence are not overridden or falsified by RBW; they do not require presentism or libertarian free will to be true. It is customary to assume that in a block universe dynamical determinism must be true, but in the block universe of RBW the question of dynamical determinism is moot. First, the claim that given certain initial and boundary conditions, and given a certain dynamical law, there is only one unique outcome or evolution of the state possible, presupposes that dynamical explanation is fundamental. In RBW adynamical global constraints are fundamental and dynamical laws are merely descriptions of regularities from the ant's-eye perspective. So what does the thesis of determinism even mean in such a world? While it is true that RBW is a block universe, it isn't deterministic as defined dynamically—there is no dynamical or causal "glue" in RBW. The mere reality of the future doesn't entail dynamical determinism. Furthermore, at least for so-called QM phenomena, RBW has direct action as given by the AGC, so there need not be contiguous dynamical processes connecting events at all. Given that discrete spacetimesource elements with direct action are fundamental in RBW, there is really no sense of a continuous dynamical deterministic evolution/trajectory in space and time, and nothing to rule out true ontological stochasticity or indeterminism in the traditional sense in certain natural and experimental contexts. The totality is given by the AGC but the sequence of events, as viewed dynamically, is under determined.

The point is that RBW is compatible with spontaneity and openness, properly conceived. Why should the mere reality of the future preclude these? Notice that

RBW in no way denies the spontaneity and openness which are the phenomenological hallmark of thought/experience. And given the neutral monism of RBW, thought/experience are not distinct from the world. Imagine if we'd started this book characterizing neutral monism and then asked, what becomes of physics? Two things should be clear. First, physics (as ordinarily conceived) isn't the queen of the sciences and can't explain everything. This follows from the ontological contextuality of RBW alone. Second, physics can't be construed as everything about the world minus thought/experience and psychology can't be construed as everything about thought/experience minus the physical world. Given neutral monism, the domain and purvey of physics and psychology interpenetrate; SR is not just physics, but a theory of perception. Neither physics nor psychology is absolutely more fundamental than the other. Reality in RBW is by its very nature spontaneous and open. Indeed, even the mere fact it is "all just there" from a God's-eye view doesn't mean there couldn't be other block universes, that things couldn't be otherwise given different AGC.

Someone might object that while maybe neutral monism makes sense as regards experiences, certainly the process and phenomenology of thinking or cogitation must be distinct from the physical universe. That is a reasonable claim, but upon examination we would say it is another pragmatically useful dualism that must ultimately be abandoned. Thought and experience are ultimately non-dual. To see this, one has to abandon the idea that thought is inherently propositional, symbolic, or even imagistic. Likewise, one has to abandon the idea that experiences are inherently qualia-like. In waking consciousness, daydreams, and dream states, thought and experience transform into one another at will like matter–energy; this is the very nature of imagination. In the lucid dream state, for example, one's thought/desire for a certain event to transpire bring about the experience of that event. All of this imaginative thought/experience, even in the waking state, unfolds in the neutral basis of pure Presence. One can think of Presence as like universal attention.

One might ask why do thinking, cogitation, or inferential processes arise at all? We would say that thought arising and the subject/object split in the field of experience go hand in hand because from the ant's-eye view the sense of being a distinct subject in a separate external world necessitates inference, judgment, and explanation as driven by the need for action. This in turn leads to self-reflection and second-order thoughts as one assesses their situation. As Windt explains, there is a close link between the experience of thinking and feeling, and the sense that one is the owner, if not generator of those thoughts and feeling [Windt, 2015, p. 417].

Here is another way to make the point about the compatibility of RBW with time as experienced. Buddhism likes to stress the impermanence and the interdependent nature of reality [Yoshinori, 1991]. Its first principle says that constant change and flux are the nature of reality (i.e., Passage), and the second principle says that nothing has inherent existence and everything is interconnected and co-determining (i.e., ontological contextuality). The first principle says that, at least from the ant's-eye view, the field of experience is ever-changing. The second principle says, at least from the God's-eye view, all phenomena are empty of self-essence, meaning that things only have existence and the properties they possess in relation to everything else. These claims are meant as both phenomenology and ontology. Needless to say, these two principles are deeply related. Interdependent arising is typically given in terms of causation and dynamics in the Buddhist literature, but it should be obvious that the principle can equally and more deeply be construed in terms of adynamical global constraints from the God's-eye view; the past, present, and future co-determine each other, and not just causally, seen from the ant's-eye view. For example, it is hard to see how interdependent arising could be true given dynamical presentism wherein only the Global Now exists. The point is that both these principles remain fundamental truths in RBW. Nothing about the block universe as such negates these two principles, indeed, it affirms them.

As regards free will, if the reality of the future (however it got there) is sufficient to make one conclude that free will is an illusion or some such, then so be it. We don't see why this is any worse than a world where dynamical presentism and dynamical determinism are true. So what must really bother people is that the future isn't open from a God's-eye view, regardless of why this might be the case. But obviously, as Oaklander says, from the ant's-eye perspective, "to say that we cannot change the future is not to say that we do not have a hand in making it or that our choices do not bring it about" [Oaklander, 2004, p. 349]. So as many have pointed out, it is one thing to say that I can't *change* the future and quite another to say that I can't *affect* the future. This is especially true given the neutral monism in RBW just described. There is no causal closure of the physical, no ontological reductionism, no causal exclusion, no physicalism, etc., to get in the way of denying efficacy to human agents. Indeed, there isn't even any hard and fast distinction between essentially mental and essentially physical causal processes.

From a God's-eye perspective the future is already there in some sense, but unless one adopts the cosmic time frame as uniquely preferred, there is no frame-independent meaning to "the future." Admittedly, in RBW all talk of causation, mental or otherwise, is perspectival. If that violates one's conception of agency and free will then so be it. We would we say it doesn't so much violate the "common sense," libertarian-like conception of free will so much as it just renders it inherently perspectival, which, as Ismael says, isn't the same as rendering it false.

Le Poidevin writes:

> If the future is real—that future facts obtain in just as concrete way as facts about the present obtain—then this affects our view of causation. In particular, we can no longer view causation as a matter of *making something real*. We must settle for a watered-down conception of causation. And this has consequences for our intuitive conception of freedom, since our exercise of apparent freedom is a causal process. This picture of the future as real is one associated with the B-series of time, which sees our distinction between past, present and future as purely perspectival, and therefore as not reflecting deep ontological asymmetries.
>
> [Le Poidevin, 2013, p. 545]

Le Poidevin calls this conception of causation or agency "watered-down," which sounds pejorative. But it is only like that, however, if one insists that the ant's-eye view is fundamental, but what if it isn't? It is like finding out that solid objects are mostly space and then deciding that therefore there just aren't any solid objects. Let it go. After all, adopting the God's-eye view, as it were, has some amazing explanatory and ontological advantages, as well as uniquely beautiful vistas.

In a relational block universe with spatiotemporal ontological contextuality and neutral monism, talk about causal and dynamical processes, both for conscious agents and so-called physical processes, must be ant's-eye only; there is no point in asking for more. So what is interesting in RBW is that what is really spurious about the idea of libertarian free will (that a person's choices and actions have so many degrees of freedom regardless of what the rest of the universe is doing in the past, present, or future), isn't dynamical determinism. It is the spatiotemporal interdependence of all phenomena.

If one thinks of trying to subtract events or narratives from the block universe, as in the Capra film, *It's a Wonderful Life*, it would be like pulling on a thread from a vast interwoven tapestry: the whole thing would unravel and the picture depicted disappear. However, matters are even worse in the relational block universe because the threads are 4D and they don't even exist independently of the other threads. Jorge Luis Borges tried to convey this idea when he wrote that "[t]ime's march is a web of causes and effects, and asking for any gift of mercy, however tiny it might be, is to ask that a link be broken in that web of iron, ask that it be already broken. No one deserves such a miracle" (*A Prayer*). He goes even further in this regard in *The Aleph*: "Yes, the place where, without admixture or confusion, all the places and times of the world, seen from every angle, coexist":

> On the back part of the step, toward the right, I saw a small iridescent sphere of almost unbearable brilliance. At first I thought it was revolving; then I realised that this movement was an illusion created by the dizzying world it bounded. The Aleph's diameter was probably little more than an inch, but all space was there, actual and undiminished. Each thing (a mirror's face, let us say) was infinite things, since I distinctly saw it from every angle of the universe. I saw the teeming sea; I saw daybreak and nightfall; I saw the multitudes of America; I saw a silvery cobweb in the center of a black pyramid; I saw a splintered labyrinth (it was London) . . . I saw your face; and I felt dizzy and wept, for my eyes had seen that secret and conjectured object whose name is common to all men but which no man has looked upon—the unimaginable universe.

> [Borges, 1945, p. 38]

Borges is trying to get the reader to imagine this "God's-eye view," where that experience is not an "external view" as if outside the block universe looking in from some preferred frame of reference. Rather, the point is that there exist only all the perspectives (or possible perspectives) from within the universe, so to speak, but in this case seen all at once: "In that unbounded moment, I saw millions of delightful and horrible acts; none amazed me so much as the fact that all occupied the same point, without superposition and without transparency. What my eyes saw was simultaneous; what I shall write is successive, because language is successive" [Borges, 1945, p. 37]. The character in Borges' story is seeing the entire block universe as one moment from various different perspectives or frames of reference, as perhaps the Tralfamadorians or heptapods might.

So, just as with physical systems, the histories and narratives of individual lives are bounded by their future boundary conditions and those of others, especially those with whom they interact.[2] So while there is no story-of-stories in a block universe because of the relativity of simultaneity, all the stories or lives are inextricably spatiotemporally interconnected. While these are, of course, normative aesthetic judgments, we admit to being awestruck in the face of a universe that is a non-decomposable amalgam of stories that are all interdependently bounded by an adynamical global constraint. Recall the concerns about dynamical presentism from chapter 2 and 7. Even if dynamical presentism is a coherent idea, we would happily trade it to be a part of the cosmic web so described. For an excellent overview of the many problems of dynamical presentism see Frischhut [2017].

Is libertarian free will a coherent notion? Recall our earlier critiques of all the A-series alternatives from chapter 7. Is libertarian free will, for example, obviously any better off in any of them? Certainly not in the moving spotlight picture where the future is also "just there." How about dynamical presentism? We think it is interesting that those craving libertarian free will find dynamical presentism a comforting and coherent notion, especially, by hypothesis, in the absence of any kind of dynamical determinism.

So the idea is that frame-independent individual point-like Nows or Global Nows go in and out of existence; the past and present literally cease to exist (with respect to what meta-time frame?), and the future is literally completely open. The next Global Now could be willy-nilly any number of different events depending in part on (but not determined by) the presumably completely causally/statistically independent free decisions of goodness knows how many willful agents on the past Now slice that itself ceases to exist before the future indeterministically pops into being. Is this really coherent? As Le Poidevin points out, any version of the A-series

> must embrace the notion that there is something causally privileged about the present moment: that this is where events can exert their causal influence. But . . . the [the external] present can only be instantaneous, we have a serious problem for freedom. For any conscious state is essentially temporally extended and temporally structured: it cannot be instantaneous . . . In rationalizing the actions which appear to be the highest expressions of our freedom, we put our conscious desires and intentions center stage, but if causality is confined to the instant, the fundamental causes are hidden ones.
>
> [Le Poidevin, 2013, p. 546]

---

[2] The novel, *Cloud Atlas*, by David Mitchell and the film version directed by the Wachowskis and Tom Tykwer capture this idea nicely.

So, as Le Poidevin writes:

> And so we face a dilemma . . . That is, either the A-theory, in one or other of its versions, or the B-theory must be the correct theory of time. But both present a challenge to human freedom. If we opt for the B-theory, then we can no longer say that the real was previously unreal. Causation is not, after all, a dynamic process [from a God's-eye view], and this then reduces the significance of our own causal contribution. If we opt for a version of the A-theory which avoids this consequence, then we make the causal contribution of our conscious states wholly derivative.
>
> My own view is that the second horn of this dilemma is much more serious than the first. Granted, the B-theory requires some shift in our intuitive understanding of causation. But is is one that applies quite generally, not just to our agency. Our actions can be causal in just as real a sense as anything else can be a cause.
>
> [Le Poidevin, 2013, p. 546]

We concur.

Following the insights of Husserl and others, as we have already discussed, in terms of the Extended specious present, coherent temporal experience seems to require that a unit of experience includes retention of the past and anticipation of the future, otherwise, how would we even perceive motion and change. If the claim is that this isn't a problem for dynamical presentism because each new global Now will have agents who retain memories of the last iteration, then what guarantees this fact—this trans-temporal identity—if there is no determinism at work here?

Much of this is an expression of the problem of transtemporal causation as characterized by John Bigelow in what follows:

> Causation is existence symmetric: if an event exists and it is a cause of some other event, then that other event exists; and if an event exists and it is caused by some other event, then that other event exists. Some present events are caused by events which are not present. And some present events are the causes of other events which are not present. Therefore things exist which are not present.
>
> [Bigelow, 1996, p. 40]

Essentially, the point is how can the past cause or bring about the present if the past ceases to exist before the present comes in to being.

Perhaps this is just a cosmological version of the standard concern about libertarian free will. Is libertarianism a logically consistent or coherent set of beliefs? For example, how can my actions be under my control *and* be undetermined? Is it logically possible? Several related logical worries arise about "deep freedom" such as:

1. How could an action be uncaused but under the control of an agent?
2. If the answer is that the causal chains begin with the agents themselves, then our actions (and by regression, will, characters, desires, beliefs, etc.) are unmoved prime movers or self-causing, which most of us find incoherent.

3. Libertarianism implies same past and different possible futures, but this
   gets unfathomable in the context of the relationship between reasons and
   action because an individual could reason exactly the same way up to time t
   but then reach opposite conclusions about action. This makes deliberation
   totally irrational and takes away from agent control.

So, as strange as it may sound to some, for reasons that would not be lost on
Nagarjuna or any of his students, we think agent causation (or any other kind
of causation) actually requires something like spatiotemporal contextuality or a
block universe. In any case the bottom line is that Bergson gets literally everything
he wants in RBW except the openness of the future, which turns out not be so
important for his global vision.

8.7   The neutral presence: Reconciling the ant's-eye perspective with a view from the God's-eye perspective

If you thought this chapter was on the verge of veering into too much speculation
thus far, you had better buckle up. In the words of Hunter S. Thompson, "[w]hen
the going gets weird, the weird turn pro." Here we speculate how the ant's-eye
view and the God's-eye view might relate perspectively with respect to subjective
experience. We have sketched out how one might explain Passage.

   We have posited that Presence is universal and fundamental. Presence, that
is, "being," in the West is typically thought of as either bracketed experience
in the phenomenological tradition or merely a qualitative experience to ultimately be explained by neuroscience. There are important exceptions in the West
such as those posited by Martin Heidegger, William James, Henri Bergson, and
Maurice Merleau-Ponty. In the traditions of existentialism, pragmatism, and phenomenology, one can find expression of the idea that the "lived present," "lived
experience" or "living present" are among the most fundamental aspects of
reality. Also, as alluded to earlier, in some Hindu and Buddhist texts Presence
is neither bracketed nor just a brain state; it is fundamental. When all qualitative and intentional states have ceased what remains is Presence (*nirvikalpa
samadhi*). Given that Presence is fundamental, it cannot be defined in terms of
other concepts, of either a material or mental nature:

> Advaita Vedanta centrally posits the existence of a permanent "self" (atman). This
> self is characterized as the "witness" (saksin) of the experiences, that is as that
> which is conscious of them—yet not in the sense of some substantial entity that
> performs the witnessing, but rather as nothing but the taking place of witnessing
> (consciousness) itself. This chapter argues for the plausibility of this notion of a
> witness consciousness, interpreted as the abiding experiencing of the ever-changing
> experiences. Synchronically and diachronically, manifold experiences are presented

in one and the same consciousness, whose oneness is not reducible to some unifying relations between the experiences, but rather forms the dimension in which
they, together with all their interrelations, have their existence in the first place.
This presence-dimension is [our emphasis] . . . what is called atman (qua witness) in
Advaita.

[Fasching, 2010, p. 20]

Once again, don't let the terminology put you off the deeper phenomenological
point: this is not awareness of or consciousness of presence, this is Presence itself.
Perceiver (subject) and perceived (object) co-exist in a self-consistent fashion as
a single non-dual aspect of Presence. Presence or "Nowness" is fundamental in
the sense of being irreducible. This is not to say that there is a preferred spatial
or temporal frame but rather that Presence is a *universal* and *neutral* base of reality, decidedly not a moving spotlight given by individual consciousness as Weyl
suggests, nor some special physical marker added to physics as Smolin posits.
To paraphrase John Wheeler, Presence is what puts the fire in the equations. It
is why there is something rather than nothing, and always has been, so to speak.
This is exactly the sort of strangeness Seager warned us about once we abandon
the physical as fundamental in favor of neutral monism.

When we say Presence is fundamental and universal, it is easy to assume here
that Presence is like a field that pervades all of spacetime, but Presence (i.e.,
the "witness" consciousness, "being itself," "witnessing itself," "pure awareness,"
etc.) is not in space or time. Rather, it is the neutral ground of a subject in space
and time. In the Advaita Vedanta tradition the idea is expressed that fundamental
reality is neither subjective nor objective, neither mind nor matter, neither time
nor space. These divisions need somebody to whom it happens, a self or subject.

It is also easy to assume that somehow conscious beings "emerge" from or
generated by Presence in some temporal or causal sense. But keep in mind that
from the God's-eye perspective, from the perspective of Presence (if there is
one, and more on this shortly), no such thing is occurring. The very idea of an
essential dualism between the external world and subject (the ant's-eye perspective) is a kind of cognitive illusion (from the God's-eye perspective), a partition
brought about perhaps by thought or conceptualization, not some emergent temporal process or a spontaneous popping into existence. According to this view,
thought/conceptualization/imagination isn't within time but generates the form of
time. Neither time nor thought originates in the subject. Rather, the subject/object
split is a result (adynamically speaking) of thought and conceptualization.

In both the Buddhist and Hindu traditions, this idea is often expressed as the
"I am" thought arises as the result of the bare witness state which is then reified
leading to thoughts like "I am in pain," etc. As Sri Nisargadatta Maharaj states,
"The sense 'I am a person in time and space' is the poison . . . Whatever you are
engrossed in you take to be real . . . There can be no universe without the witness,
there can be no witness without the universe . . . The witness that is enmeshed in
what he perceives is the person" [Maharaj, 1973, pp. 350–1]. This is the ant's-eye
view. As we said earlier, if one assumes physicalism or ontological reductionism,

then the ant's-eye perspective seems the only possible one; we have a bunch of separate brains running around out there each with their own neurally generated internal virtual reality show. But in the neutral monism of RBW, however, it is a whole new world. Pure awareness or Presence is ultimately the only subject but it has multiple foci of experience; to quote the Purusa Sukta hymn: "The Universal Being has infinite heads, unnumbered eyes, and unnumbered feet." Again, the idea here is that which experiences is not a *thing* or *entity* but a spatiotemporally localized *perspective* that comes to have a narrative identity and an entire worldview which adds to that perspective.

And again, the self is the very subjectivity of experience, not something separate from the experiential flow. Yet, this process we call self comes to conceive itself as a distinct entity in time and space, embodied, the bearer of qualia, with free will or agency, etc. Sri Nisargadatta Maharaj writes that, "a reflection of the watcher [the witness] in the mind creates the sense of 'I' and the person acquires an apparently independent existence. In reality there is no person, only the watcher identifying himself with the 'I' and the 'mine'" [Maharaj, 1973, p. 343], and "[t]he moment you say: 'I am', the entire universe comes into being along with its creator" [Maharaj, 1973, p. 363].

As we wrote earlier, in this new world of neutral monism it isn't crazy to wonder if, as James offers the "timemask" can be removed. According to science fiction and various Asian traditions such as Buddhism and Hinduism, the answer is "yes." This is just another way of asking what perspectives are possible in principle and in practice beyond those of the ant's-eye view. We have already encountered several speculations along these lines from literature. There are many other examples, of course. In the recent Christopher Nolan film, *Interstellar*, for example, there are the "bulk beings" of the future, a super-civilization formed by a race evolved from humanity who occupy the fifth dimension (whatever that means), who can observe the 4D block universe (past, present, and future) in the same way the 3D beings in the story *Flatland* view the 2D beings that occupy it. There are the aliens known as the Prophets in the TV program "Star Trek: Deep Space Nine," who somehow are 4D and live in wormholes. Here is a dialog between one of the Prophets (the Emissary) and a human character where the former tries to explain serial human existence to the latter while playing baseball:

Sisko: In the end, it comes down to throwing one pitch after another, and seeing what happens. With each new consequence, the game begins to take shape.

Alien Batter: And you have no idea what that shape is until it is completed?

Sisko: That's right. In fact, the game wouldn't be worth playing if we knew what was going to happen.

Jake Prophet: You value your ignorance of what is to come?

Sisko: That may be the most important thing to understand about humans. It is the unknown that defines our existence. We are constantly searching, not just for answers to our questions, but for new questions. We are explorers. We explore our lives day by day, and we explore the galaxy, trying to expand the boundaries of our knowledge. And that is why I am here. Not to conquer you with weapons, or with ideas. But to coexist . . . and learn.

[Roddenberry, 1993]

This is certainly a nice existential explanation of the ant's-eye view, which greatly puzzles the Prophets. Often, in these stories, the spatiotemporal God's-eye view is achieved by some technological or scientific means that is vaguely described. The way Dr. Manhattan describes his 4D experiences suggests he is experiencing his entire worldtube from birth to death (assuming he can die) as one moment. But is that a coherent notion? Is he somehow witnessing it all at once from some external preferred perspective or having every internal perspective at once as in the Borges story? In the traditions of Hinduism and Buddhism, the claim is that meditation and other techniques can in principle, for some, eventually lead to the removal of the "timemask."

The phenomenological description of perceiving from the bare-witness perspective (as opposed to the ant's-eye human perspective) is that "you" are Presence or 'witnessing itself' (not a person or agent) and what is being witnessed includes not only the usual "external environment" such as clouds and cats but also so-called mental-internal phenomena/processes such as thoughts and feelings. As Sri Nisargadatta Maharaj states, "Destroy the wall that separates, the 'I-am-the-body' idea, and the inner and outer will become one" [Maharaj, 1973, p. 346]. If these experiences are veridical, this would be a kind of immediate vindication of neutral monism. Contrast this with the deep meditative state in which there is neither self nor world but only pure Presence. But suppose we ask what spatial/temporal perspective the witness witnessing takes? Throughout this chapter we have seen some possibilities. Sri Nisargadatta Maharaj writes that, "[y]ou are the changeless background against which changes are perceived . . . The dreams flow before you the immutable witness." But from what angle? For example, we can imagine viewing every frame of a film or a graphic novel at once, but when we imagine that, we are looking at it from some one specific spatial angle and viewing it from beginning to end as in the cosmic time frame, from the first frame to the last. Does this analogy hold for the witness consciousness? An added wrinkle, given relativity, is that the universe can be multiply foliated into many different event-orderings; past, present, and future are relativistic frame-dependent. This is most unlike viewing a film reel.

Admittedly, we are not sure how to answer these questions, but perhaps lucid dreaming as an analogy can be somehow helpful here. As Thompson writes:

We need to go back to the threefold framework of the witnessing aspect of awareness, the changeable contents of awareness and ways of experiencing particular contents of awareness as the self. What marks a strong lucid dream is the felt presence of witnessing awareness, which can observe or witness the dream precisely as a dream. Its presence can inhibit the automatic identification with the dream ego that characterizes dreaming. The same witnessing awareness can be felt in the waking states in moments of heightened mindfulness; its presence can inhibit the automatic identification with the 'I–me–mine' that characterizes the waking state.

[Thompson, 2015, p. 161]

Thompson says that to see dreams in this way is to view them as acts of imagination, and not "false perception" or "pseudo-perceptions" [Thompson, 2015, p. 179]. Recall that lucid dreams are subject to "voluntary guidance" by the witness and with practice and due focus/attention can be manipulated at will. And, of course by extension all this is true of the waking state as well: "Perception, therefore, isn't online hallucination; it's sensorimotor engagement with the world. Dreaming isn't offline hallucination; it's spontaneous imagination during sleep. We aren't dreaming machines but imaginative beings. We don't hallucinate at the world; we imaginatively perceive it" [Thompson, 2015, p. 188].

The idea is that if neutral monism is true and therefore imagination is in some sense fundamental, then perhaps "our" spatiotemporal perspective is ultimately only limited by what can be imagined by what Advaita Vedanta calls "the witness consciousness," or what we are calling Presence. For example, we can all easily imagine a good lucid dreamer being able to adopt several different spatial and temporal perspectives in their dreams at will, perhaps even simultaneously.

What is interesting about deep meditation, other related techniques, and deep sleep is that they potentially suspend the conditions of embodiment and spatiotemporal situatedness that underwrite the account of subjectivity and time as experienced that we have been discussing here. So perhaps these various spatiotemporal perspectives can be had in the waking state as well. As Thompson notes:

Out-of-body experiences reveal something crucial about the sense of self: you locate yourself as an experiential subject wherever your attentional perspective feels located . . . In other words, your sense of who you are and where you're located goes with your self-as-subject and not your body-as-object.

[Thompson, 2015, p. 211]

As Thompson notes, this has all been confirmed several times by psychologists who artificially induce both out-of-body experiences and, using recording and virtual reality techniques, body-swap experiences wherein people identify with the spatiotemporal perspective of manikins and the like [Thompson, 2015, chap. 7]. So there is ample evidence that we are not stuck in the ant's-eye view of our own bodies, at least. As Thompson states:

In sum, bodily self-awareness includes feelings of self-identification, self-location, having a first-person perspective, and body ownership. All of which depend on the way our body's sensory and motor systems converse with our brain. Using virtual reality, we can alter this conversation in systematic ways on the sensory and motor side, and thereby create corresponding alterations to all these aspects of our bodily sense of self.

[Thompson, 2015, p. 218]

Obviously, these induced experiences in the laboratory are a far cry from the literary examples we discussed or the claims of Swami's etc., for one thing they are spatial and not spatiotemporal, but it certainly suggests these possibilities are worth exploring, and that the relationship between the experiential aspects of space/time and the brain/body and the physical are malleable. Part of the point here is that once we adopt RBW and neutral monism, give up mind–brain supervenience and the identity theory, then perhaps it is worth exploring the possibility, that just as some people have claimed over the years in a number of different meditative traditions, it is possible to have relatively God's-eye experiences.

8.8 Conclusion

We hope this chapter, while speculative, goes a long way to assuage concerns over the missing Passage, Presence, and Direction in a block universe approach to physics—at least the RBW approach. There is a price, of course. One has to let go physicalism and ontological reductionism, and all the baggage accompanying them. One has to accept neutral monism instead. Accepting neutral monism as described here means accepting there is one thing that isn't relational or contextual, one thing that is intrinsic and fundamental, and that is Presence. Presence is neither physical nor mental as these are typically characterized. It isn't a thing, substance, entity, property, law, qualia, event, or representation. Presence doesn't exist in space and time, it is the adynamical basis for them. Everything is grounded in Presence. The bottom line is we agree with Einstein "that the experience of the Now means something special for man, something essentially different from the past and the future, but that this important difference does not and cannot occur within physics." We understand neither those inclined toward dualism or physicalism (or both) will find our approach very palatable. The dynamical/mechanical worldview has enjoyed many successes and it is therefore a hard habit to break. But while the reader may disagree with our adynamical worldview, it certainly cannot be faulted for a lack of inclusivity; that is, we really have tried in earnest to explain everything from our God's-eye vantage point.

Coda for Ants

To my mind there must be at the bottom of it all an utterly, not equation, not an utterly simple equation, but an utterly simple idea. And to me that idea when we finally discover it will be so compelling, so inevitable, so beautiful that we will all say to each other, "Oh how could it have been otherwise."

John Wheeler in Ferris, 1985

While it is true that relativity and quantum mechanics in their dynamical forms certainly added many strange wrinkles and "epicycles" to mechanical science, those theories still conform to the dynamical/mechanical philosophy. As Carroll pointed out, "[e]ver since Newton, the paradigm for fundamental physics has been the same, and . . . This way of thinking is just as true for [quantum mechanics] or general relativity or quantum field theory as it was for Newtonian mechanics or Maxwell's electrodynamics" [Carroll, 2012]. In this book we have argued that both relativity and quantum mechanics are telling us that we need to move even further beyond the dynamical universe paradigm; that is, beyond the ant's-eye view to the God's-eye view wherein the universe is not ultimately a machine like a computer. We have proposed overturning every major tenet of the dynamical universe, at least at the fundamental level:

1. We replaced time-evolved objects with a web of 4D relations distributed contextually per disordered locality via direct action.
2. We replaced dynamical laws with 4D adynamical global constraints.
3. We replaced the dualism of time-evolved consciousness (Newton's ghost in the machine) supervening on and generated by time-evolved objects (such as brains) with neutral monism.

From the God's-eye perspective, there is no machine and there are no ghosts. This enormous violation of our dynamical bias had an equally large payoff in its resolution of the puzzles, paradoxes, problems, and conundrums of the dynamical universe, and its new way of resolving the problems of dark matter, dark energy, unification, and quantum gravity. This is not to mention having deflated the hard problem of consciousness and all the related problems of mental causation that plagued that famous mechanical philosopher, Descartes. Since, as we showed, our proposed modified lattice gauge theory and modified Regge calculus make correspondence with quantum field theory and general relativity, respectively, we don't have to reproduce their successes, we only have to resolve their problems,

Beyond the Dynamical Universe. Michael Silberstein, W.M. Stuckey and Timothy McDevitt, Oxford University Press (2018). © Michael Silberstein, W.M. Stuckey and Timothy McDevitt. DOI 10.1093/oso/9780198807087.001.0001

and we do. That is just the beginning, of course, because concordant new physics must now be produced.

Our spatiotemporal ontological contextuality gives us an enormous advantage over methodological and ontological reductionism in pursuing new physics because it makes quantum gravity empirically accessible. Specifically, we are using type Ia supernova data to provide clues to our metric perturbations for Relational Blockworld (RBW) quantum gravity, and we're using contextuality already inherent in GR to provide fits of galactic rotation curves, galactic cluster mass profiles, and the CMB angular power spectrum, as we showed in chapter 6. With these fits in hand, we are now seeking a graphical action for Regge calculus as modified per disordered locality, and an ansatz to complete a cosmological model without a cosmological constant (dark energy) or non-baryonic dark matter. Modified Regge calculus will also provide the appropriate classical context for modified lattice gauge theory which is RBW quantum gravity (contextuality!). Issues that modified lattice gauge theory in the context of modified Regge calculus must address include the tunneling rate from black holes and possible new particles arising when the energy density becomes large enough that probability amplitudes for quantum exchanges must take into account the affect that the exchange has on its classical spacetime context (the graphical lattice structure of modified Regge calculus).

Finally, and closely related to the study of RBW quantum gravity, we must study modified lattice gauge theory in the context of modified Regge calculus when quantum exchanges don't appreciably change the classical spacetime context. This individual "click-by-click analysis" must make correspondence with the Standard Model of particle physics and its fundamental spacetime worldlines (particle trajectories). Understanding the spacetime trajectories of particle physics per their composite individual detector clicks (individual quantum exchanges) means we are modeling all possible information in our direct action approach: there is no counterfactual definiteness and there are no hidden variables to account for, just the individual data points. Thus, per RBW, this is where empirical science has exhausted all possible data (information has bottomed out), so our accounting of particle trajectories via individual "click-by-click analysis" should explain the otherwise inexplicable number of free parameters, particle masses, and hierarchy problem in the Standard Model of particle physics via discrete graph theory.

If our modified lattice gauge theory and modified Regge calculus fail to accomplish any of these tasks, they will have to be discarded and we will have to find another approach to physics per RBW. Both integrity and coherence are demanded by physics, or what we want to return to calling "natural philosophy." We hope that even if the reader has no taste for our particular account, then this project can serve as a model for the foundations of physics and foundational physics communities about how philosophy and physics should interact to produce alternative physical models that go beyond mere metaphysics and interpretation.

It would seem that we're living in very exciting times, scientifically speaking. There is something in the dynamical universe paradigm many believe to be true with empirical certainty, but we will come to believe is absolutely false in the new worldview. For the Aristotelian scholar, it was an empirical fact that the natural state for objects is that of rest. If an object is put in motion without an independent source of propulsion, it will eventually come to rest. You have no doubt done this experiment yourself many times and can attest to its empirical veracity. Thus, an "empirically verified" law of motion in the Aristotelian universe is that an object in motion tends to come to rest, so an object in motion must have a mover to remain in motion. For an Aristotelian scholar, the idea that Earth orbits the Sun is absurd. What would be moving the entire Earth? In our current paradigm of the dynamical universe, we believe instead that an object in motion tends to remain in motion, so an object in motion must be forced to come to rest. For the Newtonian scholar, it is not at all ridiculous that Earth orbits the Sun. What would stop it? Ascending from Aristotle's teleological universe to Newton's mechanical universe resulted in a radical change in our method of explanation. Likewise, there is something that has been "empirically verified" in the dynamical/mechanical universe that will be deemed false in the forthcoming scientific worldview. What will it be?

If we're right and you're a millennial, your grandchildren will think you were crazy for believing the primary tenets of the dynamical/mechanical paradigm. Maybe this new way of thinking is the first step to becoming like the future posthumans in *Interstellar* whose conscious experiences are not limited to the ant's-eye view. In the old dynamical/mechanical paradigm with its views about the virtual-ghost-in-the-brain, such an idea is surely destined to remain science fiction, but RBW is a new world that needs a new kind of consciousness studies to go with it. If we are right, the fundamental physics and psychology of the future (and the past and present) for human beings is that of the heptapods.

Let us end by honoring our hero and guiding light, John Wheeler. He gave us five clues about looking for a theory of quantum gravity decades ago [Wheeler, 1994, pp. 302–9]:

1. "The boundary of a boundary is zero." [See Philosophy of Physics for Chapter 6.]

2. "No bit-level question, no bit-level answer." Reality is a game of 20 questions: "The contribution of the environment becomes overwhelming." What is the relevant environment? To paraphrase: "the whole show is wired up together." "It is wrong to regard this or that physical quantity as sitting 'out there' with this or that numerical value in default of question asked and answer obtained by way of appropriate observing device."

3. "The super-Copernican principle." This principle rejects now-centeredness in any account of existence as firmly as Copernicus repudiated here-centeredness. Try the more-oft quoted, "the past has no

evidence except as it is recorded in the present. The photon that we are going to register tonight from that four-billion-year-old quasar cannot be said to have had an existence 'out there' three billion years ago, or two (when it passed an intervening gravitational lens) or one, or even a day ago. Not until we have fixed arrangements at our telescope do we register tonight's quantum as having passed to the left (or right) of the lens or by *both* routes."

4. "Consciousness." He means here that "from the past through the billeniums to come, so many observer-participants, so any bits, so much exchange of information, as to build what we call existence."

5. "More is different." He is referring to P. W. Anderson here and follows with his own twist on this ode against naive reductionism: "Will we someday understand time and 'space' and all the other features that distinguish physics—and existence itself—as the similarly self-generated organs of a self-synthesized information system?"

At first glance, the reader, especially the reader familiar with the works of John Wheeler, might be surprised that we invoke him in the name of RBW. After all, he seems to reify quantum mechanics, suggests that the past isn't fixed, gives conscious observers a role in physics, and argues for an irreducible plurality in science and existence. While we agree that Wheeler had not completely transcended the dynamical/mechanical worldview, he came very, very close, and from our perspective he was prescient.

Start with clue number 1. As we showed in chapters 4, 5, and 6, our adynamical global constraint (AGC) is based in the boundary of a boundary principle. Spatiotemporal contextual emergence is our version of clue number 2. While clue number 3 might be read as some sort of dynamical backward causation, non-existence of the past as in presentism, or even re-writeability of the past, that is because it is being looked at dynamically. RBW, with its adynamical global constraint, is a profound example of Wheeler's central idea that the arrow of determination in the universe is not exclusively past to present to future. Once one fully transcends the dynamical and mechanical bias, and adopts the AGC, one can see how true this is without any residual dynamical/causal spookiness. Our take on clue number 4 is embodied in chapter 8. Consciousness is fundamental not to explaining the outcomes of physical processes per se but to time as experienced and more generally the world as experienced. The universe really is "one being with many eyes" and precisely what relativity does is tell us how to relate all those perspectives to create a world of shared experience. Finally, as for clue number 5, while RBW does provide a unified account of physics and experience, it certainly doesn't reduce everyday objects of experience or experience itself to fundamental physical entities as dreamed of in the mechanical/dynamical paradigm. Most importantly of all, while we stressed the fundamentality of the adynamical global perspective for certain explanatory purposes, the truth is that the God's-eye view and the ant's-eye view are complementary and in very much a kind of complementarity relation with each other. They are both real, can't be reduced to the other, and possess non-commuting properties.

All this brings us to Wheeler's quotation at the beginning of this Coda. What is the "utterly simple idea" that motivated this book? Adynamical explanation (via adynamical global constraints) in the block universe is a simpler and more comprehensive explanatory system than dynamical explanation (via dynamical laws) in the mechanical universe. Wilczek's challenge is Wheeler's desideratum. Oh how could it have been otherwise?

So dear ants, we end this story with a bow to the maestro, John Archibald Wheeler, whose direction we hope we've been following.

References

Abbott, B., Abbott, R., Abbott, T., and Abernathy, M.R., et al. (2016). Observation of gravitational waves from a binary black hole merger. *Physical Review Letters*, 116:061102.

Ackerman, L., Buckley, M., Carroll, S., and Kamionkowski, M. (2009). Dark matter and dark radiation. *Physical Review D*, 79:023519.

Aczel, A. (2002). *Entanglement: The Greatest Mystery in Physics*. Raincoast Books, Vancouver.

Aharonov, Y., Bergmann, P., and Lebowitz, J. (1964). Time symmetry in the quantum process of measurement. *Physical Review*, 134(6B):1410–16.

Aharonov, Y., Cohen, E., Elitzur, A., and Smolin, L. (2016). Interaction-free effects between distant atoms, https://arxiv.org/abs/1610.07169.

Aharonov, Y. and Vaidman, L. (1990). Properties of a quantum system during the time interval between two measurements. *Physical Review A*, 41:11–20.

Albert, D. (2012). On the origin of everything: "a universe from nothing," by Lawrence M. Krauss. *New York Times Book Review*, 3, http://www.nytimes.com/2012/03/25/books/review/a-universe-from-nothing-by-lawrence-m-krauss.html?_r=4.

Albrecht, A. and Iglesias, A. (2012). The clock ambiguity: Implications and new developments. In Mersini-Houghton, L., and Vaas, R., editors, *The Arrows of Time: A Debate in Cosmology*, pages 53–68. Springer, Berlin.

Almheiri, A., Marolf, D., Polchinski, J., and Sully, J. (2013). Black holes: complementarity or firewalls? *Journal of High Energy Physics*, **2013**(2):62.

Ambjorn, J., Goerlich, A., Jurkiewicz, J., and Loll, R. (2013). Quantum gravity via causal dynamical triangulations, http://arxiv.org/abs/1302.2173.

Anandan, J. (2003). Laws, symmetries, and reality. *International Journal of Theoretical Physics*, 42:1943–55.

Angus, G. (2009). Are sterile neutrinos consistent with clusters, the CMB and mond? *Monthly Notices of the Royal Astronomical Society*, 394(1):527–32.

Arntzenius, F. and Maudlin, T. (2013). Time travel and modern physics, Zalta, E.N., *The Stanford Encyclopedia of Philosophy* (Winter 2013 Edition), http://plato.stanford.edu/archives/win2013/entries/time-travel-phys/.

Arstila, V. and Lloyd, D. (2014). *Subjective Time*. MIT Press, Cambridge, MA.

Ashby, N. (2002). Relativity and the global positioning system. *Physics Today*, 55(5):41–7.

Atmanspacher, H. and Kronz, F. (1999). Relative onticity. In Atmanspacher, H., Amann, A., and Müller, H., editors, *On Quanta, Mind and Matter: Hans Primas in Context*, pages 273–94. Kluwer, Dordrecht.

Auffèves, A. and Grangier, P. (2014). Contexts, systems and modalities: a new ontology for quantum mechanics. *Foundations of Physics*, pages 1–17. Springer.

Baggott, J. (2013). *Farewell to Reality: How Modern Physics Has Betrayed the Search for Scientific Truth*. Pegasus Books, New York, NY.

Ballentine, L. E. (1998). *Quantum Mechanics: A Modern Development*. World Scientific. Singapore.

Barausse, E, Sotiriou, T.P., and Miller, J.C. (2008). A no-go theorem for polytropic spheres in palatini f(r) gravity. *Classical and Quantum Gravity*, 25:062001.

Barbour, J. (1999). *The End of Time: The Next Revolution in Physics*. Oxford University Press, New York, NY.

Barbour, J. (2012). Reductionist doubts, http://fqxi.org/community/forum/topic/1495, Questioning the Foundations Essay Contest.

Bardon, A. (2012). *The Future of the Philosophy of Time*. Routledge, New York, NY.

Barrett, J. (1987). The geometry of classical regge calculus. *Classical and Quantum Gravity*, 4(6):1565–76.

Becker, K. (2015). Is gravity time's archer? FQXi Blogs, 3.

Beisbart, C. (2016). Philosophy and cosmology. pages 817–35. Oxford University Press, Oxford.

Bekenstein, J. D. (2004). Relativistic gravitation theory for the modified newtonian dynamics paradigm. *Physical Review D*, 70(8):083509.

Bell, J. (1964). On the Einstein–Podolsky–Rosen paradox. *Physics*, 1:195–200.

Bell, J. (1987). *Speakable and Unspeakable*. Cambridge University Press, Cambridge

Bergson, H. (1896). *Matter and Memory*. Zone Books, London.

Bergson, H. (1913). *Time and Free Will: An Essay on the Immediate Data of Consciousness*. University of Michigan Library, Manufactured in the United States by Courier Corporation.

Bergson, H. (2002). *Henri Bergson: Key Writings*. Continuum Press, New York, NY.

Bergstrom, L. (2000). Non-baryonic dark matter: observational evidence and detection methods. *Reports on Progress in Physics*, 63:793.

Bernal, T., Capozziello, S., Hidalgo, J., and Mendoza, S. (2011). Recovering mond from extended metric theories of gravity. *The European Physical Journal C*, 71(11):1794.

Bianchi, E., Rovelli, C., and Kolb, R. (2010). Is dark energy really a mystery? *Nature*, 466:321–22.

Bigelow, J. (1996). Presentism and properties. *Nous*, 30:35-52.

Bishop, R. (2005). Patching physics and chemistry together. *Philosophy of Science*, 77: 710–22.

Bishop, R. (2010a). Downward causation in fluid convection. *Synthese*, 160:229–48.

Bishop, R. (2010b). Whence chemistry? *Studies in the History and Philosophy of Modern Physics*, 41:171–77.

Bishop, R. (2012). Fluid convection, constraint and causation. *Interface Focus*, 2:4–12.

Blanchet, L. and Le Tiec, A. (2008). Model of dark matter and dark energy based on gravitational polarization. *Physical Review D*, 78(2):024031.

Bohr, A. and Ulfbeck, O. (1995). Primary manifestation of symmetry. origin of quantal indeterminacy. *Reviews of Modern Physics*, 67:1–35.

Bohr, N. (1954). *The Philosophical Writings of Niels Bohr, Volume II: Essays 1932–1957 on Atomic Physics and Human Knowledge*. Ox Bow Press, Woodbridge, Connecticut.

Borges, J.L. (1945). El aleph, http://web.mit.edu/allanmc/www/borgesaleph.pdf.

Born, M., Einstein, A., and Born, I. (1971). *The Born Einstein Letters: Correspondence between Albert Einstein and Max and Hedwig Born from 1916 to 1955 with Commentaries by Max Born. Translated by Irene Born*. Macmillan Press, London.

Bourne, C. (2007). *A Future for Presentism*. Oxford University Press, New York, NY.

Brennan, A. (2011). Necessary and sufficient conditions, Zalta, E.N., *The Stanford Encyclopedia of Philosophy* (Winter 2012 Edition), https://plato.stanford.edu/entries/necessary-sufficient/.

Brewin, L. (2000). Is the Regge calculus a consistent approximation to general relativity? *General Relativity and Gravitation*, 32(5):897–918.

Brewin, L. C. and Gentle, A. P. (2001). On the convergence of Regge calculus to general relativity. *Classical and Quantum Gravity*, 18(3):517–26.

Briceno, S. and Mumford, S. (2016). *Relations all the Way Down? Against Ontic Structural Realism. The Metaphysics of Relations.* Oxford University Press, Oxford.

Brown, H. (2005a). *Physical Relativity: Space-time Structure from a Dynamical Perspective.* Oxford University Press, Oxford.

Brown, H. and Pooley, O. (2006). Minkowski space-time: a glorious non-entity. In Dieks, D., editor, *The Ontology of Spacetime*, pages 67–89. Elsevier, Utercht.

Brown, L. (2005b). *Feynman's Thesis: A New Approach to Quantum Theory.* World Scientific Press, New Jersey.

Brownstein, J. (2009). *Modified gravity and the phantom of dark matter.* PhD thesis, University of Waterloo.

Brownstein, J. and Moffat, J. (2006a). Galaxy cluster masses without non-baryonic dark matter. *Monthly Notices of the Royal Astronomical Society*, 367(2):527–40.

Brownstein, J. and Moffat, J. (2006b). Galaxy rotation curves without nonbaryonic dark matter. *The Astrophysical Journal*, 636(2):721–41.

Bub, J. (2009). Quantum mechanics is about quantum information. *Foundations of Physics*, 3(4):541–60.

Bub, J. and Pitowsky, I. (2010). Two dogmas about quantum mechanics. In Saunders, S., Wallace, B. J. K. A., and Wallace, D., editors, *Many Worlds?* pages 433–59. Oxford University Press, Oxford.

Buchert, T., Larena, J., and Alimi, J. (2006). Correspondence between kinematical backreaction and scalar field cosmologies and the scalar 'morphon field', https://arxiv.org/abs/gr-qc/0606020

Bull, P., Akrami, Y., Adamek, J., Baker, T., et al. (2016). Beyond ΛCDM: problems, solutions, and the road ahead. *Physics of the Dark Universe*, 12:56–99, https://arxiv.org/abs/1512.05356.

Buonomano, D. (2017). *Your Brain Is a Time Machine: The Neuroscience and Physics of Time.* Norton and Company, New York, NY.

Burdick, A. (2017). *Why Time Flies.* Simon and Schuster, New York, NY.

Butterfield, J. (2014). Dark matter and dark radiation. *Studies in the History and Philosophy of Modern Physics*, 46:57–69.

Butterfield, J. and Isham, C. (1999). *On the Emergence of Time in Quantum Gravity. In the Arguments of Time.* Oxford University Press, Oxford.

Callender, C. (2010a). Is time an illusion? *Scientific American*, 302, 58–65.

Callender, C. (2010b). Stephen Hawking says there's no theory of everything, *New Scientist*, https://www.newscientist.com/blogs/culturelab/2010/09/stephen-hawking-says-theres-no-theory-of-everything.html, 9.

Callender, C. (2012). Is time an illusion? *Scientific American*, 306(1s).

Callender, C. (2016). Thermodynamic asymmetry in time, Zalta, E.N., *The Stanford Encyclopedia of Philosophy* (Winter 2016 Edition), https://plato.stanford.edu/archives/win2016/entries/time-thermo/>.

Canal, J. (2015). *The Physicist and the Philosopher: Einstein, Bergson, and the Debate That Changed Our Understanding of Time.* Princeton University Press, Princeton, NJ.

Cao, T. (1999). Introduction: conceptual issues in quantum field theory. In Cao, T., editor, *Conceptual Foundations of Quantum Field Theory*, volume 1, pages 1–27. Cambridge University Press, Cambridge.

Capozziello, S. (2002). Curvature quintessence. *International Journal of Modern Physics D*, 11(04):483–91.

Capozziello, S. and Francaviglia, M. (2008). Extended theories of gravity and their cosmological and astrophysical applications. *General Relativity and Gravitation*, 40(2):357–420.

Caravelli, F. and Markopoulou, F. (2012). Disordered locality and lorentz dispersion relations: an explicit model of quantum foam. *Physical Review D*, 86(2):024019.

Carlini, A., Frolov, V., Mensky, M., Novikov, I., et al. (1995). Time machines: the principle of self-consistency as a consequence of the principle of minimal action. *International Journal of Modern Physics D*, 4(5):557–80.

Carnap, R. (1963). Carnap's intellectual biography. In Schilpp, P., editor, *The Philosophy of Rudolf Carnap*, pages 3–84. Open Court, La Salle, IL.

Carrick, J. and Cooperstock, F. (2012). General relativistic dynamics applied to the rotation curves of galaxies. *Astrophysics and Space Science*, 337.

Carroll, R. (2013). Kip Thorne: physicist studying time travel tapped for Hollywood film, *The Guardian*, https://www.theguardian.com/science/2013/jun/21/kip-thorne-time-travel-scientist-film, 6.

Carroll, S. (2001). The cosmological constant. *Living Reviews in Relativity*, 4(1).

Carroll, S. (2010a). The arrow of time, *Engineering & Science*, http://calteches.library.caltech.edu/705/2/Time.pdf.

Carroll, S. (2010b). *From Eternity to Here*. Penguin Group, New York, NY.

Carroll, S. (2011). Dark matter: just fine, thanks, http://www.preposterousuniverse.com/blog/2011/02/26/dark-matter-just-fine-thanks.

Carroll, S. (2012). A universe from nothing. *Discover Magazine Online (Apr, 28 2012)*.

Carroll, S. (2014). 2014 : what scientific idea is ready for retirement? *Edge*, https://www.edge.org/response-detail/25322.

Carroll, S. (2015). Why is there dark matter? http://www.preposterousuniverse.com/blog/2015/07/07/why-is-there-dark-matter/, July.

Carroll, S. (2016). *The Big Picture: On the Origins of Life, Meaning, and the Universe Itself*. Dutton, Penguin Random House, New York, NY.

Carroll S. What Does "Happy New Year" Even Really Mean? http://smithsonianmag.com/science-nature/what-does-happy-new-year-even-really-mean-180953633

Carruth, A. (2008). *Molnar, Intrinsicality and Iterated Powers*. PhD thesis, Durham University.

Chalmers, D. (1996). *The Conscious Mind: In Search of a Fundamental Theory*. Oxford University Press, Oxford.

Chalmers, D. (2016). The combination problem for panpsychism. In Bruntrup, G. and Jaskolla, L., editors, *Panpsychism*, pages 179–214. Oxford University Press, Oxford.

Chemero, A. and Silberstein, M. (2008). After the philosophy of mind: from scholasticism to science. *Philosophy of Science*, 75:1–27.

Chiang, T. (2002). *Stories of Your Life and Others*. ORB Press, New York, NY.

Chodos, A. (2011). Nobels honor discoveries of accelerating universe, quasicrystals. APS News 20(10), 1 and 5.

Choi, S. and Fara, M. (2016). Dispositions, Zalta, E.N., *The Stanford Encyclopedia of Philosophy* (Winter 2013 Edition), https://plato.stanford.edu/archives/spr2016/entries/dispositions/.

Cifone, M. (2004). Undergraduate thesis.

Clarkson, C. and Maartens, R. (2010). Inhomogeneity and the foundations of concordance cosmology. *Classical and Quantum Gravity*, 27(12):124008, IOP Publishing.

Clifton, T., Ferreira, P., Padilla, A., and Skordis, C. (2012). Modified gravity and cosmology. *Physics Reports*, 513(1).

Clowe, D., Bradac, M., Gonzalez, A., Markevitch, M., et al. (2006). A direct empirical proof of the existence of dark matter. *The Astrophysical Journal Letters*, 648: L109–L113, http://arxiv.org/abs/astro-ph/0608407.

Colosi, D. and Rovelli, C. (2009). What is a particle? *Classical and Quantum Gravity*, 26(2):025002, http://arxiv.org/abs/gr-qc/0409054.

Cooper, R. (1993). *Heidegger and Whitehead*. Ohio University Press, Athens, OH

Cooperstock, F. and Tieu, S. (2007). Galactic dynamics via general relativity—a compilation and new results. *International Journal of Modern Physics A*, 13.

Corry, R. (2015). Retrocausal models for EPR. *Studies in History and Philosophy of Modern Physics*, 49:1–9.

Craig, W. and Smith, Q. (2007). *Einstein, Relativity and Absolute Simultaneity*. Routledge, London.

Craig, W. L. (2001). *Time and Eternity: Exploring God's Relationship to Time*. Crossway Books, Wheaton, IL.

Cramer, J. (1986). The transactional interpretation of quantum mechanics. *Reviews of Modern Physics*, 58(3):647–87.

Crick, F. (1994). *The Astonishing Hypothesis: The Scientific Search for the Soul*. Simon & Schuster, New York, NY.

Crick, F. and Koch, C. (1990). Toward a neurobiological theory of consciousness. *Seminars in the Neurosciences*, 2:263–75.

Cross, D. J. (2006). Comments on the Cooperstock–Tieu galaxy model, https://arxiv.org/abs/astro-ph/0601191.

Croswell, K. (2001). *The Universe at Midnight*. The Free Press, New York.

Crowther, K. (2016). *Effective Spacetime: Understanding Emergence in Effective Field Theory and Quantum Gravity*. Springer, Switzerland.

Cushing, J. (1988). Foundational problems in and methodological lessons from quantum field theory. In Brown, H. and Harre, R., editors, *Philosophical Foundations of Quantum Field Theory*, pages 25–39. Oxford University Press, Oxford.

Dainton, B. (2001). *Time and Space*. McGill-Queen's University Press, Montreal.

Dainton, B. (2011). Time, passage and immediate experience. In Callender, C., editor, *The Oxford Handbook of the Philosophy of Time*, pages 382–419. Oxford University Press, Oxford.

Dainton, B. (2012). Time and temporal experience. In Bardon, A., editor, *The Future of the Philosophy of Time*, pages 123–48. Routledge, New York, NY.

Dainton, B. (2016). Some cosmological implications of temporal experience. *Cosmological and Pyschological Time*, 23:75–105.

Dainton, B. (2017). Bergson on temporal experience and duree reelle. In Phillips, I., editor, *The Routledge Handbook of Philosophy of Temporal Experience*, pages 93–106. Routledge, London.

Davies, P. (1971). Extension of Wheeler–Feynman quantum theory to the relativistic domain. I. scattering processes. *Journal of Physics A: General Physics*, 4(6):836–45.

Davies, P. (1972). Extension of Wheeler–Feynman quantum theory to the relativistic domain. II. emission processes. *Journal of Physics A: General Physics*, 5(7):1025–36.

Davies, P. (2006). *The Goldilocks Enigma: Why Is the Universe Just Right for Life?* Penguin Books, London.

de Blok, W. (2010). The core–cusp problem. *Advances in Astronomy*, 2010:789293.

De Felice, A. and Fabri, E. (2000). The Friedmann universe of dust by Regge calculus: study of its ending point, http://arxiv.org/pdf/gr-qc/0009093.pdf, Sep 27, 2000.

Dehlinger, D. and Mitchell, M. (2002). Entangled photons, nonlocality, and bell inequalities in the undergraduate laboratory. *American Journal of Physics*, 70(9):903–10.

DeWitt, R. (2004). *Worldviews: An Introduction to the History and Philosophy of Science*. Blackwell Publishing, Printed and bound in the United Kingdom by TJ International, Padstow, Cornwall.

Dicker, G. (2013). Transcendental arguments and temporal experience. In Bardon, A. and Dyke, H., editors, *A Companion to the Philosophy of Time*. John Wiley & Sons, Inc Chichester, UK.

Dieks, D. (2016). Physical time and experienced time. *Cosmological and Pyschological Time*, 23:3–20.

Dirac, P. (1933). The Lagrangian in quantum mechanics. *Physikalische Zeitschrift der Sowjetunion 3*, pages 64–72.

Dirac, P. (1978). *Directions in Physics*. John Wiley & Sons, New York.

Dolev, Y. (2007). *Time and Realism: Metaphysical and anti-Metaphysical Perspectives*. MIT Press, Cambridge, MA.

Dolev, Y. (2016). Relativity, global tense and phenomenology. *Cosmological and Pyschological Time*, 23:21–40.

Dolev, Y. and Roubach, M. (2016). Boston studies in the philosophy and history of science. *Cosmological and Pyschological Time*, 285:1–218.

Dorato, M. (2006a). Absolute becoming, relational becoming and the arrow of time: some non-conventional remarks on the relationship between physics and metaphysics. *Studies in the History and Philosophy of Modern Physics*, 37:559–76.

Dorato, M. (2006b). The irrelevance of the presentist/eternalist debate for the ontology of minkowski spacetime? In Dieks, D., editor, *The Ontology of Spacetime*, pages 93–109. Elsevier, Utrecht.

Dorato, M. (2016). *Rovelli's Relational Quantum Mechanics, Anti-Monism, and Quantum Becoming. The Metaphysics of Relations*. Oxford University Press, Oxford.

Dowker, F. (2014). Are there premonitions in quantum measure theory? Presented at "Free Will and Retrocausality in a Quantum World," 7, Trinity College, Cambridge.

Dreyfus, H. (2015). *Being-in-the-World: A Commentary on Heidegger's Being and Time*. MIT Press. Cambridge, MA

Dunlap, L. (2015). On the common structure of the primitive ontology approach and the information-theoretic interpretation of quantum theory. *Topoi*, 34(2):359–67.

Earman, J. (1974). An attempt to add a little direction to the problem of the direction of time. *Philosophy of Science*, 41:15–47.

Earman, J. (2003). Tracking down gauge: an ode to the constrained hamiltonian formalism. In Brading, K. and Castellani, E., editors, *Symmetries in Physics: Philosophical Reflections*, pages 140–62. Cambridge University Press, Cambridge.

Earman, J. and Fraser, D. (2006). Haag's theorem and its implications for the foundations of quantum field theory. *Erkenntnis*, 64(3):305–44.

Earman, J., Wüthrich, C., and Manchak, J. (2016). Time machines, Zalta, E.N., *The Stanford Encyclopedia of Philosophy* (Summer 2016 Edition), http://plato.stanford.edu/archives/sum2016/entries/time-machine/.

Echeverria, F., Klinkhammer, G., and Thorne, K. (1991). Billiard balls in wormhole space-times with closed timelike curves: classical theory. *Physical Review D*, 44(4):1077–99.

Einstein, A. (1920). *Einstein Papers, The Berlin Years: Writings, 1918–1921*, 7(31):265.

Einstein, A. (1948). Quanten-mechanik und wirklichkeit. *Dialectica*, 2(3–4):320–324.

Eisenstein, D. J., Seo, H.-J., and White, M. (2007). On the robustness of the acoustic scale in the low-redshift clustering of matter. *The Astrophysical Journal*, 664(2):660–74.

Elitzur, A. and Vaidman, L. (1993). Quantum mechanical interaction-free measurements. *Foundations of Physics*, 23:987.

Ellis, G. (2011). Inhomogeneity effects in cosmology. *Classical and Quantum Gravity*, 28(16):164001.

Ellis, G. (2014). Flow of time. *Annals of the New York Academy of Sciences*, 1326:26–41.

Esfeld, M. (2013). Ontic structural realism and the interpretation of quantum mechanics. *European Journal for Philosophy of Science*, 3(1):19–32.

Evans, P. (2015). Retrocausality at no extra cost. *Synthese*, 192(4):1139–55.

Evans, P. W., Gryb, S., and Thébault, K. (2016). Psi-epistemic quantum cosmology? pages 10–11, https://arxiv.org/abs/1606.07265.

Fasching, W. (2010). I am of the nature of seeing: phenomenological reflections on the Indian notion of witness-consciousness. In Siderits, M., T. E. and Zahavi, D., editors, *Self, No Self? Perspectives from Analytical, Phenomenological, and Indian Traditions*, pages 193–216. Oxford University Press, Oxford.

Feinberg, G., Friedberg, R., Lee, T., and Ren, H. (1984). Lattice gravity near the continuum limit. *Nuclear Physics B*, 245:343–68.

Felline, L. and Bacciagaluppi, G. (2013). Locality and mentality in everett interpretations: Albert and Loewer's many minds. *Mind and Matter*, 11(2).

Feng, J. (2010). Dark matter candidates from particle physics and methods of detection. *Annual Reviews of Astronomy and Astrophysics*, 48:495.

Fernow, R. (1986). *Introduction to Experimental Particle Physics*. Cambridge University Press, Cambridge.

Ferris, T. (1985). The creation of the universe, Northstar Productions.

Feynman, R. (1964). Electromagnetic mass, http://www.feynmanlectures.caltech.edu/II_28.html.

Feynman, R., Leighton, R., and Sands, M. (1963). The Feynman lectures on physics volume 1, Addison-Wesley, Reading, MA.

Flanagan, E.E. (2004). The conformal frame freedom in theories of gravitation. *Classical and Quantum Gravity*, 21:3817.

Flatow, I. (2016). Science Friday: in the quantum world, physics gets philosophical, http://www.sciencefriday.com/segments/in-the-quantum-world-physics-gets-philosophical/.

Folger, T. (2002). Does the universe exist if we're not looking? *Discover*, 23(6):44.

French, S. and Krause, D. (2006). *Identity in Physics: A Historical, Philosophical and Formal Account*. Oxford University Press, Oxford.

Frisch, H. (1993). Pattern recognition at the Fermilab collider and superconducting supercollide. *Proceedings of the National Academy of Sciences*, 90(11):9754–57.

Frischhut, A. (2017). Presentism and temporal experience. In Phillips, I., editor, *The Routledge Handbook of Philosophy of Temporal Experience*, pages 249–61. Routledge, London.

Fuchs, B. and Phleps, S. (2006). Comment on "general relativity resolves galactic rotation without exotic dark matter" by F.I. Cooperstock and S. Tieu. *New Astronomy*, 11(2):608–10.

Fulling, S. (2005). Review of some recent work on acceleration radiation. *Journal of Modern Optics*, 52:2207–13.

Gallagher, S. and Zahavi, D. (2008). *The Phenomenological Mind*. Routledge, London.

Gallagher, S. and Zahavi, D. (2014). Primal impression and enactive perception. In Arsila, V. and Lloyd, D., editors, *Subjective Time: The Philosophy, Psychology and Neuroscience of Temporality*. MIT Press, Cambridge, MA.

Gao, S. (2017). *The Meaning of the Wave Function: In Search of the Ontology of Quantum Mechanics*. Cambridge University Press.

Garfinkle, D. (2006). Inhomogeneous spacetimes as a dark energy model. *Classical and Quantum Gravity*, 23(15):4811–18.

Garrett, K. and Duda, G. (2011). Dark matter: a primer. *Advances in Astronomy*.

Gentile, G., Burkert, A., Salucci, P., Klein, U., et al. (2005). The dwarf galaxy DDO 47 as a dark matter laboratory: testing cusps hiding in triaxial halos. *The Astrophysical Journal Letters*, 634(2).

Gentile, G., Famaey, B., and de Blok, W. (2011). Things about mond. *Astronomy & Astrophysics*, 527:A76.

Geroch, R. (1978). *General Relativity from A to B*. University of Chicago Press, Chicago, IL.

Gibson, J. (1979). *The Ecological Approach to Visual Perception*. Houghton Mifflin, Boston.

Gleick, J. (2016). *Time Travel*. Pantheon Books, New York, NY.

Glymour, C. (1977) Indistinguishable spacetimes and the fundamental group. In Earman, J., and Glymour, C., and Stachel, J., editors, *Foundations of Spacetime Theories: Minnesota Studies in Philosophy of Science*, volume 8, pages 50–60. University of Minnesota Press.

Goff, P. (2013). Non-compositional panpsychism, www.youtube.com/watch?v=bCNCR2niqQY.

Goff, P. (2017). *Consciousness and Fundamental Reality*. Oxford University Press, Oxford.

Goodstein, D. (1985). The mechanical universe ... and beyond, Caltech and INTELECOM Intelligent Telecommunications.

Gott, J. (1991). Closed timelike curves produced by pairs of moving cosmic strings: exact solutions. *Physical Review Letters*, 66(9):1126–29.

Greene, B. (2011). *The Hidden Reality*. Alfred A. Knopf, New York, NY.

Gross, D. (1999). *Conceptual Foundations of Quantum Field Theory*, pages 56–67. Cambridge University Press, Cambridge.

Gryb, S. (2014). A shape dynamics tutorial, http://hef.ru.nl/~sgryb/research/shape_dynamics.html.

Gupta, B. (1998). *The Disinterested Witness: A Fragment of Advaita Vedanta Phenomenology*. Northwestern University Press, Evanston.

Hawking, S. (1965). On the Hoyle–Narlikar theory of gravitation. In *Proceedings of the Royal Society of London A: Mathematical, Physical and Engineering Sciences*, volume 286, pages 313–20. The Royal Society.

Hawking, S. (1992). Chronology protection conjecture. *Physical Review D*, 46(2):603–611.

Hawking, S. (1996). Why gauge? http://www.hawking.org.uk/the-beginning-of-time.html.

Hawking, S. and Mlodinow, L. (2010). *The Beginning of Time*. Bantam Books, New York, NY.

Hawthorne, J. and Silberstein, M. (1995). For whom the bell arguments toll. *Synthese*, 102:99–138.

Healey, R. (2002). Can physics coherently deny the reality of time? In Callender, C., editor, *Time, Reality, and Experience*, pages 293–316. Cambridge University Press, Cambridge.

Healey, R. (2004). Gauge theories and holisms. *Studies in History and Philosophy of Modern Physics*, 35:619–42.

Healey, R. (2007). *Gauging What's Real: The Conceptual Foundations of Gauge Theories*. Oxford University Press, Oxford.

Healey, R. (2016). Holism and nonseparability in physics, Zalta, E.N., *The Stanford Encyclopedia of Philosophy* (Spring 2016 Edition), http://plato.stanford.edu/ entries/physics-holism/#Aharonov-Bohm.

Healey, R. and Uffink, J. (2013). Part and whole in physics: an introduction. *Studies in History and Philosophy of Science Part B*, 44(1):20–21.

Heidegger, M. (1925). *Supplements: From the Earliest Essays to Being and Time and Beyond*. State University of New York Press, Albany, NY.

Heidegger, M. (1996). *Being and Time*. Harper and Row, New York, NY.

Held, C. (2013). The Kochen–Specker theorem, Zalta, E.N., *The Stanford Encyclopedia of Philosophy* (Winter 2014 Edition), http://plato.stanford.edu/archives/ win2014/entries/kochen-specker/.

Hoefer, C. (2011). Time and chance propensities. In Callender, C., editor, *The Oxford Handbook of Philosophy of Time*, pages 68–90. Oxford University Press, Oxford.

Hoefer, C. and Smeenk, C. (2016). Philosophy of the Physical Sciences. pages 115–136. Oxford University Press, The Oxford Handbook of Philosophy of Science, Humphreys, P.

Holt, J. (2012). Physicists, stop the churlishness, http://www.nytimes.com/2012/06/10/ opinion/sunday/what-physics-learns-from-philosophy.html?_r=0, 6, NY Times.

Horgan, J. (2010). Cosmic clowning: Stephen Hawking's "new" theory of everything is the same old crap, *Scientific American*, http://blogs.scientificamerican.com/cross-check/ cosmic-clowning-stephen-hawkings-new-theory-of-everything-is-the-same-old-crap/, 9.

Horgan, J. (2014). Physicist slams cosmic theory he helped conceive. *Scientific American Blog*, https://blogs.scientificamerican.com/cross-check/physicist-slams-cosmic-theory-he-helped-conceive/.

Horgan, J. (2015). Can integrated information theory explain consciousness, *Scientific American*, https://blogs.scientificamerican.com/cross-check/can-integrated-information-theory-explain-consciousness/.

Horst, S. (2007). *Beyond Reduction*. Oxford University Press, Oxford.

Howard, D. A. (2015). Einstein's philosophy of science, Zalta, E.N., *The Stanford Encyclopedia of Philosophy* (Winter 2015 Edition), http://plato.stanford.edu/ archives/win2015/entries/einstein-philscience.

Hoy, D. (2009). *The Time of Our Lives: A Critical History of Temporality*. MIT Press, Cambridge, MA.

Heraclitus and parmenides. In Bardon, A. and Dyke, H., editors, *A Companion to the Philosophy of Time*, pages 9–29. John Wiley & Sons, Inc, Chichester, UK.

Hoyle, F. and Narlikar, J. V. (1995). Cosmology and action-at-a-distance electrodynamics. *Reviews of Modern Physics*, 67(1):113–55.

Hu, W. (1995). *Wandering in the Background: A Cosmic Microwave Background Explorer.* PhD thesis, University of California at Berkeley.

Hu, W. and Sugiyama, N. (1995). Anisotropies in the cosmic microwave background: an analytic approach. *Astrophysical Journal,* 444(2):489–506.

Hu, W. and Sugiyama, N. (1996). Small scale cosmological perturbations: an analytic approach. *Astrophysical Journal,* 471:542–70.

Huggett, N., Vistarini, T., and Wüthrich, C. (2013). Time in quantum gravity. In Dyke, H. and Bardon, A., editors, *A Companion to the Philosophy of Time,* pages 242–261. John Wiley & Sons, Ltd, Chichester.

Huggett, N. and Wüthrich, C. (2012). Emergent spacetime and empirical incoherence, https://arxiv.org/abs/1206.6290.

Humphreys, P. (1997). How properties emerge. *Philosophy of Science,* 64:1–17.

Hurley, S. (2002). *Consciousness in Action.* Harvard University Press, Harvard.

Hurley, S. and Noë, A. (2003). Neural plasticity and consciousness. *Biology and Philosophy,* 18:131–68.

Husserl, E. (2001). *Logical Investigations, 3 Volumes.* Routledge, London.

Hutto, D. and Myin, E. (2013). *Radicalizing Enactivism: Basic Minds without Content.* MIT Press, London.

Ijjas, A., Steinhardt, P., and Loeb, A. (2017). Pop goes the universe. *Scientific American,* 316(2):32–9.

Isham, C. (1992). Canonical quantum gravity and the problem of time, https://cds.cern.ch/record/241615/files/9210011.pdf.

Ismael, J. (2007). *The Situated Self.* Oxford University Press, Oxford.

Ismael, J. (2016). From physical time to human time. *Cosmological and Pyschological Time,* 23:107–24.

Israel, N. and Moffat, J. (2016). The train wreck cluster and bullet cluster explained by modified gravity without dark matter, Israel, N.S., and Moffat, J.W., https://arxiv.org/abs/1606.09128.

Jackiw, R. (1999). The unreasonable effectiveness of quantum field theory. In Cao, T., editor, *Conceptual Foundations of Quantum Field Theory,* volume 1, pages 148–160. Cambridge University Press, Cambridge.

James, W. (1898). Human immorality, William James: Writings 1878–1899, Library of American, New York.

James, W. (1904). Does consciousness exist? *Journal of Philosophy, Psychology and Scientific Methods,* 1(18).

James, W. (1905a). The notion of consciousness, Sciousness, Eirini Press, 89–111.

James, W. (1905b). The place of affectional facts in a world of pure experience, William James: Writings 1902–1910, Library of American, New York.

James, W. (1912). *Essays in Radical Empiricism.* Longmans, Green and Co., New York, NY.

James, W. (1988). Manuscript lectures, Harvard University Press, Cambridge.

Jaroszkiewicz, G. (2016). *Images of Time: Mind, Science, Reality.* Oxford University Press, Oxford.

Kaiser, G. (1981). Phase-space approach to relativistic quantum mechanics iii: quantization, relativity, localization and gauge freedom. *Journal of Mathematical Physics,* 22.

Kant, I. (1998). *The Critique of Pure Reason.* Cambridge University Press, Cambridge.

Kardar, M. and Zee, A. (2002). Information optimization in coupled audio-visual cortical maps. *Proceedings of the National Academy 2003,* 99(25):15894–97.

Kastner, R. (2013). *The Transactional Interpretation of Quantum Mechanics: The Reality of Possibility*. Cambridge University Press, Cambridge.

Kastner, R. (2016). Personal Correspondence, 5.

Käuffer, S. and Chemero, A. (2014). *Phenomenology: An Introduction*. Polity.

Kefalis, C. (2015). Steven Weinberg's attack on philosophy, http://dissidentvoice. org/2015/09/steven-weinbergs-attack-on-philosophy/.

Khavari, P. and Dyer, C. (2009). Aspects of causality in parallelisable implicit evolution scheme, http://arxiv.org/pdf/0809.1815.pdf, 2 Apr 2009.

Kiefer, C. (2011). Time in quantum gravity. In Callender, C., editor, *The Oxford Handbook of the Philosophy of Time*, pages 663–78. Oxford University Press, Oxford.

Kirchhoff, M. (2015). Experiential fantasies, prediction and enactive minds. *Journal of Consciousness Studies on Embodied and Extended Theories of Conscious Experience*, 22(3-4):68–92.

Kiverstein, J. (2016). The interdependence of embodied cognition and consciousness. *Journal of Consciousness Studies*, 23(5-6):105–37.

Kiverstein, J. and Arstila, V. (2013). Time in mind. In Bardon, A. and Dyke, H., editors, *A Companion to the Philosophy of Time*. John Wiley & Sons, Inc, Chichester, UK.

Kleinert, H. and Schmidt, H. (2002). Cosmology with curvature-saturated gravitational lagrangian $r/\sqrt{1 + l^4 R^2}$. *General Relativity and Gravitation*, 34:1295–18.

Knee, G., Kakuyanagi, K., Yeh, M., Matsuzaki, Y., et al. (2016). A strict experimental test of macroscopic realism in a superconducting flux qubit. *Nature Communications*, 7:13253.

Koch, C. (2004). *The Quest for Consciousness: A Neurobiological Approach*. Roberts & Company Publishers Englewood, CO.

Koch, C. (2014). A 'complex' theory of consciousness: is complexity the secret to sentience, to a panpsychic view of consciousness? *Scientific American*, https://www.scientificamerican.com/article/a-theory-of-consciousness/.

Korzynski, M. (2005). Singular disk of matter in the Cooperstock–Tieu galaxy model, https://arxiv.org/pdf/astro-ph/0508377.pdf.

Krauss, L. (2012). *A Universe from Nothing: Why There Is Something Rather than Nothing*. Free Press, New York, NY.

Kuhlmann, M. (2009). *Quantum Field Theory*, Zalta, E.N., *The Stanford Encyclopedia of Philosophy* (Spring 2009 Edition), http://plato.stanford. edu/archives/spr2009/entries/quantum-field-theory/.

Kuhlmann, M. (2013). What is real? *Scientific American*, 309(2):40–47.

Kupczynski, M. (2016). What do we learn from computer simulations of bell experiments? https://arxiv.org/abs/1611.03444.

Kutach, D. (2013). *Causation and Its Basis in Fundamental Physics*. Oxford University Press, New York, NY.

Ladyman, J. (2007). *Structural Realism*, Zalta, E.N., *The Stanford Encyclopedia of Philosophy* (Summer 2009 Edition), http://plato.stanford.edu/archives/ sum2009/entries/structural-realism.

Ladyman, J. and Ross, D. (2007). *Everything Must Go: Metaphysics Naturalized*. Oxford University Press, Oxford.

Lancaster, T. and Blundell, S. (2014). *Quantum Field Theory for the Gifted Amateur*. Oxford University Press, Oxford.

Le Poidevin, R. (2013). Time and freedom. In Bardon, A. and Dyke, H., editors, *A Companion to the Philosophy of Time*, page 535. John Wiley & Sons, Inc, https://www.scientificamerican.com/article/a-theory-of-consciousness/.

Leibniz, G. (1991). Monadology. In *G.W. Leibniz's Monadology*. University of Pittsburgh Press, Pittsburgh, Pennsylvania.

Leifer, M. (2014). Is the quantum state real? An extended review of ψ-ontology theorems. *Quanta*, 3(1):67–155.

Leinaas, M. (1991). Hawking radiation, the unruh effect and the polarization of electrons. *Europhysics News*, 22:78–80.

Lewis, P. (2016). *Quantum Ontology: A Guide to the Metaphysics of Quantum Mechanics*. Oxford University Press, Oxford.

Lewis, S. (1982). Two cosmological solutions of regge calculus. *Physical Review D*, 25(2):306–12.

Lisi, A. G. (2006). Quantum mechanics from a universal action reservoir, https://arxiv.org/abs/physics/0605068.

Lloyd, S. (2006). *Programming the Universe: A Quantum Computer Scientist Takes On the Cosmos*. Knopf, New York, NY.

Lobo, F. S. N. (2008). Closed timelike curves and causality violation. In Frignanni, V. R., editor, *Classical and Quantum Gravity: Theory and Applications*, Nova Science Publishers, https://arxiv.org/abs/1008.1127.

Magalhaes, N. S. and Cooperstock, F. I. (2015). Galactic mapping with general relativity and the observed rotation curves, http://arxiv.org/abs/1508.07491.

Maharaj, N. (1973). *I Am That: Talks with Sri Nisargadatta Maharaj*. The Acorn Press, Durham, NC.

Malament, D. (1977). Observationally indistinguishable spacetimes. In Earman, J. Glymour, C., and Stachel, J., editors, *Foundations of Spacetime Theories: Minnesota Studies in Philosophy of Science volume 8*, pages 61–80. University of Minnesota Press, Minneapolis, MN.

Maldacena, J. (2016). Black holes, wormholes and the secrets of quantum spacetime. *Scientific American*, 315(5):26–31.

Manchak, J. (2009). Can we know the global structure of spacetime? *Studies in the History and Philosophy of Modern Physics*, 40:53–6.

Manchak, J., (2011) What is a physically reasonable spacetime? *Philosophy of Science*, 78:410–420,

Marra, V. and Notari, A. (2011). Observational constraints on inhomogeneous cosmological models without dark energy. *Classical and Quantum Gravity*, 28:164004.

Martila, D. (2015). Cooperstock is wrong: the dark matter is necessary, http://www.academia.edu/10991940/Cooperstock is wrong The Dark Matter is necessary.

Mashhoon, B. (2015). Nonlocal general relativity. *Galaxies*, 3:1–17.

Maudlin, T. (2002). Remarks on the passing of time. *Proceedings of the Aristotelian Society*, 102:237–52.

Maudlin, T. (2006). On the passing of time, http://philocosmology.rutgers.edu/images/uploads/TimDavidClass/05-maudlin-chap04.pdf.

Maudlin, T. (2007). A modest proposal concerning laws, counterfactuals, and explanations. *The Metaphysics Within Physics*, Oxford University Press, Oxford.

Maudlin, T. (2007). Completeness, supervenience and ontology. *Journal of Physics A: Mathematical and Theoretical*, 40(12):3151–3171, http://stacks.iop.org/1751-8121/40/i=12/a=S16,2007

Maudlin, T. (2011). Existence: does time exist? A discussion on whether time and its arrow are 'illusions' or fundamentally real, https://www.youtube.com/watch?v=xF-L_iWeAHo&t=15s, Foundational Questions Institute Conference: Setting Time Aright: Investigating the Nature of Time.

McCabe, G. (2007). *The Structure and Interpretation of the Standard Model.* Elsevier, Amsterdam.

McGaugh, S. (2015a). A tale of two paradigms: the mutual incommensurability of ΛCDM and MOND. *Canadian Journal of Physics*, 93(2):250–59.

McGaugh, S. (2015b). A tale of two paradigms: the mutual incommensurability of ΛCDM and MOND. *Canadian Journal of Physics*, 93:250, http://arxiv.org/abs/1404.7525.

McGaugh, S. (2016). Personal communication.

McGaugh, S., Lelli, F., and Schombert, J. (2016). The radial acceleration relation in rotationally supported galaxies. *Physical Review Letters*.

Meir, H. and Orly, S. (2016). The arrow of time. *Cosmological and Pyschological Time*, 23(3-20):155–64.

Menzies, D. and Mathews, G. J. (2006). General relativistic galaxy rotation curves: implications for dark matter distribution, https://arxiv.org/abs/gr-qc/0604092.

Merali, Z. (2013). Astrophysics: Fire in the hole! *Nature*, 496(7443).

Merleau-Ponty, M. (1962). *Phenomenology of Perception.* Routledge, London.

Mermin, N. (1981). Bringing home the atomic world: quantum mysteries for anybody. *American Journal of Physics*, 49(10):940–43.

Mermin, N. (1994). Quantum mysteries refined. *American Journal of Physics*, 62(10): 880–87.

Mermin, N. (1998). What is quantum mechanics trying to tell us? *American Journal of Physics*, 66:753–67.

Merritt, D. (2016). In rotating galaxies, distribution of normal matter precisely determines gravitational acceleration. *Case Western Reserve University Science Daily*, www.sciencedaily.com/releases/2016/09/160921085052.htm.

Metzinger, T. (2003). *Being No One: The Self-Model Theory of Subjectivity.* Bradford Book, printed and bound in the United States of America.

Milgrom, M. (1983). A modification of the newtonian dynamics as a possible alternative to the hidden mass hypothesis. *The Astrophysical Journal*, 270:365–70.

Milgrom, M. (2015). Mond theory. *Canadian Journal of Physics*, 93(2):107–18.

Miller, K. (2013). Presentism, eternalism, and the growing block. In Bardon, A. and Dyke, H., editors, *A Companion to the Philosophy of Time.* John Wiley & Sons, Inc, Chichester, UK.

Miller, M. A. (1995). Regge calculus as a fourth-order method in numerical relativity. *Classical and Quantum Gravity*, 12:3037–51.

Misner, C., Thorne, K., and Wheeler, J. (1973). *Gravitation.* W.H. Freeman, San Francisco, CA.

Moffat, J. (2005). Gravitational theory, galaxy rotation curves and cosmology without dark matter. *Journal of Cosmology and Astroparticle Physics*, 2005:003.

Moffat, J. (2006). Scalar–tensor–vector gravity theory. *Journal of Cosmology and Astroparticle Physics*, 0603:004.

Moffat, J. (2008). A modified gravity and its consequences for the solar system, astrophysics and cosmology. *International Journal of Modern Physics D*, 16:2075–90.

Moffat, J. and Rahvar, S. (2013). The mog weak field approximation and observational test of galaxy rotation curves. *Monthly Notices of the Royal Astronomical Society*, 436: 1439–51.

Moffat, J. and Rahvar, S. (2014). The mog weak field approximation ii. observational test of chandra x-ray clusters. *Monthly Notices of the Royal Astronomical Society*, 441:3724–32.

Moffat, J. and Toth, V. (2013). Cosmological observations in a modified theory of gravity (mog). *Galaxies*, 1(1):65–82.

Montero, B. (2010). A Russellian response to the structural argument against physicalism. *Journal of Consciousness Studies*, 17(3–4): 70–83.

Monton, B. (2006). Presentism and quantum gravity. In Dieks, D., editor, *The Ontology of Spacetime*, pages 263–280. Elsevier, Utrecht.

Moore, A. (2016). *Jerusalem.* Liveright, New York, NY.

Moore, A. and Gibbons, D. (1986). *Watchmen.* DC Comics, New York, NY.

Morris, M. S., Thorne, K. S., and Yurtsever, U. (1988). Wormholes, time machines, and the weak energy condition. *Physical Review Letters*, 61(13):1446.

Mott, N. (1929). The wave mechanics of alpha-ray tracks. *Proceedings of the Royal Society of London A: Mathematical, Physical and Engineering Sciences*, 126:79–84.

Mukhanov, V. (2005). *Physical Foundations of Cosmology.* Cambridge University Press, Chicago, IL.

Munoz, C. (2004). Dark matter detection in the light of recent experimental results. *International Journal of Modern Physics A*, 19:3093–170.

Nagel, T. (1986). *The View from Nowhere.* Oxford University Press, Oxford.

Narlikar, J. (2003). Action at a distance and cosmology: a historical perspective. *Annual Review of Astronomy and Astrophysics*, 41(1):169–89.

Nobel Prize (2011). The Nobel Prize in Physics 2011.

NobelPrize.org (2016). Relativity, http://www.nobelprize.org/educational/physics/relativity/postulates-1.html.

Noë, A. (2004). *Action in Perception.* MIT Press, Cambridge.

Noë, A. (2009). *Out of Our Heads.* Hill and Wang, New York.

Nojiri, S. and Odintsov, S. (2011). Unified cosmic history in modified gravity: from theory to lorentz non-invariant models. *Physics Reports*, 505(2–4):59–144.

Novikov, I. (1989). An analysis of the operation of a time machine. *Soviet Phyics JETP*, 68:439–43.

Oaklander, L. (2004). Freedom and the new theory of time. In Le Poidevin, R., editor, *Questions of Time and Tense*, pages 185–206. Oxford University Press, Oxford.

O'Connor, T. (2014). Free will and metaphysics. In Palmer, D., editor, *Libertarian Free Will*, pages 27–36. Oxford University Press, Oxford.

O'Connor, T. and Wong, H. (2015). Emergent properties, Zalta, E.N., *The Stanford Encyclopedia of Philosophy* (Summer 2015 Edition), https://plato.stanford.edu/archives/sum2015/entries/properties-emergent/.

Oerter, R. (2006). *The Theory of Almost Everything: The Standard Model, the Unsung Triumph of Modern Physics.* Pi Press, New York, NY.

Olmo, G. (2011). Palatini approach to modified gravity: f(r) theories and beyond. *International Journal of Modern Physics D*, 20(04):413–62.

Oort, J. (1932). The force exerted by the stellar system in the direction perpendicular to the galactic plane and some related problems. *Bulletin of the Astronomical Institutes of the Netherlands*, 6:249–87.

O'Regan, J. (2011). *Why Red Doesn't Sound Like a Bell: Understanding the Feel of Consciousness*. Oxford University Press, Oxford.

O'Regan, J. and Noë, A. (2001). A sensorimotor account of vision and visual consciousness. *Behavioral and Brain Sciences*, 24:883–917.

Oriti, D. (2013). Disappearance and emergence of space and time in quantum gravity, http://arxiv.org/abs/1302.2849.

Ouellette, J. (2012). Alice and bob meet the wall of fire. *Quanta Magazine*, https://www.quantamagazine.org/20121221-alice-and-bob-meet-the-wall-of-fire/.

Padmanabhan, T. (2004). From gravitons to gravity: myths and reality, http://arxiv.org/pdf/gr-qc/0409089v1.pdf.

Padmanabhan, T. (2016). Do we really understand the cosmos?, https://arxiv.org/pdf/1611.03505.pdf.

Paranjape, A. and Singh, T. (2006). The possibility of cosmic acceleration via spatial averaging in Lemaître Tolman–Bondi models. *Classical and Quantum Gravity*, 23:6955–69.

Paul, L. and Healy, K. (2016). Transformative treatments. *Nous*, Early View, 11, 10.1111/nous.12180, Wiley-Blackwell.

Peres, A. (1997). Critique of the Wheeler–Dewitt equation. In Harvey, A., editor, *On Einstein's Path*, pages 367–79. Springer, http://arxiv.org/abs/gr-qc/9704061.

Perlmutter, S. (2003). Supernovae, dark energy, and the accelerating universe. *Physics Today*, 56:53–60.

Perlmutter, S., Aldering, G., Goldhaber, G., Knop, R., et al. (1999). Measurements of ω and λ from 42 high-redshift supernovae. *The Astrophysical Journal*, 517(2):565–86.

Peterson, D. and Silberstein, M. (2010). Relativity of simultaneity and eternalism: in defense of the block universe. In Petkov, V., editor, *Space, Time, and Spacetime—Physical and Philosophical Implications of Minkowski's Unification of Space and Time*, pages 209–37. Springer, Berlin.

Petkov, V. (2005). *Relativity and the Nature of Spacetime*. Springer, Berlin.

Petkov, V. (2006). Is there an alternative to the block universe view? In Dieks, D., editor, *The Ontology of Spacetime*, pages 207–28. Elsevier, Utrecht.

Petkov, V. (2007). On the reality of minkowski space. *Foundational Physics*, 37: 1499–1502.

Pigliucci, M. (2012). Lawrence krauss: another physicist with an anti-philosophy complex, http://rationallyspeaking.blogspot.com/2012/04/lawrence-krauss-another-physicist-with.html.

Pillsbury, M., Orland, H., and Zee, A. (2005). A steepest descent calculation of RNA pseudoknots. *Physical Review E*, 72, 011911.

Planck Collaboration (2016). Planck 2015 results. XIII. Cosmological parameters. *Astronomy & Astrophysics*, 594:A13.

Planck Collaboration (2016a). COM PowerSpect COM PowerSpect CMB-base-plikHM-TT-lowTEB-minimum-theory R2.02.txt, http://pla.esac.esa.int/pla/.

Planck Collaboration (2016b). COM PowerSpect CMB R2.02.fits, http://irsa.ipac.caltech.edu/data/Planck/release 2/ancillary-data/HFI Products.html.

Prescod-Weinstein, C. and Smolin, L. (2009). Disordered locality as an explanation for the dark energy. *Physical Review D*, 80(6):063505.

Price, H. (1996). *Time's Arrow and Archimedes Point: New Directions for the Physics of Time*. Oxford University Press, Oxford.

Price, H. (2005). *Causal Perspectivalism*, http://philsci-archive.pitt.edu/4475/1/TimeAsymmetryOfCausation.pdf.

Price, H. (2008). Toy models for retrocausality. *Studies in History and Philosophy of Science Part B: Studies in History and Philosophy of Modern Physics*, 39(4):752–61.

Price, H. (2011). The flow of time. In Callender, C., editor, *The Oxford Handbook of Philosophy of Time*, pages 276–311. Oxford University Press, Oxford.

Price, H. (2013). Huw Price reviews Lee Smolin's time reborn, http://www.3quarksdaily.com/3quarksdaily/2013/08/huw-price-reviews-lee-smolins-time-reborn.html#sthash.qlXSkNmi.dpuf.

Price, H. and Weslake, B. (2008). *The Time-Asymmetry of Causation*, http://philsci-archive.pitt.edu/4475/1/TimeAsymmetryOfCausation.pdf.

Price, H. and Wharton, K. (2013). Dispelling the quantum spooks—a clue that Einstein missed? *arXiv preprint arXiv:1307.7744*.

Prosser, S. (2013). The passage of time. In Bardon, A. and Dyke, H., editors, *A Companion to the Philosophy of Time*, pages 315–327. John Wiley & Sons, Inc, Chichester, UK.

Putnam, H. (1967). Time and physical geometry. *Journal of Philosophy*, 64:240–47.

Redhead, M. (1990). Explanation. In *Explanation and Its Limits*, pages 135–54. Cambridge University Press, Cambridge.

Regge, T. (1961). General relativity without coordinates. *Nuovo Cimento*, 19(3):558–71.

Reiprich, T.H. (2001). *Cosmological Implications and Physical Properties of an X-Ray Flux-Limited Sample of Galaxy Clusters*. PhD thesis, https://arxiv.org/abs/astro-ph/0308137.

Reiprich, T. and Böhringer, H. (2002). The mass function of an X-ray flux-limited sample of galaxy clusters. *The Astrophysical Journal*, 567(2):716–40.

Rescher, N. (1970). *Scientific Explanation*. Free Press, New York, NY.

Revonsuo, A. (2006). MIT Press.

Revonsuo, A. (2016). Hard to see the problem? *Journal of Consciousness Studies*, 22: 52–67.

Rickles, D. (2012). Time, observables, and structure. In *Structural Realism*, pages 135–45. Springer, Dordrecht.

Rickles, D. and Bloom, J. (2015). Things ain't what they used to be. Physics without objects. In Tomasz, B. and Wüthrich, C., editors, *Metaphysics in Contemporary Physics*, pages 101–21. Brill, Leiden.

Riess, A., Filippenko, A., Challis, P., Clocchiattia, A., et al. (1998). Observational evidence from supernovae for an accelerating universe and a cosmological constant. *Astronomical Journal*, 116:1009–38.

Rietdijk, C. (1966). A rigorous proof of determinism derived from the special theory of relativity. *Philosophy of Science*, 33:341–44.

Robertson, T. and Atkins, P. (2016). Essential vs. accidental properties, Zalta, E.N., *The Stanford Encyclopedia of Philosophy* (Summer 2016 Edition), http://plato.stanford.edu/archives/sum2016/entries/essential-accidental/.

Roddenberry, G. (1993). Star trek: deep space nine, Emissary.

Rodrigues, D.C., Piattella, O.F., Fabris, J.C., and Shapiro, I.L. (2012). Renormalization group approach to gravity: the running of g and λ inside galaxies and additional details on the elliptical NGC 4494. *Proceedings of Science, VIII International Workshop on the Dark Side of the Universe*, https://arxiv.org/abs/1301.4148.

Roos, M. (2000). Shear-dependent pressure in a Lemaître–Tolman–Bondi metric, https://arxiv.org/abs/1107.3028.

Rothe, H. (1992). *Lattice Gauge Theories: An Introduction*. World Scientific, Singapore.

Rovelli, C. (1996). Relational quantum mechanics. *International Journal of Theoretical Physics*, 35:1637–78.

Rovelli, C. (1999). "Localization" in quantum field theory: how much of QFT is compatible with what we know about space-time? In *Conceptual Foundations of Quantum Field Theory*, pages 207–32. Cambridge University Press, Cambridge.

Rovelli, C. (2007). *Quantum Gravity*. Cambridge University Press, Cambridge.

Rovelli, C. (2013). Why gauge? quant-ph/1308.5599.

Rovelli, C. (2016). An argument against the realistic interpretation of the wave function. *Foundations of Physics*, 46:1229–1237.

Rovelli, C. (2017). *Reality Is Not What It Seems: The Journey to Quantum Gravity*. Riverhead Books, New York, NY.

Rubin, V. and Ford Jr, K. (1970). Rotation of the andromeda nebula from a spectroscopic survey of emission regions. *The Astrophysical Journal*, 159:379–403.

Russell, B. (1913). *Theory of Knowledge: The 1913 Manuscript*. Routledge, London.

Russell, B. (1915). On the experience of time. *The Monist*, 25(2):211–33.

Saint-Ours, A. (2015). Does time differ from change? philosophical appraisal of the problem of time in quantum gravity and in physics. *Studies in the History and Philosophy of Modern Physics*, 52:48–54.

Salmon, W. (1989). *Four Decades of Scientific Explanation*. University of Pittsburgh Press, Pittsburgh, PA.

Sanders, R. (2005). A tensor–vector–scalar framework for modified dynamics and cosmic dark matter. *Monthly Notices of the Royal Astronomical Society*, 363(2):459–68.

Sanders, R. H. and McGaugh, S. S. (2002). Modified newtonian dynamics as an alternative to dark matter. *Annual Reviews of Astronomy & Astrophysics*, 40:263–317.

Saunders, S. (2002). How relativity contradicts presentism. In Callender, C., editor, *Time, Reality, and Experience*, pages 277–92. Cambridge University Press, Cambridge.

Savitt, S. (2006a). Being and becoming in modern physics, Zalta, E.N., *The Stanford Encyclopedia of Philosophy* (Fall 2007 Edition), http://plato.stanford.edu/archives/fall2007/entries/spacetime-bebecome.

Savitt, S. (2006b). Presentism and eternalism in perspective. In Dieks, D., editor, *The Ontology of Spacetime*, pages 111–27. Elsevier, Utrecht.

Schaffer, J. (2010). Monism: the priority of the whole. *Philosophical Review*, 119(1).

Schopenhauer, A. (1969). *The World As Will and Representation*. Dover, New York, NY.

Seager, W. (2007). A brief history of the philosophical problem of consciousness. In Zelazo, P., Moscovitch, M., and Thompson, E., editors, *The Cambridge Handbook of Consciousness*, pages 9–34. Cambridge University Press, Cambridge.

Seager, W. (2009). Dual aspect theories. In T. Bayne, T., Cleeremans, A., and Wilken, P., editors, *Oxford Companion to Consciousness*, pages 243–244. Oxford University Press, Oxford.

Seager, W. (2012). Emergentist panpsychism. *Journal of Consciousness Studies*, 19(9–10):19–39.

Seager, W. (2013). Monism and models, Manuscript.

Seager, W. (2016a). Panpsychist infusion. In Bruntrup, G. and Jaskolla, L., editors, *Panpsychism*, pages 229–246. Oxford University Press, Oxford.

Seager, W. (2016b). *Theories of Consciousness: An Introduction and Assessment*. Routledge, London.

Searle, J. (2017). Biological naturalism. In Schneider, S. and Velmans, M., editors, *The Blackwell Companion to Consciousness*. John Wiley & Sons, Inc., Malden, MA

Serway, R. and Jewett, J. (2014). *Physics for Scientists and Engineers, 9th Edition*. Brooks/Cole, Boston, MA.

Shankar, R. (1994). *Principles of Quantum Mechanics, 2nd edition.* Plenum Press, New York, NY.

Sider, T. (2001). *Four-Dimensionalism.* Oxford University Press, Oxford.

Siegfried, T. (2008). It's likely that times are changing. *Science News,* 174(6):26–8.

Silberstein, M. (2002). Reduction, emergence, and explanation. In Machamer, P. and Silberstein, M., editors, *The Blackwell Guide to the Philosophy of Science,* pages 203–26.

Silberstein, M. (2006). In defense of ontological emergence and mental causation. In Davies, P., editor, *The Re-emergence of Emergence,* pages 203–226. Oxford University Press, Oxford.

Silberstein, M. (2007). On finding 'real' time in quantum mechanics. In Craig, W. and Smith, Q., editors, *Absolute Simultaneity,* pages 254–57. Oxford University Press, Oxford.

Silberstein, M. (2009). Emergence. In T. Bayne, T., Cleeremans, A., and Wilken, P., editors, *Oxford Companion to Consciousness,* pages 254–57. Oxford University Press, Oxford.

Silberstein, M. (2010). Why neutral monism is superior to panpsychism. *Mind and Matter,* 7:239–48.

Silberstein, M. (2012). Emergence and reduction in context: philosophy of science and/or analytic metaphysics. *Metascience,* 21:627–42.

Silberstein, M. (2013). Constraints on localization and decomposition as explanatory strategies in the biological sciences. *Philosophy of Science,* 80:958–70.

Silberstein, M. (2014). Experience unbound: neutral monism, contextual emergence and extended cognitive science. *Mind and Matter,* 12(2):289–340.

Silberstein, M. (2016). The implications of neural reuse for the future of cognitive neuroscience and the future of folk psychology. *Brain and Behavioral Sciences,* 120:1–45.

Silberstein, M. (2017a). Strong emergence no, contextual emergence yes. Special issue of Philosophica on emergence.

Silberstein, M. (2017b). Neutral monism reborn: breaking the gridlock between emergent versus inherent. In Seager, W., editor, *Routledge Companion to Panpsychism.* Routledge, London.

Silberstein, M. and Chemero, A. (2011a). After the philosophy of mind: from scholasticism to science. *Journal of Philosophical Studies,* 15:1–19.

Silberstein, M. and Chemero, A. (2011b). Dynamics, agency and intentional action. *Journal of Philosophical Studies,* 15:1–19.

Silberstein, M. and Chemero, A. (2012). Complexity and extended phenomenological-cognitive systems. *Topics in Cognitive Science,* 4(1):35–50.

Silberstein, M. and Chemero, A. (2015). Extending neutral monism to the hard problem. *Journal of Consciousness Studies on Embodied and Extended Theories of Conscious Experience,* 22(3–4):181–94.

Silberstein, M., Cifone, M., and Stuckey, W. (2008). Why quantum mechanics favors adynamical and acausal interpretations such as relational blockworld over backwardly causal and time-symmetric rivals. *Studies in History and Philosophy of Modern Physics,* 39(4):736–51.

Silberstein, M., Stuckey, W., and Cifone, M. (2007). An argument for 4d blockworld from a geometric interpretation of non-relativistic quantum mechanics. In Petkov, V., editor, *Relativity and the Dimensionality of the World,* pages 197–216. Springer-Verlag, Dordrecht.

Silberstein, M., Stuckey, W., and McDevitt, T. (2013). Being, becoming and the undivided universe: a dialogue between relational blockworld and the implicate order concerning the unification of relativity and quantum theory. *Foundations of Physics*, 43(4):502–32.

Sinha, S. and Sorkin, R. (1991). A sum-over-histories account of an epr(b) experiment. *Foundations of Physics Letters*, 4(4):303–35.

Skordis, C. (2009). The tensor–vector–scalar theory and its cosmology. *Classical & Quantum Gravity*, 26:143001.

Skordis, C. (2016). Personal communication.

Skordis, C., Mota, D., Ferreira, P., and Boehm, C. (2006). Large scale structure in bekenstein's theory of relativistic modified newtonian dynamics. *Physical Review Letters*, 96:011301.

Smeenk, C. (2014). Predictability crisis in early universe cosmology. *Studies in the History and Philosophy of Modern Physics*, 46:122–33.

Smolin, L. (2006). *The Trouble with Physics*. Houghton Mifflin, Boston, MA.

Smolin, L. (2009). The unique universe: against the timeless multiverse. *Physics World*, 22(6):21–6.

Smolin, L. (2013). *Time Reborn: From the Crisis in Physics to the Future of the Universe*. Mariner Books, Houghton Mifflin Harcourt, Boston, MA.

Sorkin, R. (2007a). Quantum dynamics without the wavefunction. *Journal of Physics A: Mathematical and Theoretical*, 40(12):3207–222.

Sorkin, R. (2007b). Relativity theory does not imply that the future already exists: a counterexample. In *Relativity and the Dimensionality of the World*, pages 153–61. Springer, Dordrecht.

Stachel, J. (2002). *Einstein from 'B' to 'Z': Einstein Studies, Volume 9. Albert Einstein, Lecture in Kyoto on December 14, 1922*. Birkhäuser, Boston, MA.

Stein, H. (1968). On Einstein–Minkowski spacetime. *Journal of Philosophy*, 65:5–23.

Stein, H. (1991). On relativity theory and openness of the future. *Philosophy of Science*, 58:147–67.

Steinhardt, P. (2011). The inflation debate: is the theory at the heart of modern cosmology deeply flawed? *Scientific American*, 304(4):36–43.

Storage, B. (2013). You're so wrong, Richard Feynman, https://themultidisciplinarian.com/2013/11/02/youre-so-wrong-richard-feynman/.

Strawson, G. (2008). *Real Materialism and Other Essays*. Oxford University Press, New York, NY.

Strawson, G. (2009). Realistic monism: Why physicalism entails panpsychism. In Skrbina, D., editor, *Mind that Abides: Panpsychism in the New Millenium*, pages 33-66. John Benjamins, Amsterdam.

Stubenberg, L. (2009). Neutral monism, *Stanford Encyclopedia of Philosophy*, https://plato.stanford.edu/archives/sum2016/entries/neutral-monism/.

Stuckey, W. (1993). The Schwarzschild black hole as a gravitational mirror. *American Journal of Physics*, 61(5):448–56.

Stuckey, W. (1994). The observable universe inside a black hole. *American Journal of Physics*, 62(9):788–95.

Stuckey, W., McDevitt, T., and Silberstein, M. (2012a). Explaining the supernova data without accelerating expansion. *International Journal of Modern Physics D*, 21(11):1242021.

Stuckey, W., McDevitt, T., and Silberstein, M. (2012b). Modified Regge calculus as an explanation of dark energy. *Classical and Quantum Gravity*, 29(5):055015.

Stuckey, W.M., McDevitt, T., Sten, A.K., and Silberstein, M. (2016a). Dark matter as a metric perturbation, https://arxiv.org/abs/1509.09288.

Stuckey, W., McDevitt, T., Sten, A., and Silberstein, M. (2016b). End of a dark age? *International Journal of Modern Physics D*, 25(12):1644004.

Stuckey, W.M., Silberstein, M., and Cifone, M. (2005). Reversing the arrow of explanation in the relational blockworld: why temporal becoming, the dynamical brain and the external world are in the mind. In Buccheri, R., Elitzur, A., and Saniga, M., editors *Endophysics, Time, Quantum and the Subjective*, pages 293–316. World Scientific, Singapore.

Stuckey, W.M., Silberstein, M., and Cifone, M. (2006). Deflating quantum mysteries via the relational blockworld. *Physics Essays*, 19(2):269–283.

Stuckey, W., Silberstein, M., and Cifone, M. (2008). Reconciling spacetime and the quantum: relational blockworld and the quantum liar paradox. *Foundations of Physics*, 38(4):348–83.

Stuckey, W., Silberstein, M., and McDevitt, T. (2015). Relational blockworld: providing a realist psi-epistemic account of quantum mechanics. *International Journal of Quantum Foundations*, 1:123–70.

Stuckey, W., Silberstein, M., and McDevitt, T. (2016c). An adynamical, graphical approach to quantum gravity and unification. In Licata, I., editor, *Beyond Peaceful Coexistence: The Emergence of Space, Time and Quantum*, pages 499–544. Imperial College Press, London.

Stuckey, W., Silberstein, M., and McDevitt, T. (2016d). Underwriting information-theoretic accounts of quantum mechanics with a realist, psi-epistemic model. *International Journal of Quantum Information*, 14(1):1640007.

Sullivan, J.W.N. (1931a). Interviews with great scientists IV. Prof. Schrödinger. *The Observer*, pages 15–16, 11 January, London.

Sullivan, J.W.N. (1931b). Interviews with great scientists VI. Max Planck. *The Observer*, page 17, 25 January, London.

Suzuki, N., Rubin, D., Lidman, C., Aldering, G., et al. (2012). The hubble space telescope cluster supernova survey. v. improving the dark-energy constraints above z > 1 and building an early-type-hosted supernova sample. *The Astrophysical Journal*, 746(1):85.

't Hooft, G. (1997). *In Search of the Ultimate Building Blocks*. Cambridge University Press, Cambridge.

't Hooft, G. (2007). The conceptual basis of quantum field theory. In Butterfield, J. and Earman, J., editors, *Philosophy of Physics Part A*, pages 661–729. Elsevier, Amsterdam.

't Hooft, G. (2016). Foreword. In Licata, I., editor, *Beyond Peaceful Coexistence: The Emergence of Space, Time and Quantum*, pages ix–x. Imperial College Press, London.

Tanimoto, M. and Nambu, Y. (2007). Luminosity distance–redshift relation for the ltb solution near the center. *Classical and Quantum Gravity*, 24(15):3843–857.

Tegmark, M. and Wheeler, J. (2001). 100 years of quantum mysteries. *Scientific American*, 284:68–75.

Teller, P. (1997). *An Interpretive Introduction to Quantum Field Theory*. Princeton University Press, Princeton, NY.

Thalos, M. (2013). *Without Hierarchy: The Scale Freedom of the Universe*. Oxford University Press, Oxford.

Thompson, E. (2007). *Mind in Life: Biology, Phenomenology and the Sciences of Mind*. Harvard University Press, Cambridge, MA.

Thompson, E. (2015). *Waking, Dreaming, Being: Self and Consciousness in Neuroscience, Meditation, and Philosophy*. Columbia University Press, New York, NY.

Thompson, E. and Stapleton, M. (2009). Making sense of sense-making: reflections on enactive and extended mind theories. *Topoi*, 28(1):23–30.

Toffoli, T. (2003). What is the Lagrangian counting? *International Journal of Theoretical Physics*, 42(2):363–81.

Tononi, G. and Koch, C. (2015). Consciousness: here, there and everywhere? *Philosophical Transactions of the Royal Society B*, 370:20140167.

Unger, R. and Smolin, L. (2015). *The Singular Universe and the Reality of Time*. Cambridge University Press, Cambridge.

Varela, F., Thompson, E., and Rosch, E. (1991). *The Embodied Mind*. MIT Press, Cambridge, MA.

Veltman, M. (1994). *Diagrammatica: The Path to Feynman Rules*. Cambridge University Press, Cambridge.

Vilenkin, A. and Tegmark, M. (2011). The case for parallel universes. *Scientific American Blog*, https://www.scientificamerican.com/article/multiverse-the-case-for-parallel-universe/.

Vollick, D. (2004). On the viability of the palatini form of 1/r gravity. *Classical and Quantum Gravity*, 21:3813.

Vonnegut, K. (1969). *Slaughter House Five*. Dell Publishing, New York, NY.

Wald, R. (1984). *General Relativity*. University of Chicago Press, Chicago, IL.

Wallace, D. (2006). In defence of naivete: the conceptual status of lagrangian quantum field theory. *Synthese*, 151:33–80.

Walter, F., Brinks, E., de Blok, W., Bigiel, F., Kennicutt Jr, R., Thornley, M., and Leroy, A. (2008). Things: the hi nearby galaxy survey. *The Astronomical Journal*, 136(6):2563–647.

Wang, L., Caldwell, R., Ostriker, J., and Steinhardt, P. (2000). Cosmic concordance and quintessence. *The Astrophysical Journal*, 530:17–35.

Warman, M. (2011). Stephen Hawking tells Google 'philosophy is dead', http://www.telegraph.co.uk/technology/google/8520033/Stephen-Hawking-tells-Google-philosophy-is-dead.html.

Weinberg, S. (1972). *Gravitation and Cosmology: Principles and Applications of the General Theory of Relativity*. John Wiley & Sons, New York.

Weinberg, S. (1999). Panel discussion, chapter 26. In Cao, T., editor, *Conceptual foundations of quantum field theory*, volume 1, pages 368–86. Cambridge University Press, Cambridge.

Weinberg, S. (2000). The cosmological constant problems, https://arxiv.org/abs/astro-ph/0005265.

Weinberg, S. (2009). The quantum theory of fields: effective or fundamental? http://cdsweb.cern.ch/record/1188567, 7.

Weinstein, S. (2015). Patterns in the fabric of nature. In *Questioning the Foundations of Physics*, pages 139–50. Springer, Heidelberg.

Weinstein, S. and Rickles, D. (2016). Quantum gravity, Zalta, E.N., *The Stanford Encyclopedia of Philosophy* (Winter 2016 Edition), https://plato.stanford.edu/archives/win2016/entries/quantum-gravity/.

Wharton, K. (2014). Quantum states as ordinary information. *Information*, 5(1):190–208.

Wharton, K. (2015). The universe is not a computer. In Aguirre, A., Foster, B., and Merali, Z., editors, *Questioning the Foundations of Physics*, pages 177–90. Springer, Heidelberg.

Wharton, K., Miller, D., and Price, H. (2011). Action duality: a constructive principle for quantum foundations. *Symmetry*, 3(3):524–40.

Wheeler, J.A. (1990). Information, physics, quantum: the search for links. In Ezawa, H., Kobayashi, S.I., and Murayama, Y., editors, *Proceedings of the 3rd International Symposium on Foundations of Quantum Mechanics in the Light of New Technology*. Physical Society of Japan, https://jawarchive.files.wordpress.com/2012/03/informationquantumphysics.pdf.

Wheeler, J. (1994). *At Home in the Universe*. The American Institute of Physics, New York, NY.

Wheeler, J. and Feynman, R. (1949). Classical electrodynamics in terms of direct inter-particle action. *Reviews of Modern Physics*, 21(3):425–33.

Wilczek, F. (2015). How physics will change—and change the world—in 100 years.

Wilczek, F. (2016). Physics in 100 years. *Physics Today*, 69(4):32–9.

Williams, R. M. and Tuckey, P. A. (1992). Regge calculus: a brief review and bibliography. *Classical and Quantum Gravity*, 9(5):1409–22.

Windt, J. (2015). *Dreaming: A Conceptual Framework for Philosophy of Mind and Empirical Research*. MIT Press, Cambridge, MA.

Wise, D. (2006). *p*-form electromagnetism on discrete spacetimes. *Classical and Quantum Gravity*, 23(17):5129–76.

Wittmann, M. (2015). Modulations of the experience of self and time. *Consciousness and Cognition*, 38(12):172–81.

Wittmann, M. (2016). *Felt Time: The Psychology of How We Perceive Time*. MIT Press, Cambridge, MA.

Woit, P. (2012). Linde on inflation and the multiverse, http://www.math.columbia.edu/~woit/wordpress/?p=5076, 8.

Wolchover, N. (2015). A fight for the soul of science, *Quanta Magazine*, https://www.quantamagazine.org/20151216-physicists-and-philosophers-debate-the-boundaries-of-science/, 12.

Wong, C. (1971). Applications of regge calculus to the Schwarzschild and Reissner-nordstrom geometries at the moment of time symmetry. *Journal of Mathematical Physics*, 12.

Wong, H. (2006). Emergents from fusion. *Philosophy of Science*, 73:345–67.

Woo, M. (2017). Why the multiverse isn't just madness, https://sciencefriday.com/articles/why-the-multiverse-isnt-just-madness/. Science Friday.

Wüthrich, C. (2012). The fate of presentism in modern physics, https://arxiv.org/abs/1207.1490.

Wyse, R. (2017). The cosmological context of the milky way galaxy, Franklin & Marshall College, 4.

Yinon, D. (2016). Change's order: on Deleuze's notion of time. In Dolev, Y. and Roubach, M., editors, *Cosmological and Psychological Time*, pages 203–218, Springer, Cham, Switzerland.

Yoshinori, T. (1991). *Heart of Buddhism: In Search of the Timeless Spirit of Primitive Buddhism*. Crossroad Publishing Company.

Young, M. (2016). No dark matter from lux experiment, Sky & Telescope, http://www.skyandtelescope.com/astronomy-news/no-dark-matter-from-lux-experiment/, 7.

Yourgrau, P. (2005). *A World Without Time*. Basic Books, Cambridge, MA.

Zahavi, D. (2005). *Subjectivity and Selfhood: Investigating the First-Person Perspective*. MIT Press, Cambridge, MA.

Zahavi, D. (2014). *Self & Other Exploring Subjectivity, Empathy, and Shame*. Oxford University Press, Oxford.

Zee, A. (2003). *Quantum Field Theory in a Nutshell*. Princeton University Press, Princeton, NJ.

Zhao, H. and Li, B. (2010). Dark fluid: a unified framework for modified newtonian dynamics, dark matter, and dark energy. *The Astrophysical Journal*, 712(1):130–41.

Zibin, J., Moss, A., and Scott, D. (2008). Can we avoid dark energy? *Physical Review Letters*, 101:8251303.

Zimmerman, D. (2011). Presentism and the space-time manifold. In Callender, C., editor, *The Oxford Handbook of Philosophy of Time*, pages 163–246. Oxford University Press, Oxford.

Zlatev, I., Wang, L., and Steinhardt, P. (1999). Quintessence, cosmic coincidence, and the cosmological constant. *Physical Review Letters*, 82:896–99.

Zlosnik, T., Ferreira, P., and Starkman, G. D. (2007). Modifying gravity with the aether: an alternative to dark matter. *Physical Review D*, 75(4):044017.

Zwicky, F. (1933). Spectral displacement of extra galactic nebulae. *Helvetica Physica Acta*, 6:110–27.

Zwicky, F. (1937). On the masses of nebulae and of clusters of nebulae. *The Astrophysical Journal*, 86:217–46.

Index